Advanced Structured Materials

Volume 233

Series Editors

Common engineering materials are reaching their limits in many applications, and new developments are required to meet the increasing demands on engineering materials. The performance of materials can be improved by combining different materials to achieve better properties than with a single constituent, or by shaping the material or constituents into a specific structure. The interaction between material and structure can occur at different length scales, such as the micro, meso, or macro scale, and offers potential applications in very different fields.

This book series addresses the fundamental relationships between materials and their structure on overall properties (e.g., mechanical, thermal, chemical, electrical, or magnetic properties, etc.). Experimental data and procedures are presented, as well as methods for modeling structures and materials using numerical and analytical approaches. In addition, the series shows how these materials engineering and design processes are implemented and how new technologies can be used to optimize materials and processes.

Advanced Structured Materials is indexed in Google Scholar and Scopus.

Viacheslav L. Bogdanov · Alexander Ya. Grigorenko · Ihor B. Chepkov · Ihor V. Odnoralov · Andrii V. Kuchynskyi · Valerii V. Kremenytskyi · Svitlana O. Sperkach

Identification of Special-Purpose Structures by Their Fragments Based on Scanning Electron Microscopy

Springer

Viacheslav L. Bogdanov
S. P. Timoshenko Institute of Mechanics
National Academy of Sciences of Ukraine
Kyiv, Ukraine

Alexander Ya. Grigorenko
S. P. Timoshenko Institute of Mechanics
National Academy of Sciences of Ukraine
Kyiv, Ukraine

Ihor B. Chepkov
Central Research Institute of Armaments and Military Equipment of Armed Forces of Ukraine
Kyiv, Ukraine

Ihor V. Odnoralov
Central Research Institute of Armaments and Military Equipment of Armed Forces of Ukraine
Kyiv, Ukraine

Andrii V. Kuchynskyi
Central Research Institute of Armaments and Military Equipment of Armed Forces of Ukraine
Kyiv, Ukraine

Valerii V. Kremenytskyi
Technical Center
National Academy of Sciences of Ukraine
Kyiv, Ukraine

Svitlana O. Sperkach
Technical Center
National Academy of Sciences of Ukraine
Kyiv, Ukraine

ISSN 1869-8433 ISSN 1869-8441 (electronic)
Advanced Structured Materials
ISBN 978-3-031-88826-7 ISBN 978-3-031-88827-4 (eBook)
https://doi.org/10.1007/978-3-031-88827-4

This Springer imprint is published by the registered company Springer Nature Switzerland AG
The registered company address is: Gewerbestrasse 11, 6330 Cham, Switzerland

Preface

In recent years, a significant challenge has emerged in international relations. Democratic governments and transnational organizations must establish effective export controls. They need to ensure rapid and scientifically accurate identification of destructive means based on their fragments, considering the accepted classifications and countries of origin. Additionally, it is essential to develop measures to prevent the uncontrolled or "grey" transfer of special-purpose structures as well as military or dual-use technologies and materials.

This challenge gained the utmost importance after the beginning of the unprovoked full-scale aggression of the Russian Federation against Ukraine. The specific feature of the abovementioned armed confrontation is the mass use of a large variety of ground-, aircraft- and sea-launched cruise, ballistic and aeroballistic missiles, including hypersonic ones, along with reconnaissance and strike unmanned aerial vehicles, loitering munitions.

Another peculiarity of the present moment is the covert mass transfer to the Russian Federation of the means of destruction manufactured by its allies (Iran and North Korea) [1, 2]. It should be noted that this can only be the first step in the shadow arming of the aggressor with weaponry also made in other authoritarian countries. This situation further enhances the importance of providing scientifically based identification of means of destruction by their fragments. Even a superficial glance at the events in Ukraine allows one to conclude that the Russian Federation employs in this war quite a powerful arsenal of high-tech weapons. It is a comprehensive analysis of the structural materials used in fabricating them that is provided in this work.

Chapter 1 provides a detailed analysis of the available experimental techniques of investigating physico-mechanical characteristics of metallic and composite materials used in the structural elements of today's special-purpose products. These techniques are subdivided into two large groups, depending on whether the specimen under study is destroyed during investigation or not.

The chapter presents the main physical principles of employing the above techniques; their limitations are described. A comparative analysis of those techniques is made.

The destructive techniques of investigating the physicochemical structure of solid materials have a common limitation from the viewpoint of examining the structure and chemical composition of metallic and composite materials. These methods require that a sample of the material be taken for further atomization, ionization, etc., so they do not provide information on the structure of the specimen surface and the concentration of chemical elements on it.

It has been shown that X-ray microanalysis is a non-destructive technique. It allows the chemical analysis of microinclusions to be performed, and images of the studied specimen areas to be produced. In addition, it permits the concentration of chemical elements to be measured not only at points and rectangular areas but along scanning lines as well. Besides, this technique can be employed to investigate specimens of unknown chemical compositions.

The conclusion is made that X-ray microanalysis is the most suitable technique of studying metallic and composite materials, including those intended for special purposes.

In Chap. 2, a comprehensive analysis of present-day metals and alloys used in manufacturing special-purpose structures is performed.

It is shown that high-quality, high-strength alloyed steels and alloys that contain special additives of chemical elements are used. Non-destructive research methods are mainly employed to investigate the structures and chemical compositions of those objects.

The studies have been carried out with reliance on raster electron microscopy—the most universal technique of non-destructive research and analysis of the microstructural characteristics of solids. Specific features of preparing original metal and alloy specimens for examining their chemical composition, structure and surface topography have been presented. The general stages in the experimental studies of the fragments of special-purpose structures made of metals and alloys are described.

The physico-chemical composition, structure and surface topography of a wide class of special-purpose products manufactured from metals and alloys have been determined.

Differences in the chemical composition and structure of various specimens of the components of the means of destruction employed by the Russian Federation have been shown.

Chapter 3 provides a comprehensive consideration of various composite materials that are widely used in manufacturing current special-purpose structures (means of destruction) since those materials are characterized by high strength, heat resistance, low density, thermal and electric conductivity, along with other special properties.

Composite materials can be classified in terms of their structure, molding methods, matrix and filler materials, reinforcement techniques, purpose, fields of application, etc. Professional literature provides a detailed analysis of the advantages and limitations of current composite materials from the viewpoint of their application for manufacturing elements of special-purpose structures.

The research into microstructural characteristics of the fragments of special-purpose structures (means of destruction), whose results are presented in this chapter, was carried out with the most versatile non-destructive scanning electron microscopy technique and the method of X-ray microanalysis.

To examine the structure, chemical composition and elemental distribution on the surfaces and in the bulks of the fragments, as well as measure the dimensions of different layers of composite materials, sections of the structural elements of means of destruction used by Russian troops during hostilities in Ukraine were produced.

Physico-chemical composition, structure and surface topography of the elements of a broad class of fragments of special-purpose means made of composite materials were determined.

The results obtained indicate that in manufacturing special-purpose structures (means of destruction) quite well-known composite materials were mostly used, namely: carbon fibers, fiberglass, organoplastics, boron-containing plastic composites and Kevlar-based hybrid composite materials. The abovementioned materials are used both for manufacturing structural elements and for producing coatings. The main purpose of using the abovementioned composite materials in manufacturing special-purpose structures (means of destruction) is to increase their strength and heat resistance and reduce their weight and radar signature (cross-section).

The use of scanning electron microscopy permitted a most accurate determination of the chemical composition, structure and topology of the elements of special-purpose means (means of destruction) that were manufactured using composite materials.

Chapter 4 presents a review of the methods of measuring materials' electrophysical parameters in the UHF range. It considers theoretical foundations of electromagnetic radiation interaction with composite materials and shows the application of analytical expressions obtained with reliance on well-known theoretical approaches to calculating the parameters of RF absorbent materials.

An approach is developed for computing the parameters of frequency characteristics of RF absorbent coatings and materials of the components of special-purpose structures. It is based on using the results of theoretical analytical research into the interaction of electromagnetic waves with composite RF absorbent materials and on specific experimental data obtained with electron microscopy and X-ray microanalysis techniques.

The approach proposed allows a highly accurate study of the internal structure, including geometric parameters, of fillers and coating layers, and the quantitative chemical compositions of coating materials, which, in turn, makes it possible to model the physical processes occurring during the interaction of radio waves and the material of RF absorbent coating in a wide frequency range.

Examples of calculating the operation range width of RF absorbent coatings and the limit wavelengths of that range obtained with the application of the proposed approach are demonstrated.

Chapter 5 is concerned with a comparative analysis of the existing approaches to identifying special-purpose structures and estimates the advantages and limitations of each of them. The tasks to be solved in the identification of special-purpose structures (means of destruction) are established. It is accepted in the current practice to distinguish three classes of tasks addressed in the identification of special-purpose structures: identifying, classifying and diagnostic ones. Characteristics of each class of tasks are described.

Special emphasis is placed on using standards and technical normative documentation in the process of identifying special-purpose structures.

A general scheme of the identification study, which includes four main stages, namely, preparatory, analytical, comparative and summarizing (synthesizing) ones is proposed. Each stage is analyzed in detail.

An important role in identifying means of destruction belongs to the analysis of the chemical composition and structure of their fragments. The results of the comprehensive analysis allow for determining the material composition and quantitative characteristics of the compounds it consists of. These results are used to identify metal or alloy grades and determine the compositions of composite materials along with the technologies used in their production.

An interactive knowledge base for monitoring the materials of special-purpose structures in the framework of the transdisciplinary cluster of information support to the development of special-purpose structures has been compiled.

The principles of creating a transdisciplinary cluster for information support to the development of means of destruction (TCIS DMD) and monitoring of materials and technologies for special-purpose structures have been formulated, and corresponding artificial neural networks have been constructed.

Supplement presents a general review of high-tech means of destruction that are employed by the Russian Federation in the war against Ukraine, along with their technical characteristics.

The book could be interesting to a wide circle of professionals in mechanical engineering, materials science, as well as those of defense industry sector.

Kyiv, Ukraine

Viacheslav L. Bogdanov
Alexander Ya. Grigorenko
Ihor B. Chepkov
Ihor V. Odnoralov
Andrii V. Kuchynskyi
Valerii V. Kremenytskyi
Svitlana O. Sperkach

References

1. Bogdanov V, Grigorenko A, Chepkov I, Odnoralov I, Kremenytskyi V, Kuchinskyy A, Sperkach S (2023) Identification of the country of origin of unmanned aerial vehicles based on the analysis of the element base by scanning electron microscopy. Weapons and military equipment: scientific journal 1(37): 57–62
2. Bogdanov V, Grigorenko A, Chepkov I, Odnoralov I, Kremenytskyi V, Kuchinskyy A, Sperkach S (2024) Application of the method of scanning electron microscopy in determining the chemical composition of metal fragments of promptly-tactical missiles cruise KH-23 made in North Korea and aircraft missiles manufactured by the Russian Federation. Weapons and military equipment: scientific journal 4(44): 33–43

Contents

Chapter 1
Analysis of Available Methods of Investigating Metallic and Composite Materials, and the Choice of Optimal Research Methods

Analysis of Existing Methods of Research of Metallic and Composite Materials; Selection of Optimal Research Tools.

The experimental techniques of studying the physico-chemical characteristics of a solid materials can be divided into two large groups, depending on whether the sample examined is destroyed or not during the study.

The first group embraces the techniques which require that the investigated sample should be destroyed. They include atomic absorption spectroscopy analysis, atomic emission spectroscopy (including laser-induced spark breakdown spectrometry as one of the types of spectral analysis), mass spectrometry, and electron paramagnetic resonance (EPR) spectroscopy.

The other group involves non-destructive research techniques. Those of them that are most in demand and most widely used include optical microscopy, atomic force microscopy, X-ray diffraction analysis, X-ray fluorescence spectroscopy, scanning electron microscopy, and X-ray microanalysis.

1.1 Atomic Absorption Spectroscopy

Atomic absorption spectroscopy is the analytical technique based on selective light absorption by atoms of a substance transformed into atomic gaseous state. The emission from the light source, passing through the vaporized substance at the frequencies coinciding with the frequency of electron transition from the ground level to the closest one, is absorbed (the resonance line); by the degree of weakening of the spectral lines of the element studied, its concentration in the sample is determined (Price 1972; Haswell 1991).

The intensity of light absorption in the technique of atomic absorption is determined by the well-known Bouguer–Lambert–Beer law:

V. L. Bogdanov et al., *Identification of Special-Purpose Structures by Their Fragments Based on Scanning Electron Microscopy*, Advanced Structured Materials 233,
https://doi.org/10.1007/978-3-031-88827-4_1

$$D = \ln(I_0/I) = klC,$$

where D is the optical density (absorption), I_0 is the initial intensity of the excited light, I is intensity of the light that passed through the sample, k is the absorption coefficient, l is the absorption layer thickness, and C is the concentration of the element identified. The absorption coefficient k is proportional to resonance transition probability and does not depend on temperature.

To atomize the sample, the temperature of ~2000–3000 °C is necessary. In this temperature range, flame and electrothermal atomizers are used, as well as HF and UHF discharges; the flame atomizers operating on acetylene-nitrogen oxide are the most common. In the abovementioned temperature range, over 90% of atoms are not in the excited state, hence, other atoms and molecules cannot influence the absorption coefficient. This fact, along with a small number of absorption lines, results in a high selectivity of the technique. The light source emits a linear spectrum containing the required line of the element to be identified. Hollow cathode lamps, electrodeless discharge lamps, and tunable lasers are used as radiation sources. A considerable limitation of atomic absorption technique as compared to atomic emission spectroscopy is the impossibility of simultaneous detection of several elements in the sample, so their successive identification is needed. Yet, the technique of atomic absorption analysis is much simpler in comparison with other methods of atomic spectral analysis, and it permits the identification of up to 70 elements with the sensitivity of 10–5 mass percent of both low and high concentrations in the specimens. As of today, atomic absorption spectroscopy is considered to be one of the most selective, fast, effective, precise, and, at the same time, relatively inexpensive technique (Britzke and Savelyeva 1970; Slavin 1971).

A unit for atomic absorption analysis generally consists of

(1) source of monochromatic light;
(2) atomization source;
(3) monochromator; and
(4) light detector (photomultiplier).

The principal requirements concerning light sources in the atomic absorption analysis are high brightness and stability of resonance lines, as well as the ease and safety of their handling. There are hollow-cathode lamps and ball-type high-frequency electrodeless discharge lamps. Lately, frequency-tunable lasers are used as light sources, their advantage being narrow-line light generation (Khavezov and Tsalev 1983).

Along with frequently used flame atomizers, electrothermal atomizers are also employed effectively. They consist of miniature graphite tubes heated in an inert gas atmosphere by a powerful electric arc (PE pyro-coated graphite furnace with integrated Lvov platform) or by electric current passed through its walls (King and Massmann furnaces) to evaporate the samples fed into atomizers as solutions or powders.

Photometers and spectrophotometers are employed as monochromators. In photometers, the spectral line is isolated with a light filter. Photometers can detect few elements: potassium, sodium, lithium, calcium, and other alkali and alkaline-earth metals. Photometers have low resolution and allow the analysis of solutions simple in composition. In spectrophotometers, light is split with a prism or a diffraction grating in order to isolate the line of the element to be identified and separate it from the entire radiation of the atomizer.

Light detectors—photomultipliers—amplify the photocurrent generated by a light-receiving photocell.

The following factors cause deviations from the basic law of atomic absorption emission (Bouguer–Lambert–Beer law) that lead to measurement errors in quantitative analysis: unstable functioning of device components; non-monochromatic emission lines; self-absorption (re-absorption) at high concentrations of the element identified; ionization of metal atoms at high temperatures; anion effect (in the presence of some anions, compounds form that are hard to evaporate, hard to dissociate, and hard to atomize, which results in underestimating the element content in comparison with its actual amount); and cation effect (the effect of silicon and aluminum is the most common).

It should be emphasized that in current atomic absorption spectrometers many measuring processes are automated and all operations are checked and controlled by a personal computer with compatible software (Volynsky 2005). The following options are used: automatic regulation of the speed of flammable gas stream to optimize the combustion effectiveness in elemental analysis; automated adjustment of the burner height to optimize the conditions for analyses; a turret with an automatic choice of lamps; automated scanning of wavelengths and search for peaks; and automatic switching of spectral slit width.

1.2 Atomic Emission Spectroscopy

Emission spectral analysis relies on examining the emission spectrum of the studied sample atoms in the wavelength range of 150–800 nm and achieves the limit of determining the content of elements of 10–5% of their mass (Török et al. 1978; Thomas et al. 1984). The abovementioned emission is due to the processes taking place in the atoms excited by thermal or electrical impact, or due to absorption of electromagnetic waves. The processes of atomic excitations involve electrons of outer electron shells. If an atom does not lose its excess energy by colliding with other particles, it can become excited. The excited state of an atom is characterized with one or several of its electrons being on the orbits with higher energies. According to the quantum theory, a transition of an electron to a higher energy level happens when the excess energy received by the atom is equal to or higher than the energy difference of the levels between which the transition occurs. The minimal energy necessary for electron transition from the ground level to a higher energy level is referred to as excitation energy. An atom can only stay in an excited state for a limited time (10^{-7}–10^{-8}s). If during that time the atom does not give off its excess energy by collision, it emits a photon. With that, electron from a higher energy level goes to one of the lower energy levels or to the ground level. The emitted photon energy is equal to the energy difference of the levels between which the transition occurs. Since the wavelengths emitted by the atom depend on its structure and do not depend on the number of atoms, the spectrum qualitative parameter is the wavelengths in the specimen spectrum. The intensity of a certain wavelength line depends on the number of the atoms of the corresponding element in the sample and is used as the quantitative parameter of an analytical signal.

Each device for emission spectral analysis consists of three necessary parts: an excitation source (gas flame, electric arc discharge, gas discharge, HF or UHF discharge); a dispersing element (prism, diffraction gratings); and emission detector. The task of the excitation source is to transform a part of the sample into atomic vapor (atomization) and provide atoms with additional excitation energy due to which they emit a line spectrum. The dispersing element splits the emission generated in the excitation source by wavelengths. Thus, the equipment for atomic emission spectroscopy is similar to that employed in atomic absorption spectroscopy. However, the additional source of light (lamps, etc.) is absent in the case of atomic emission spectroscopy, since this technique detects the intensities of the emission lines of excited vapor. In the atomic absorption spectroscopy, the intensities of the emission lines of an additional light source are detected, which were partially absorbed in the vapors of the specimen.

A promising type of excitation sources in atomic emission spectroscopy is laser-induced spark spectrometry, in which spectra of laser sparkle breakdown (laser spark) spectra are used to analyze solid samples. Laser sparkle (plasma) is formed by focusing pulsed laser beam on the sample surface. At present, the methods of laser spark emission spectrometry are being advanced in connection with the development of multi-purpose emission analyzers that are capable of analyzing any types of samples (including microscopical ones, due to laser beam focusing) simultaneously for all elements, with obtaining high spatial resolution. However, it should be stressed that laser spark forms very hot torch plasma. And the torch plasma produced by absolutely different samples often has similar characteristics. The utilization of femtosecond laser pulses (their wavelengths being shorter than 1000 fs) partially improves the accuracy and reliability of the analysis results. It should be noted that the quantitative atomic emission spectroscopy requires the use of a complex of standards (reference specimens, materials for comparison) of known chemical compositions.

1.3 Mass Spectrometry

Mass spectrometer is a vacuum device that relies on the physical laws of charged particle motion in magnetic and electrical fields and is necessary for obtaining mass spectrum (Herman and Lemasters 1993). In mass spectrometer, the sample to be examined is ionized before the analysis. Mass spectrum is the dependence of ion current intensity on the mass-to-charge ratio. A significant difference between mass spectrometry and other analytical physico-chemical methods is that optical, X-ray, and some other techniques detect energy emission or absorption by molecules or atoms, whereas mass spectrometry directly detects substance particles themselves. The atoms of chemical elements have specific masses. Hence, the precise measuring of the analyzed molecule mass permits the determination of its elemental composition. Mass spectrometry also permits isotopic analysis to be carried out.

The scheme of mass spectrometer includes three principal elements—ion source, analyzer, and detector. When various methods of substance ionization in the ion source are used, the beams of either positively or negatively charged ions are generated, depending on the task assigned, and sometimes both are generated simultaneously. The mass spectrometry technique is based on the ability of chemical compounds to get ionized under the influence of various factors that lead to electron detachment. The ionizing factors can be accelerated electrons, photons, heating, etc. Besides, electrons moving in mass spectrometer are not to collide with gas molecules, since such collisions change the trajectories of ion motion and cause peaks broadening in mass spectrum. To prevent such a situation, the entire working volume of the mass spectrometer has to be pumped down to provide high vacuum (the content of foreign gases should not exceed 10^{-7} Pa).

Another important component of mass spectrometer is mass analyzer. Most often this is a tube of irregular cross section placed between magnet poles. The part of the tube located in the magnetic field is bent in such way that ion motion takes place as far as possible from the tube walls. One end of this tube is joined to the ion source, and the other to the registration device. The radius of the curvature of ions motion in magnetic and electron fields is determined by the formula:

$$R = \sqrt{\frac{2Em}{H^2 z}},$$

where H is the magnetic field intensity, E is the electric field intensity, and m and z are the ion mass and charge. This formula implies that for ions with arbitrary m/z ratio, such intensities of the electric and magnetic fields can be selected that those ions would be moving along the curve of a predetermined radius. At the end of their way, they are to get on the recording device. Each mass spectrometer has its own scan system, which makes it possible to consequentially direct ions with different m/z ratios in the given range to the recording device. Usually, the current in the analyzer magnet is varied for that purpose, which results in magnetic field changing within the necessary limits (Ishchenko et al. 2018).

When dealing with mass spectrometry techniques, secondary ion mass spectrometry (SIMS) should be considered (Feldman and Mayer 1986; Woodruff and Delchar 1994). The secondary ion mass spectrometry is a method of investigating elemental composition, which takes a special place in science. This method is characterized with a low sensitivity level (10–12 g), high depth resolution (of nanometer order), it is able to distinguish the isotopes of elements. In addition to that, this technique permits not only determining the elemental composition of the object under study and presenting distribution of elements in the sample volume, but it also reveals the concentration of their components. Today, SIMS is one of the most widely used methods of examining materials (Behrisch 1981). So, we will consider it in more detail.

In the process of atomizing, surface layers are damaged, so the measurement of the relative content of atomized products is a direct method of determining the composition of the damaged layer. Atomized particles are released as neutral atoms that are in various excited states—in the form of positive and negative ions singly or multiply charged, or as clusters. The ratio of the number of ions to that of neutral atoms can vary by several orders of magnitude in the same specimen, depending on the state of its surface. The analysis of atomization products is the most sensitive method of studying the surface. It is usually employed to measure small atom concentrations in solids. Atomization is most widely used with the purpose of further registration and analyzing the secondary ions—ionized products of atomization. Secondary ions first get into the filter, which is a common electrostatic analyzer, and then they are collected in the mass spectrometer. That gave the name to the method—mass spectrometry of secondary ions (SIMS). All SIMS devices permit surface and volumetric analyses of element concentrations. In one of the operation modes, the ion beam moves across the sample surface on which it produces a crater. To prevent the registration of ions from crater walls, the detection system is provided with the electron gate that passes ion signals from the central part of the crater. There are also direct display devices—ion microscopes in which secondary ions from certain specimen microareas are registered in such a way that, as a result, a visual image of surface composition is formed on the display. The spectra of both positive and negative secondary ions have complex structures since they are formed by not only singly and multiply charged ions, but by all other ionized clusters as well. The mass spectrum acquired due to ion bombardment is emitted not only by once but by twice or three times ionized atoms, along with clusters of two, three, and four atoms. However, singly ionized atoms prevail in the majority of cases.

1.4 EPR Spectroscopy

The techniques of studying solids that rely on magnetic resonance phenomena are widely used. On the basis of electron paramagnetic resonance (EPR), a powerful spectroscopic method of studying the properties of materials has been developed. This method is based on studying the absorption of electromagnetic waves in the microwave range (from 1 to 1000 GHz) resulting from electron transitions between energy levels, which are caused by the interaction of electron spins with the external magnetic field. The EPR phenomenon emerges if the substance contains unpaired electrons whose carriers can be ions of transition elements or radicals of chemical compounds. EPR can also be observed when atoms or molecules of a substance have paired electrons only. However, in this case, they should be in a high-spin state. EPR is an extremely sensitive method (10–16 spin/T, magnetic field resolution is 0.006 mT). By analyzing EPR spectra, it is possible to identify paramagnetic particles in any state of matter and determine their concentration (Russel 1992; Abragam and Bleaney 2012).

Degeneracy of electron spin states increases if an unpaired electron is placed in a magnetic field, which creates two spin states $m_s = \pm 1/2$, where $m_s = -1/2$ is the lower energy state that aligns with the magnetic field. Electron spin state can be reversed when electromagnetic radiation is used. In the case of electron spin transitions, this corresponds to radiation in the microwave range.

The energy difference between two states is given by the equation:

$$\Delta E = E_+ - E_- = h\nu = g\beta B$$

where h is the Plank constant, ν is the radiation frequency, β is the Bohr magneton, B is the strength of magnetic flux density, and g is g-factor (dimensionless quantity of internal magnetic moment of electron, its value for free electron is 2.0023). The value of g can be determined by the values of magnetic field and spectrometer frequency.

EPR spectrometers function by generating microwaves from a source (e.g., from a klystron), sending them through an attenuator onto the specimen which is placed in the microwave cavity. Microwaves reflected back from the cavity are directed to a detector diode, and the signal goes out in the form of reduced current on the detector, similar to microwave absorption by the specimen. Specimens for EPR can be gases, single crystals, solutions, powders, and frozen solutions. For solutions, solvents with high dielectric constants are not advisable, since they would absorb microwaves. For frozen solutions, solvents that vitrify during freezing are the best. Good glass is formed from low-symmetry solvents and the solvents that do not interact with hydrogen (Banwell 1983).

For the ease of interpretation, EPR spectra are usually presented as the first derivative of absorption spectra.

EPR experiments are carried out either by varying the frequency of microwave radiation incident on the specimen, with magnetic field remaining constant, or by varying the field at a constant frequency. In practice, as a rule, the latter procedure is employed. The frequency and the magnetic field intensity are to be determined with high precision. A reference sample with the well-known g-factor is to be used.

As electromagnetic radiation sources, today's EPR spectrometers make use of microwave generators that have a number of set frequencies and provide good monochromaticity, frequency stability, and power. Magnetic field with the induction up to 16 T is generated by a superconducting solenoid whose geometric dimensions ensure high magnetic field uniformity at the site of sample location. To achieve the superconducting state, the solenoid is cooled with liquid helium. The rate of magnetic field change is 0.7 T/min in the field range 0–7 T and 0.5 T/min in the range 7–16 T. The chamber with the specimen is located at the center of the solenoid. As the detector for registering UHF emission, an InSb bolometer could be employed. Detectors of this type have high sensitivity and quick response (~1 μs). To operate properly, the detector must be cooled with liquid helium (T = 4.2 K). The spectrometer permits conducting experiments in the temperature range 1.5–300 K. Recording of the absorption/transmission spectrum takes place at the set frequency of UHF signal with magnetic field sweep. To increase sensitivity, modulation measurement technique is employed. Two types of signal modulation are used. Magnetic field modulation results in a better signal/noise ratio for narrower lines, radiation power modulation with using a breaker (optical modulation) allows direct measurement of absorption or transmission and provides a better sensitivity for wider resonance lines. A synchronous phase detector is used for signal detection. The specimen is placed inside a waveguide. The chamber has the shape facilitating radiation focusing on the specimen.

Lately, free-electron lasers are quite often used for EPR spectroscopy in strong magnetic fields as sources of electromagnetic radiation. Such fields induce spin polarization and increase both sensitivity and resolution of the spectrometer. Besides, in spin-echo experiments, the intense pulse of terahertz radiation generated by a free-electron laser can cause a 90° spin rotation in several nanoseconds, which is faster by an order of magnitude than in the case of using low-frequency solid-state sources. That permits the detection of weak signals from specimens with low concentrations of paramagnetic centers, which occur quite frequently in examining complex-composition solid materials (Böcker 1997).

The following advantages of EPR spectroscopy can be emphasized: high sensitivity (small concentrations of paramagnetic substances are sufficient to obtain EPR signal); acquiring detailed and very precise information about the properties of the paramagnetic ion and its nearest environment; and spin Hamiltonian parameters can be determined with high precision. The technique provides an estimate of the spin–spin interaction between paramagnetic ions and the evaluation of spin–lattice relaxation time.

If EPR spectroscopy and NMR spectroscopy are compared, it should be noted that electron paramagnetic resonance spectroscopy is not employed so widely as nuclear magnetic resonance spectroscopy. The reason for this is that EPR spectroscopy requires unpaired electrons, while electron pairing is usually energetically beneficial. Hence, only a small part of pure substances generates EPR signals, while NMR spectroscopy can be used for nearly any compound. On the other hand, since electron pairing is the basis of chemical bonds, unpaired electrons are related to reactivity. Correspondingly, EPR spectroscopy is a very important technique for understanding radical reactions, electron transport processes, and transition metal catalysis, which are associated with "unpaired electron reactivity". Some materials with unpaired electrons are chemically stable and can be used as spin probes for studying systems where NMR spectroscopy works within resolution limits or cannot provide information sufficient for complete characteristic of structure and dynamics. It should also be emphasized that EPR requires the magnetic field intensity which is 660 times as low as that necessary for NMR with the same operating frequency, since electron mass is significantly less than proton mass.

All the methods of studying solid materials (including metals) that were considered above are destructive (i.e., they require specimen sampling for further atomization, ionization, etc.); therefore, they do not provide information on the structure of the specimen surface, which in many cases can prevent their use.

Another class of research methods are non-destructive techniques of studying materials. Those include optical microscopy, scanning electron microscopy, X-ray microanalysis, atomic force microscopy, fluorescence microscopy, and X-ray diffraction analysis.

1.5 Optical Microscopy

It is commonly known that microscopes are the instruments which produce enlarged images of small objects in the way that those could be seen as convenient for observation and analysis. Such devices include optical microscope, which consists of lens and eyepiece. It was first described in the sixteenth century and was constantly improved for increasing its magnification factor and improving image quality. The image in optical microscope is formed due to focusing, diffraction, and reflection of visible-range electromagnetic waves in the instrument.

In its classical (simplest) design, optical microscope consists of two parts: the lens and the eyepiece. The object to be studied is located directly behind the front focus of the lens. In this case, the image will be real, inversed, enlarged, and will be situated behind the double focus of the lens. For comfortable and convenient work with a microscope, it is desirable for the eyepiece to be situated in such a way that the plane of image is located in the focal plane of the eyepiece or between its front focus and the eyepiece itself, but in close proximity to the focus.

The most important parameter of optical microscope (like all types of microscopes) is its resolution (Böcker 1997). Resolution is determined by the resolution limit Z_{min}—the minimum distance between two light dots of an object that are seen separately in optical microscope. The formula for the resolution limit is as follows:

$$Z_{min} = \lambda/2n \sin\theta,$$

where $n \sin\theta$ is the numerical aperture, n is refractive index (for the liquid located between the object and the lens in the case of immersion lens), θ is the aperture angle (i.e., the angle between the extreme rays of the light beam that get onto the lens), and λ is the light wavelength. For the maximum numerical aperture with $\theta_{max} = 70$, $n = 1.5$, $\lambda = 0.555$ µm (the human eye is the most sensitive to this wavelength), for the immersion microscope we obtain $Z_{min} = 0.2$µm. It should be emphasized that it is not reasonable to reduce the value of Z_{min} indefinitely, since the details of the object are to be resolved by the eye as well. Therefore, the real magnification of an optical microscope with the classical (the simplest) design is determined by the relation $\Gamma_{val} = Z'/Z_{min}$, where Z' is the eye resolution limit, i.e., the size of the retina image of the object that has the size coinciding with the microscope resolution limit at 25 cm—the best eyesight distance. Since $Z' \approx 280\mu m$, for the useful magnification of the simplest optical microscope, we obtain $\Gamma_{val} \approx 1500$.

Optical microscopes differ from each other by their purpose: biological (intended to study transparent and semi-transparent objects); stereoscopic (reflected-light microscopes providing 3d image, which are intended for examining opaque objects); and metallographic (reflected-light microscopes, in which illumination comes from a built-in illuminator, intended for studying opaque objects).

Consider in more detail metallographic microscopes, since this type of instruments is commonly used to study metals and alloys. By means of metallographic microscope, one can carry our qualitative and quantitative phase analysis; determine one of the most important structure parameters that influences mechanical properties—the size of crystal grains in metals and alloys; reveal the presence of cracks, pores, non-metallic inclusions, and other structural defects; and detect changes in metal structure due to thermal, chemical-thermal and thermomechanical treatment, casting, forming, treatment, welding, cutting, etc. (Cherepin and Vasiliev 1982).

Depending on the sphere of application of metals and alloys, various materials properties are prioritized. For example, for structural materials, of primary importance are mechanical properties: characteristics of materials' behavior under the action of external loads—strength, plasticity, hardness, impact toughness, etc.; for instrumental materials, the most important properties are wear resistance and heat resistance; and electrical materials are characterized with certain physical properties—electrical, magnetic, etc. Many of these properties depend on the structure of materials, they are referred to as structure-sensitive. The structure of a material is considered as its internal organization, which is studied with special instruments, including optical metallographic microscopes. Revealing the dependence between structural elements (or a single taken structural feature) and the indicators of material's characteristics is one of the major tasks of materials science and an important stage of material's quality control.

It should be noted that the preparation of specimens for metallographic microscopy is a labor-intensive process (Brandon and Kaplan 2008), as very strict requirements to the quality of sample surface are imposed. The specimen for metallographic examination of metal structure is known as a polished section (slice). Since metal microstructure is examined in reflected light, the section surface is to meet a number of requirements: it is to have the minimal deformed layer; be flat (without "potholes"), which allows studies to be carried out at high magnification; have maximum reflectivity; be free of scratches and dimples; and different surface areas are to reflect light differently, which makes it possible to distinguish surface elements. The latter requirement is met either due to varied reflection of light (polarized or non-polarized, mono- or poly-chromatic) from the flat surface, or due to producing a relief on the surface by etching or by producing films whose thickness and, hence, color depends on the nature or orientation of the crystallites on which they formed.

Digital microscope takes a special place among optical microscopes. Professional digital microscopes have made a major breakthrough in today's instrumentation. In these devices, the optical system is coupled with a transforming matrix, which makes it possible to transform the light flow into a digital signal and transfer it to a computer for further processing. Connection of photo and video devices is provided for data recording, as well as monitors and printers for their visualization.

Today, due to a very wide use of personal computers, digital microscopes have become unified, most of them operate when connected to a standard USB port of a computer, their installed software compatible with a certain model of digital microscope. Maximum magnification in digital microscopes (nearly 5000 times) also is much better than that in optical microscopes, which opens up new prospects for their users. Digital technology in state-of-the-art microscopes often permits saving video streams. To obtain the sharpest and high-quality images, advanced methods of camera focusing and automatic adjustment are employed. In some models, the issue of additional illumination is addressed by highlighting with in-built light diodes.

The digital technology of image saving and processing facilitates data documentation in the catalogs and comparison of the objects observed. Users can save the necessary images in respective catalogs and later compare the objects examined or the state of the same object in different time periods. That way, quality control and other tests can be carried out. The saved images can be processed by means of external editors.

1.6 Atomic Force Microscopy

Atomic force microscope (AFM) is a high-tech scientific instrument enabling researchers to obtain the images of specimen surfaces with the resolution of the order of several nanometers and manipulate nanoscopic objects, e.g., individual molecules.

Before describing the principle of atomic force microscope functioning, we will consider its predecessor—scanning tunneling microscope (Binnig et al. 1982; Binnig and Rohrer 1987).

Scanning tunneling microscopy (STM) is a method of studying the detailed structure of an electrically conductive surface with atomic resolution. It relies on using the tunneling effect in the following way: to the tip of a thin (molecule-sized) needle placed above a surface a certain (very low) voltage is applied that generates a small quantum mechanical tunneling current to overcome the energy barrier between the needle and the surface. By using the value of this current, a topographic map of the surface is produced. An increase in voltage can lead to the displacement of surface atoms and/or induce a chemical reaction.

Based on the principle of STM functioning, it is clear that the study of electrically non-conductive substances becomes rather problematic if a scanning tunneling microscope is employed, since the issue of charge carrier transport through the studied specimen remains unresolved. Numerous attempts were made to implement STM experiments in systems with thin dielectric films or even with spatial dielectric specimens on condition that charge carrier transport through the studied specimens is of ballistic nature. However, even under such conditions the possibility of an experiment with STM proved to be quite limited, while a routine study of arbitrary non-conductive specimens was problematic.

That motivated researchers to develop another instrument that would use the principles of scanning and probe locality, which are characteristic of STM, and, at the same time, would have no limitations regarding conductivity of the examined specimens. Atomic force microscope (AFM) was that kind of device—almost a twin of scanning tunneling microscope. The only principal difference between them is that AFM measures not the current between the probe and the specimen but the force of the interaction between them. Since this force is measured very locally, it, in fact, is the force of interaction between specimen's individual atoms and the needle tip—that gave the name to the instrument—atomic force microscope. In comparison with STM, AFM permits routine studies of non-conductive surfaces, which, in turn, opened the way to employing this technique in most varied areas, such as materials studies, chemistry, biology, nanotechnologies, and many others.

Owing to the ease of its use, AFM is widely employed in both basic and applied science spheres, in particular, for conducting research and testing specimens directly at production facilities, while STM remains the device intended mainly for basic research. Lately, a highly important trend of employing STM and AFM not only for examining but also for modifying surfaces at the atomic level has been witnessed. There has been a noticeable transition from STM and AFM as nanodiagnostic instruments to the tools of nanotechnologies. That gives rise to the hopes of creating fundamentally new substances—man-made atomic or molecular structures that do not exist in nature. As of today, there are successful experiments of forming small clusters of such artificial substances by manipulating the location of individual atoms or molecules of a "common" substance. Such artificial formations have a number of unique properties that would make them very attractive for employing in nanoelectronics—to manufacture electronic circuits with characteristic element sizes commensurate with the sizes of individual atoms or molecules (Goryachko et al. 2012).

As mentioned above, the principle of atomic force microscope functioning is based on the registration of force interaction between the surface of the specimen examined and the probe. Used as probes are nanosized spikes situated at the edge of an elastic cantilever. The force acting on the probe from the surface causes console to bend. The appearance of protrusions and depressions under the spike leads to a change in the force acting on the probe, hence, to a change in the size of cantilever bend. Therefore, by recording the size of the bend, a conclusion concerning the surface relief can be made.

Moving above an uneven surface, the probe goes up and down, and these very small vertical displacements are detected by the laser beam incident on the upper surface of the cantilever beam with a mirror attached. Although the vertical displacements of the mirror are very small, the beam reflected from it deviates at an angle sufficient to be measured with a matrix photodetector. The signal obtained is analyzed by electronics and is transformed into a surface image.

By forces acting between the probe and the specimen, we, first of all, mean long-range van der Waals forces, which at first are attraction forces and upon further approach change into repulsive ones. In AFM, the maximum height difference during surface examination is several microns, while the maximum scan field is about $150 \times 150 \mu m$ microns at best. To provide a constant force between the surface and the probe, as well as prevent damaging, a feedback mechanism is employed (Goryachko et al. 2012).

Thus, one of the major AFM applications is the field of material science. Researches can use AFM for studying surface topography and mechanical properties of materials at the nanoscale. This information is vital for designing and optimizing materials with certain properties, such as strength, conductivity, or elasticity. AFM played an important role in developing advanced materials, including graphene, carbon nanotubes, and various nanostructured materials (Yaminsky 1997).

1.7 X-Ray Diffraction Analysis

X-ray diffraction analysis is the method of studying substance structure, which is based on the phenomenon of X-ray diffraction on 3d crystal lattices.

The method permits determining the atomic structure of a substance, including the space group of the unit cell, its dimensions and shape, as well as crystal symmetry group. Employing it, one can examine metals and their alloys, minerals, inorganic and organic compounds, polymers, amorphous materials, liquids and gases, molecules of proteins and nucleic acids, as well as other substances (Rusakov 1977). The easiest and the most successful is the application of this method in establishing the atomic structure of crystalline solids which already have strict structure periodicity and, in fact, are diffraction gratings for X-rays that are produced by nature. For the rest of substances, a crystal should be formed, which is an important and difficult aspect of X-ray diffraction method (Mudrii et al. 2017).

X-ray diffraction in crystals happens due to the scattering of X-rays by crystal atoms. In particular, X-ray scattering is the result of their reflection from atomic planes. These planes can be drawn through atom centers, which are considered to be fixed. Crystal is regarded as a group of parallel atomic planes equidistant from each other at the distance d, the number of planes is sufficiently large, and X-ray refraction does not occur.

If a beam of parallel monochromatic rays is incident at the angle θ with respect of atomic plane, the rays of the parallel beam are reflected from the planes at the same angle θ. The path difference of the rays reflected by the same plain is equal to zero. At the same time, the rays reflected from neighboring planes are able to interfere, i.e., they can weaken or enhance each other depending on the ray path difference.

If the absorption of X-rays is neglected, the rays reflected from atomic planes in one direction have the same amplitudes. Besides, the phase difference of the waves reflected from neighboring planes remains constant. Taking into account the above, interference maximum of the reflected waves is observed on the following condition:

$$2d \sin \theta = n\lambda$$

which is referred to as Bragg's law, where λ is the wavelength of X-ray radiation and n is a positive integer number.

As shown by numerous experiments, Bragg's law is valid and highly accurate, though it was obtained with reliance on many assumptions. Deviations from it are observed in high-precision experiments, which is caused by X-ray refraction. On the basis of experimentally determined angles of diffraction maximums, the use of Bragg's law allows inter-plane distances to be determined (by the known radiation wavelength): $d = \lambda/2 \sin \theta$.

To generate X-ray radiation, diffractometers use special appliances—X-ray tubes. X-ray tube is an electrovacuum device (diode) containing a cathode part for producing and focusing an electron beam and a water-cooled anode, which is the source of X-ray radiation. In most diffractometers, the role of scattered radiation detector is played by a scintillation counter that couples a luminescent crystal and a photoelectron multiplier.

Diffractometer makes it possible to carry out qualitative and quantitative phase analysis (Rusakov 1977). The phase is commonly understood as the uniform part of a thermodynamic system, whose physical properties are the same at all sites. Both one and several phases can be in the state of thermodynamic equilibrium. In crystals, different phases are characterized with distinct properties and crystal structure. By employing X-ray diffraction analysis, it is possible to identify phases in single- and multi-phase materials. Its advantages are high sensitivity and fastness. In addition, diffraction analysis technique is non-destructive. Alongside with determining quality characteristics, widely used in practice is quantitative analysis, by which it is possible to determine the volumetric content of the phases in a multi-phase specimen.

Establishing the phase composition of a substance is one of the most important tasks of physico-chemical analysis since physical properties of substances depend significantly on their phase compositions. X-ray phase analysis is widely used in studying phase equilibrium and phase transformations in composite materials. An important sphere of its practical application is non-destructive control of the technological processes of material treatment. It should be noted that X-ray phase analysis is not very sensitive to spatial inhomogeneities caused by fluctuations in the chemical compositions of phases. The presence of inhomogeneities practically does not affect the location and intensity of diffraction maximums and, therefore, does not reduce the credibility of phase analysis results (Mudrii et al. 2017). The purpose of quantitative phase analysis is to determine the percentage of specific phase content in the specimen examined. The accuracy of determining the phase volume fraction is about 1%. Quantitative analysis sensitivity depends on the nature, chemical composition, absorption coefficient, and the type of phase crystal lattice. When examining nanocrystalline materials with grain sizes smaller than 100 nanometers, the analysis sensitivity also depends on dispersity of the material, the heterogeneity of its chemical composition, and the presence of residual stresses. These factors cause broadening of diffraction peaks of various phases, their possible superimposition, and, hence, reduce quantitative analysis accuracy (Mudrii et al. 2017).

X-ray methods of examining materials have for a long time remained the principal ones in studying the internal structure of matter. Along with classical techniques, novel methods of X-ray diffraction analysis for in-depth studies of atomic and molecular structure, analysis of various types of defects, dynamics of atomic vibrations, etc. are employed today. Scientific and technological achievements have contributed to automating the processes of obtaining diffraction data and quantitative interpretation of measurement results. Solids and liquids in a broad range of temperatures are examined with methods of X-ray diffraction analysis. The effective operation of modern X-ray diffractometers has been significantly facilitated by upgraded software and the opportunity to use various structure models for interpreting diffraction spectra of multi-component systems.

1.8 Scanning Electron Microscopy

The method of scanning electron microscopy makes it possible to acquire specimen surface images with resolution of several nanometers and magnification of several hundred times. The images produced by scanning electron microscope look like 3d ones and are convenient for examining surface structure.

In scanning electron microscope, a focused electron beam runs across a rectangular area of a specimen, which results in the emission of secondary and backscattered reflected electrons from its surface. The signals of those beams are detected and directed onto the synchronized image scan on the monitor, generating surface image in various beam modes. The width of the scanning zone determines the image magnification. Besides secondary and backscattered scattered electrons, other signals from other detectors located in the microscope chamber are also analyzed (Goldstein et al. 2017).

Depending on the mechanism of signal registration, several operation modes are distinguished: secondary electron mode, backscattered electron mode, cathode luminescence mode, etc.

Secondary electron mode is most frequently used in examining surface structure (Engel and Klingele 1986). The physical mechanism of secondary electron image formation can be seen in the following way (Goldstein et al. 2017). Electrons with the energies from 1 to 50 keV (electrons with such energies are used in scanning electron microscopy) incident on the surface of a solid specimen behave in a rather complicated way. First of all, the high-energy electrons of the bombarding beam in a solid target are influenced by: elastic scattering (change in electron motion trajectories with an insignificant energy loss) and inelastic scattering (significant energy losses at insignificant changes of motion trajectories). Elastic scattering occurs mainly due to the interaction with substance atomic nuclei, and significant deviations of the trajectories from the initial ones occur. Inelastic scattering results from two mechanisms—inelastic interaction with atom nuclei and inelastic interaction with bound electrons in atoms.

If there are inelastic collisions between weakly bound outer atomic electrons and incident beam electrons, beam electrons lose energy, while weakly bound external electrons are emitted. These electrons have energies smaller than or equal to 50 eV and are called secondary electrons. It is secondary electrons that are used in scanning electron microscopy for examining the surface structure of the studied specimens. If secondary electrons emerge near the surface and their energy is larger than surface barrier energy (2–6 eV), there is high probability that secondary electrons will leave the surface. However, secondary electrons are absorbed heavily, and if they arise much deeper than 100 angstroms (10 nm) from the specimen surface, the probability of them leaving the specimen becomes extremely small.

The intensity of emitted secondary electrons is enhanced considerably with the increase in the angle between the incident beam and the examined specimen surface. This is accounted for by the situation that as soon as the specimen deviates at a certain angle from the normal beam location, the length of beam path part located within 10 nm from the surface increases, and a great majority of electrons leave the sample. When the image of the specimen area containing edges inclined at various angles in respect of the beam is formed in a scanning electron microscope, a larger number of electrons will leave surface areas with a larger tilt angle. That way, the topographic contrast of secondary electron images of the examined specimen surface areas is formed in electron scanning microscopy.

Another commonly used mode of scanning electron microscope functioning is the mode of producing backscattered electron images. In this case, contrast is obtained due to the fact that electron backscattering coefficient rises with an increase in atomic number. Thus, in a multi-phase object image formed by backscattered electrons, there are areas with more intense signal, which correspond to the largest atomic number, as well as the areas with weak signal that correspond to the smallest atomic number. Areas with intermediate atomic numbers produce intermediate-level signals. When signal is registered on the cathode-ray tube screen, signal amplification and other adjustments are chosen in such a way that the element with the highest atomic number should look white on the screen, the element with the lowest atomic number should look black, while those with intermediate numbers would be shown as various shades of grey (Goldstein et al. 2017).

It should be noted that the main parameter of scanning electron microscope is its spatial resolution, which is the criterion of how closely dots can be located on the specimen surface for them to be able to be distinguished on the image. Resolution is closely related to the diameter of the electron beam scanning the surface of the specimen examined and it cannot be smaller than the beam diameter (Hirsch et al. 1965).

Advanced electron optical lens systems employed in scanning electron microscopes permit an electron beam to be focused to the diameter of several nanometers, which, in turn, makes it possible to obtain corresponding spatial resolution of the order of a few nanometers.

Current electron scanning microscopes are equipped with attached energy-dispersive spectrometers that allow research into specimen chemical composition (Goldstein et al. 2017). Therefore, another effective method of studying solid materials is X-ray microanalysis. X-ray microanalysis relies on characteristic X-ray emission appearing in the specimen as a result of incident beam electrons interacting with electrons of the inner shells of atoms. Since atomic electrons are located at discrete energy levels, the emitted X-ray quanta will also have discrete energy values. For that reason, the wavelength of characteristic emission has the characteristic values determined for each atom (chemical element). The characteristic X-ray lines revealed during analysis prove that the element in question is present in the specimen. The intensity of characteristic emission is proportional to the chemical element concentration. Hence, by comparing the intensities of characteristic emissions of the specimen studied and the reference sample with a known chemical element concentration, the quantitative chemical composition of the examined object is obtained.

In the case when the electron incident on the specimen has sufficient energy, it can knock off an electron from inner K-shells, L-shells, or M-shell and send the atom into an excited or ionized state. The atom returns to its ground state as a result of electron transfer from an external shell onto a vacancy in an inner shell. When an atom returns to its ground state, it loses energy for X-ray quantum generation. In the process, electron can leave L-shell and occupy a vacancy in K-shell. Also possible is electron transfer from M-shell to K-shell or from M-shell to L-shell. Usually, all lines of K, L, or M series appear simultaneously. In the studies by means of X-ray microanalysis method, the most intense lines of characteristic emission are registered. It should be noted that when the chemical compositions of fragments of special-purpose products were examined in this work, K lines were always detected.

To measure the energy and intensity of characteristic X-ray emission, spectrometers with wave and energy dispersion are employed. Energy-dispersion X-ray spectrometers register all spectral wavelengths simultaneously. These spectrometers consist of a semiconductor detector, which transforms the energy of photons into electric pulses whose energy is proportional to photon energies.

That way, photon energy discrimination occurs. The count of pulses with corresponding voltages is carried out. The number of pulses with the same voltage is proportional to X-ray emission intensity for a certain wavelength. By comparing the intensities of characteristic emission of the examined specimen and the reference sample with a known chemical element concentration, we obtain the quantitative chemical composition of the studied object.

It should be mentioned that besides measuring the concentrations of chemical elements contained in the examined specimen, energy-dispersive and crystal-diffraction spectrometers permit producing an image of specimen surface in characteristic X-ray emission, i.e., producing the maps of chemical elements' distribution over the surface. In addition to that, it is possible to obtain dependences of concentrations of chemical elements along the scan line (the so-called profiles). These two modes of spectrometer functioning are extremely effective in studying composite materials that are most often multi-layered in the case of RF absorbent coatings.

Considering X-ray microanalysis methods, we will briefly describe the technique of X-ray fluorescence analysis. This technique in its registering aspect is analogous to crystal-diffraction X-ray microanalysis (crystallites in which Bragg's condition is met yield a peak of X-ray emission). However, characteristic X-ray spectra are excited with an electron beam in X-ray microanalysis, while in fluorescence analysis, those spectra are excited with X-ray quanta. Since the focusing of X-ray radiation is not an easy task, and electron beam diameter is 1 μm or less, the application of fluorescence analysis does not allow studying the chemical composition of microinclusions in materials, which is possible in X-ray microanalysis.

Besides the abovementioned optional equipment that can be employed in scanning electron microscopes, the attachment based on photoluminescence effect is also interesting.

When a solid is bombarded with an electron beam, besides electron scattering and X-ray emission, electromagnetic emission emerges in the visible, infrared, and ultraviolet regions of the spectrum. The interaction of high-energy primary electrons with some types of specimens leads to the formation of electron–hole pairs. When recombination of electron–hole pairs occurs, energy is released in the form of long-wave radiation. Usually, such emission is called cathode luminescence. Cathode luminescence signal can be registered with a photomultiplier and used for modulating cathode-ray tube brightness, and in that way an image of the emitting surface can be produced.

The difficulty of studying cathode luminescence excited by an electron beam in a scanning electron microscope is due to relatively low beam power and a small excited volume. Therefore, collecting light emission requires a very effective collecting system able to collect the possible maximum of luminescence emitted on the photomultiplier in a scanning electron microscope to observe a relatively small number of photons emitted per unit time. One of the techniques to provide a highly effective photon collecting is to place an ellipsoid mirror above the specimen for collecting all radiation emitted by it. The collected light is collimated and falls on a flexible light guide normally to its edge. Through this fiber optical device, the light arrives at the photomultiplier.

It should be noted that when the electron beam scans over the raster of 200 μm or more in size, the intensity at the edges of vision drops noticeably. To achieve sufficient intensity in any case, the beam current in a scanning electron microscope should be approximately 1–10 nm. That leads to an increase in probe size to as much as 0.5 μm, and thus the spatial resolution in the mode of cathode luminescence is of the same order.

It would be worthwhile to consider another piece of additional equipment—a device for diffraction of backscattered electrons. The technique of electron backscatter diffraction (EBSD) in scanning electron microscopes is actively employed in crystallography. Since the development of computer algorithms permitted complete automation of the process of diffraction pattern analysis, a new technique was developed for scanning electron microscopy, which is known as orientation imaging microscopy (Adams et al. 1993). This technique relies on computer algorithms for the automated analysis of diffraction patterns, which consists in producing successive diffraction patterns of backscattered electrons for each point of crystal lattice of the smooth surface in the specimen inclined at a certain angle. That way, data on crystallographic orientations and phases can be obtained for each point of the specimen surface, and the quality of the diffraction patterns can be assessed. On the basis of these data, microstructure of the examined surface is determined. The maps of crystallographic orientations obtained in that way make it possible to collect detailed data on specimen properties, e.g., on phase distribution, sizes, shapes, and defects in its grains, the nature and location of grain boundaries, crystal lattice orientations, orientation distribution (texture), etc.

This method is also used to study crystal structure, first of all, symmetry and other properties of crystal lattices. Special software allows detailed investigations of any features of crystal lattices (Wilkinson et al. 2006), as well as residual stresses (Winkelmann et al. 2007). Diffraction of backscattered electrons also makes it possible to work with relatively large specimens, so the preparation of thin foils (as in the case of electron transmission microscopy) is not necessary, since the resolution of backscattered electron diffraction technique is about 50 nm (Schwartz et al. 2009; Zaefferer and Habler 2016; Thomas et al. 2003). The diffraction patterns acquired during analyzing large-sized specimens have much in common with Kikuchi lines obtained in the analysis of thin foils by means of electron transmission microscopy. Every band (Kikuchi line) corresponds to certain lattice planes, so the Miller indices of the lattices can be determined by the relative angle and width of each band. The indexed bands, in turn, allow a detailed study of crystallographic orientations and phases. To produce diffraction patterns, a detector is used, which consists of a luminescent screen and a highly sensitive video camera located near the specimen. For automatic recording and analysis of diffraction bands, specialized software is utilized.

Before the analysis, thorough and careful specimen polishing is done for achieving an ideally smooth surface without deformations. Since specimen deformation and the related impact on the subsurface layers can significantly distort the microstructure of its surface, precision polishing of the specimen is necessary to ensure appropriate analysis results. To remove a thin layer of material, a vibration polishing machine is commonly used. Properly performed vibration polishing minimizes specimen deformation, so the specimen can be used for further analysis, whose results would be reliable and reproducible, as required by the metallography standards.

1.9 Choosing Optimal Tools for Studying Metallic and Composite Materials

The techniques of studying solid materials that were considered above and which belong to the destructive ones have a common limitation from the viewpoint of studying the structure and chemical composition of metallic and composite materials. These techniques require that a specimen of the material is to be taken for further atomization, ionization, etc., and due to this they do not provide information on the surface structure of the specimen in question and concentrations of chemical elements in it. This becomes critical in the cases when it is necessary to examine metal structures or composite materials subjected to considerable mechanical loads and heating to high temperatures, which led to changes in their structure and chemical composition. After all, it is impossible to study the physical processes occurring under such loadings. Besides, composite materials have inclusions in the form of conductive and non-conductive particles of various shapes (spheres, cylinders, fibers, etc.) and often contain a large number of layers with varied structures and chemical compositions.

Naturally, such materials cannot be investigated by means of destructive methods. Comparing X-ray analyzers and the devices based on atomic absorption spectral analysis and atomic emission spectroscopy, it should be emphasized that the latter have better sensitivity and smaller measurement errors, including those for light elements. However, as mentioned above, X-ray microanalysis is a non-destructive method and permits studying the chemical composition of microinclusions, acquiring X-ray images of the examined specimen areas, i.e., producing maps of chemical element distribution over the specimen surface. Besides, it allows measuring the concentration of chemical elements not only at points and over chosen rectangular areas but also along scan lines, which is of great importance in the studies of multi-layer composition materials.

There is another advantage of X-ray microanalysis as compared to destructive methods, which should be mentioned. This concerns studying specimens of unknown chemical compositions. By employing an energy-dispersive spectrometer, one can reveal practically complete qualitative chemical composition of the specimen examined in a matter of minutes, to be followed by the quantitative one. In contrast, when using, e.g., atomic absorption analysis, it is necessary to choose the light source for illuminating the specimen with radiation lines present in the absorption lines of the studied specimen. That is, in such studies, it is desirable to have preliminary information about its chemical composition.

As mentioned above, another group of research techniques are those of non-destructive studies of materials. They include optical microscopy, scanning electron microscopy, X-ray microanalysis, atomic force microscopy, fluorescence microscopy, and X-ray diffraction analysis. These research techniques are effective in terms of some parameters but have significant limitations.

Comparing optical microscopy and scanning electron microscopy for studying specimen surface structure, it should be mentioned at once that electron microscopes provide magnifications of up to 300,000 times, with spatial resolution of several nanometers, while an advanced digital optical microscope provides 5,000-time magnification. Besides, electron microscope, due to its large focal depth, shows surface image as a three-dimensional picture, which is essential for studying specimen surface details. With the help of an optical metallographic microscope, qualitative and quantitative phase analysis of metals and alloys can be performed, but with proper specimen preparation; the same investigations can be carried out with a scanning electron microscope as well.

Taking into account all abovementioned considerations, as well as other effective parameters and operation modes of scanning electron microscope, it can be concluded that electron microscope is a more effective instrument for investigating solid specimens as compared to optical microscope.

Atomic force microscopy also has considerable limitations. In AFM, the maximum height difference in studying surfaces is a few microns, while the maximum scan field is about 150 × 150 microns at best. Taking into account that surface differences of investigated specimens of metals and composite materials are much larger than a few microns (especially in the specimens that were under the impact of high mechanical and heat loads), AFM methods, despite their better spatial resolution than that of scanning electron microscope, are inapplicable for investigations of the specimens mentioned.

As to the X-ray diffraction analysis, as noted above, the most successful is its application for determining the atomic structure of crystalline materials which have strict structure periodicity and are natural diffraction gratings for X-rays. For the rest of substances, a crystalline material should be formed, which is an important and difficult aspect of X-ray diffraction method.

X-ray fluorescence technique, as well as X-ray microanalysis, is intended for qualitative and quantitative analysis of materials' chemical composition. In X-ray microanalysis, characteristic X-ray spectra are induced with an electron beam, while in fluorescence analysis, characteristic X-ray spectra are induced by X-ray quanta. Since focusing of X-ray radiation is very difficult, while electron beam diameter is 1 μm or less, the employment of fluorescence analysis does not permit investigations of the chemical compositions of microinclusions in materials.

In choosing research methods, the following considerations should also be taken into account. Current scanning electron microscopes and X-ray analyzers are parts of one instrument. The same electron beam of a microscope is used both for forming the image of the examined specimen surface and for inducing characteristic X-ray spectra for microanalysis. Hence, that presents an opportunity for practically concurrent investigation of specimen surface and performing microanalysis, choosing areas for performing microanalysis on the specimen surface, obtaining maps of chemical element distributions over the surface. That is, this device is a powerful multi-purpose scientific instrument. Thus, scanning electron microscope is the most appropriate tool for studying metallic and composite materials, including those intended for special purposes.

References

Abragam A, Bleaney B (2012) Electron paramagnetic resonance of transition ions. Oxford University Press

Adams BL, Wright SI, Kunze K (1993) Orientation imaging: the emergence of a new microscopy. Metall Trans A 24(4):819–831. https://doi.org/10.1007/BF02656503

Banwell CN (1983) Fundamentals of molecular spectroscopy. McGraw-Hill, London

Behrisch R (1981) Sputtering by particle bombardment I: physical sputtering of single-element solids. Topics in Applied Physics, 47. Springer, Heidelberg

Binnig G, Rohrer H (1987) Scanning tunneling microscopy—from birth to adolescence. Rev Mod Phys 59(3):615–625. https://doi.org/10.1103/RevModPhys.59.615

Binnig G, Rohrer H, Gerber C, Weibel E (1982) Surface studies by scanning tunneling microscopy. Phys Rev Lett 49(1):57–61. https://doi.org/10.1103/PhysRevLett.49.57

Böcker J (1997) Spektroskopie: Instrumentelle Analytik mit Atom-und Molekülspektrometrie. Vogel, Würzburg

Brandon D, Kaplan WD (2008) Microstructural characterization of materials. John Wiley & Sons, Chichester

Britzke ME, Savelyeva AN (1970) Atomic absorption analysis in metallurgy of non-ferrous and rare metals. Tsvetmetinformatsiya, Moscow [in Russian]

Cherepin VT, Vasiliev MA (1982) Methods and tools for analyzing the surface of materials: Directory: Naukova Dumka, Kyiv [in Russian]
Engel A, Klingele G (1986) Raster electron microscopy. Destruction: The handbook. Metallurgy, Moscow [in Russian]
Feldman LC, Mayer JW (1986) Fundamentals of surface and thin film analysis. North-Holland, New York
Goldstein JI, Newbury DE, Michael JR, Ritchie NWM, Scott JHJ, Joy DC (2017) Scanning electron microscopy and x-ray microanalysis. Springer, New York
Goryachko AM, Kulyk SP, Prokopenko OV (2012) Fundamentals of scanning probe microscopy and spectroscopy (part 2): textbook. Radiophysical Faculty of Technology, Shevchenko KNU, Kyiv [in Russian]
Haswell SJ (1991) Atomic absorption spectrometry: theory, design and applications. Elsevier, Amsterdam
Herman B, Lemasters JJ (1993) Optical microscopy: emerging methods and applications. Academic Press, San Diego
Hirsch PB, Howrie A, Nicholson RB, Pashley DW, Whelan MJ (1965) Electron microscopy of thin crystals. Butterworths, London
Ishchenko OV, Gaidai SV, Beda OA (2018) Mass spectrometry. SPC Kyiv University, Kyiv
Khavezov I, Tsalev D (1983) Atomic absorption analysis. Chemistry, Leningrad [in Russian]
Mudrii SI, Kulyk YO, Yakimovich AS (2017) X-ray structural analysis in materials science I. Franka LNU, Lviv [in Ukrainian]
Price WJ (1972) Analytical atomic absorption spectroscopy. Heyden and Sons, London
Rusakov AA (1977) X-ray of metals. Atomizdat, Moscow [in Russian]
Russel SD (1992) Physical methods for chemist. Surfside Scientific Publishers, Gainesville
Schwartz AJ, Kumar M, Adams BL, Field DP (2009) Electron backscatter diffraction in materials science. Springer, New York
Slavin U (1971) Atomic absorption spectroscopy. Chemistry, Leningrad [in Russian]
Thomas JE, Thompson M, Walsh JN (1984) A handbook of inductively coupled plasma spectrometry. Glasgow and London (Blackie)
Thomas I, Zaefferer S, Friedel F, Raabe D (2003) High resolution EBSD investigation of deformed and partially recrystallised IF steel. Adv Eng Mater 5(8):566–570. https://doi.org/10.1002/adem.200300373
Török T, Mika J, Gegus E (1978) Emission spectrochemical analysis. CRC Press
Volynsky AB (2005) Modern atomic absorption spectrometry: achievements and future prospects. Ukr Chem J 71(9):25–31
Wilkinson AJ, Meaden G, Dingley DJ (2006) High-resolution elastic strain measurement from electron backscatter diffraction patterns: new levels of sensitivity. Ultramicroscopy 106:307–313
Winkelmann A, Trager-Cowan C, Sweeney F, Day AP, Parbrook P (2007) Many-beam dynamical simulation of electron backscatter diffraction patterns. Ultramicroscopy 107:414–421
Woodruff DP, Delchar TA (1994) Modern techniques of Surface science. Cambridge Solid State Science Series. Cambridge University Press
Yaminsky IV (1997) Scanning probe microscopy of biopolymers. Nauchnyi mir, Moscow
Zaefferer S, Habler G (2016) Scanning electron microscopy and electron backscatter diffraction. In: Microstructures, textures, chemical and isotopic signatures. EMU notes in mineralogy. Mineral Reaction Kinetics, pp 7–95

Chapter 2
Using Scanning Electron Microscopy for Studying Metal Fragments of Special-Purpose Structures

Despite the extensive use of non-metallic and, first of all, composite materials in manufacturing special-purpose structures, the elements made of metals and alloys still remain among the most important components of any present-day technical structure. This statement is also true for special-purpose products, means of destruction, in particular.

2.1 Analysis of Metals and Alloys Currently Used in Manufacturing Special-Purpose Structures

The right choice of materials for structural elements can considerably improve weight characteristics and tactical flight performance of special-purpose products, which include cruise, ballistic, and air-launched ballistic missiles, unmanned aerial vehicles and loitering munitions, as well as reduce the cost of their manufacturing and exploitation. However, in choosing materials for them most attention is given to providing the necessary strength and stiffness at the smallest mass, and to ensuring the weight efficiency of the material.

According to their purpose, current metals and alloys used in manufacturing special-purpose structures are subdivided into structural materials, whose determinant characteristics are mechanical properties, and materials for non-structural purposes, whose major parameters are specific physical and chemical properties. In terms of exploitation conditions, they are subdivided into high temperature, those for work at low temperatures, heat-, wear-, corrosion-, fuel-, oil-, fire-resistant materials.

Metals and alloys are an important part of today's aviation and rocket industries. Yet, the main "driver of progress" has been designers' desire to reduce the mass of flying vehicles. It is the reduction of this parameter that provides the necessary level of safety, flying vehicle efficiency, its optimal fuel consumption, and flight range.

V. L. Bogdanov et al., *Identification of Special-Purpose Structures by Their Fragments Based on Scanning Electron Microscopy*, Advanced Structured Materials 233,
https://doi.org/10.1007/978-3-031-88827-4_2

The percentages of materials used in special-purpose structures can be demonstrated with the example of today's passenger planes: composite materials make 52% of Airbus A350 weight, aluminum—20%, titanium—14%, steel—7%, and other—7%. Similar ratios can be seen in Boeing 787 Dreamliner wide-body jet: 50%—composite materials, 20%—aluminum, 15%—titanium, 10%—steel, and 5%—other (Taranenko 2014; Molyar et al. 1995).

Steel. Designers of special-purpose products highly value steel for its strength, hardness, resistance to high temperatures, and low cost. The properties mentioned make this alloy an ideal material for manufacturing landing gears, aircraft skin and body elements, hinges, cables and fastenings, along with other parts.

Incessant upgrading of aircraft designs has led to continuous enhancement of strength and specific strength (stress-to-density ratio) characteristics of steels. However, the difficulty of the problem is not so much in achieving those characteristics as in ensuring the performance of aviation structures produced of respective materials under the action of asymmetric cyclic loads. This results from the fact that increased strength of steels leads to the reduction in their ductility, toughness, crack resistance, etc. On the account of that, scientists developing new steels lead a continuous search for compromises between increasing strength and ensuring reliability.

At the current stage of engineering and technology development, the most frequently used steels in manufacturing products for special purposes are corrosion-resistant and heat-resistant steels and alloys. Chemically speaking, there are two types of corrosion:

- chemical corrosion, which occurs under the action of gases and non-electrolytes on a metal (gas corrosion);
- electrochemical corrosion, occurring due to the action of electrolytes.

Gas corrosion causes damage to turbine blades, exhaust nozzles, and aircraft skin materials around the gas turbine nozzle.

In rocketry and aviation, corrosion-resistant alloyed steels are used wherever possible, but a number of highly stressed structural parts, such as landing gears and flap tracks, are made of high-strength low-alloy steels (Taranenko 2014; Ostapchuk et al. 2015).

The steels resistant to gas corrosion at temperatures above 550 °C are referred to as scale-resistant or high-temperature (oxidation-resistant) steels.

The increase in corrosion resistance of steels is achieved by alloying them with elements that form protective films on the steel surface.

The principal alloying element of corrosion-resistant (stainless) steels is chromium. It has been found that a stainless steel is to consist at least 12.5% chromium, otherwise the protective effect is not observed. Chromium produces a protective Cr_2O_3 oxide film on the steel surface. Nickel is added into high-chromium steels to improve their anti-corrosion properties.

In terms of chemical composition, corrosion-resistant steels are subdivided into chromium and chrome-nickel steels. Chromium steels contain at least 13% chromium. Chromium steels are used for manufacturing parts of fuel equipment, aviation instruments, parts of aviation engines, fire partitions, and weaponry parts.

Increasing steel heat resistance (oxidation resistance) consists in its alloying with chromium, aluminum, and silicon. Oxide films of Cr_2O_3, SiO_2, and Al_2O_3 are formed on steel surface. They protect the metal from oxidation. By introducing 5…8% chromium into steel, its thermal resistance is raised to 700–750 °C. Adding 5.8% chromium increases steel heat stability up to 1100 °C. Heat resistance of steel can be raised to 1300 °C by its alloying with up to 25% aluminum and silicon.

Austenitic steels are not strengthened by heat treatment. They weld well, and have sufficient ductility in the hardened state. Heat-resistant (oxidation-resistant) steels are employed in manufacturing combustion chambers, flame tubes, afterburners, gas turbine blades, compressor disks, and cast rotors. These steels are supplied as sheets and castings. Standard steels, such as 20KH23N18, 12KH17, 15KH25T, and 36KH18N25S2, are used (Vambol et al. 2011).

In operation, the parts and units of current aviation engines and aircraft are heated up to 700...1000 °C, so high-temperature (heat-resistant) steels are widely used in their manufacturing. Heat resistance is the ability of a material exposed to high temperatures to resist the beginning and development of plastic deformation and fracture under the action of long-term mechanical loads. Metal strength depends on the loading time and operation temperature. The higher the temperature and longer the loading, the sooner plastic deformation develops in the material, and the earlier structure failure occurs.

Since the emergence of jet aircraft, nickel has become one of the principal materials for producing heat-resistant alloys.

Industrial nickel alloys are subdivided into high-temperature (oxidation-resistant), heat-resistant, corrosion-resistant ones, and steels with special properties. Nickel alloys are labeled either with special letters with a serial number (e.g., ZI435) or like steels—KH20N77TU.

To produce heat-resistant alloys, a metal with high melting point is usually chosen. Nickel forms a protective NiO film on its surface. Nickel-chromium system is the basis of industrial oxidation-resistant nickel steels and alloys. They are often mentioned as nichromes. Chromium content in the nickel alloy is 20 …40%. Oxidation-resistant nichromes KH10N90, KH20N80, KH30N70, and KH40N60 are used at temperatures up to 1200 °C (Taranenko 2014; Ostapchuk and Semishov 2011).

High-temperature nickel-based alloys are subdivided into wrought, cast, and dispersion-hardened (or age-hardened) ones. High-temperature wrought nickel-based alloys are referred to as nimonic alloys. They are alloyed with chromium, titanium, and aluminum. The most common makes of such alloys are ZI437 (20% chromium, 2.5% titanium, 0.75% aluminum), ZI617, and ZP742.

To manufacture hollow turbine blades and turbine disks, cast high-temperature alloys are used. The most widely used cast nickel-based alloys are ZHS, ZHS3, ZHS6, ZHS6K, ZHS6KP, ZHS6U, and ZHS6F. The principal alloying elements of those are molybdenum, chromium, tungsten, aluminum, titanium, niobium, vanadium, and cobalt. Operating temperatures of these alloys reach 900–1050 °C. Nickel-based alloys are widely used as castings and are often aluminized.

The dispersion-hardened (age-hardened) VDU-1, VDU-2, VDU-3 alloys can be produced by powder metallurgy methods. They consist of a metal matrix and dispersed oxides. Operation temperature for these alloys is 1100–1200 °C.

It should be noted that heat-resistant alloys can also include titanium heat-resistant alloys and alloys based on refractory metals. Those alloys are considered below.

In the recent two decades, new steels make for rocket and aviation industries have been developed and introduced to production in the Russian Federation. They allow the manufacture of various parts and structural elements due to their high tensile strength, workability, and reasonable price (Klimov and Kozlov 2017).

Aviation steels VKS-9 (35KHS2N3M1FA) having tensile strength about 1950 MPa and VKS-12 (38KHS3N4K2MFA) with tensile strength up to 2300 MPa have high reliability, high-quality weldability, and are used for producing power setting and other parts in the aircraft developed by Sukhoi, Antonov, Mikoyan, and Kamov design offices. Besides, these steels are substitutes to titanium-based alloys in "disposable" rocketry products, which permits a significant cost reduction for products of the same specific strength (Klimov and Kozlov 2017).

The development of fighter jets and supersonic missiles, whose speed is 2.5–3 times higher than the speed of sound, puts forward new requirements to materials. The materials employed in them are to operate above the thermal barrier—temperatures of 280–300 °C, which rules out the use of aluminum alloys. The principal material in all-welded plane compartments of MiG supersonic jets is corrosion-resistant VNS-2 (08KH15N5D2T) steel with tensile strength 1250–1400 MPa. It is supplied as sheets and ribbons to produce aircraft skin and interior set, as well as power parts—in the form of rods, forgings, etc.

Besides, VNS-5 (13KH15N4AMZ) steel, with tensile strength 1380–1600 MPa, has found a very wide application. Power parts of MiG and Su airframes are made of it.

Another "interesting" representative of this class of alloys is SN-2A (07KH16N6) steel, with tensile strength 1100–1300 MPa. It is used as a material for power parts, including fastening ones, as well as for air and oxygen balloons, with which all aircraft types, including those in naval aviation, are equipped. The most important feature of these balloons is that they do not break into pieces when hit by a bullet.

A special place in aircraft and rocket industries belongs to high-alloy steels with relatively low tensile strength but with high corrosion and heat resistance. In addition, they have high workability when bent, rolled, flared, or are under other types of mechanical processing. They are used to manufacture nozzle parts, engine manifolds, systems of hot gas communication in pipeline compressors, exhaust pipes, etc. For example, products made of VNS-53 (08KH21G11AN6ch) steel with tensile strength 980 MPa have operation temperature from −70 to +300 °C and have several times higher strength and durability than the products made of conventional 12KH18N9T steel.

All the above proves that in aviation and rocket industries, steel usually remains one of the principal structural materials used to manufacture special-purpose structures.

Aluminum-based alloys. The most characteristic property of aluminum is its low density—2.7 g/cm^3, its melting temperature being 660 °C. The tensile strength of technical aluminum is 80 MPa, its elastic modulus is 70 GPa, and tensile elongation is about 35%. Aluminum has high heat and electrical conductivity, is a chemically reactive metal, easily covered with oxide scale that protects it from interaction with various environments.

Due to its high strength and relatively low density, aluminum and its alloys are very common materials employed in special-purpose structures. Now, in aircraft and rocket industries, high-strength aluminum alloys containing magnesium, lithium, and zinc are mainly used. Yet, at high temperatures, aluminum loses its strength; therefore, pure aluminum is not used as supersonic vehicle skin.

All aluminum-based alloys are subdivided into wrought, cast, and powder-sintered ones. Bars, pipes, sheets, various cross-sectional profiles, wire, forgings, and stampings are made of them. To produce parts and semi-finished products, various methods of pressure treatment are employed: pressing, forging, hot stamping, bending, rolling, and dragging. Plastic strain is also applied for hardening aluminum alloys, since it results in anisotropy of their properties. Aluminum alloys can be welded in a variety of ways, with property anisotropy being removed in the places of welding. All strained aluminum alloys are subdivided into those strengthened and non-strengthened by thermal treatment (aging).

Chemically, wrought aluminum-based alloys are subdivided into groups in compliance with the content of principal elements contained in the chemical composition of alloys. The most common alloy group is Al-Cu-Mg alloys (duralumins). High-strength alloys are based on Al-Zn-Mg-Cu system. Alloys for forging and stamping contain Al-Mg-Si-Cu. Wrought aluminum alloys are marked with letter D, the high-strength ones with letter V, and the forged ones as AK. Wrought aluminum alloys are widely used in aircraft, rocketry, and helicopter production. They are used to manufacture frames, the majority of control and assembly elements, some aviation engine parts.

Cast aluminum alloys are put into a separate alloy class since they are united by the presence of the major properties—castability, volume, and casting shrinkage, the tendency to shrinkage cracking and liquation.

The most widely employed of aluminum cast alloys are silumins of Al-Si system. Besides silumins, alloys based on Al-Cu-Mg, Al-Cu, and other systems are used to cast products of complex shapes.

These alloys differ from the wrought alloys of the corresponding composition by a higher content of copper and magnesium, as well as refractory metals: titanium, nickel, iron, chromium, and others. Such alloys can be used as high-temperature ones.

As a rule, castings of these alloys are heat treated. Cast aluminum alloys are marked with letters AL.

At present, aviation and rocket industries widely employ complex cast aluminum alloys (Al-19, Al-20, Al-21, etc.) of Al-Cu-Mg system with small additives of Ni, Cr, Fe, Ti. They are used as high-temperature alloys for operating at temperatures up to 300 °C (Taranenko 2014; Vambol et al. 2011).

Besides, duralumins D1, D16, D18—wrought aluminum alloys—are most widely used in aviation and rocket industries. The digits after letter D denote the number of the grade and do not contain any other information. These alloys belong to Al-Cu-Mg system. All duralumins have low corrosion resistance, therefore, they are always protected either by cladding or by anodizing.

In aircraft manufacturing, high-strength Al-Zn-Mg-Cu aluminum-based alloys hardened by heat treatment, as well as medium- and high-strength Al-Mg-Cu alloys, are used extensively. They are structural materials for aircraft skin and internal power set of airplane frame (fuselage, wing, fin, etc.). Airframes, onboard systems, landing gears, propeller blades, and instruments are made of weldable aluminum-based alloys (Al-Mg, Al-Mn, Al-Cu, Al-Mg-Li, Al-Mg-Si).

To reduce aircraft mass, aluminum-lithium alloys of the basic Al-Mg-Li (1420, 1421, 1424) and Al-Cu-Li (1460, 1464, 1469) alloying systems are employed. Using high-strength aluminum-lithium alloys in welded hermetically sealed structures of fuselage load-bearing fuel tanks permits a 12–15% reduction of their own mass.

Aluminum-based alloys have the undeniable superiority in rocket-and-space technologies. The advantage of aluminum-based alloys is their performance at cryogenic temperatures in contact with liquid oxygen, hydrogen, and helium. The so-called cryogenic hardening occurs in these alloys, i.e., their strength and ductility rise with a decrease in temperature.

Magnesium-based alloys. The melting point of magnesium is 650 °C, density—1.74 g/cm^3, its elastic modulus is 45 GPa, tensile strength—180 MPa, and tensile elongation—12%.

Magnesium is a reactive metal. In the air, it oxidizes to form an oxide film. Magnesium-based alloys are welded well and easily machined. Magnesium burns in the air at 623 °C.

Magnesium forms solid solution in the entire range of concentrations only with cadmium. Iron, copper, silicon, and nickel reduce magnesium corrosion resistance. The most common alloying elements are zirconium, manganese, beryllium, aluminum, and zinc. Thorium, neodymium, and cerium improve magnesium heat resistance at temperatures 250–300 °C.

Because of its low mechanical strength and low corrosion resistance, pure magnesium is not used in aircraft manufacturing. In special-purpose structures, parts made of cast magnesium-based alloys are employed. Their main advantage as compared to other alloys is low mass since their density is 1.76–2.00 g/cm^3, which is approximately 4 times as low as that of steel, and 1.5 times as low as the density of aluminum-based alloys. Magnesium-based alloys are used to cast compressor and instrument cases, crankcases, crankcase covers, blower housings, oil pump housings, aircraft wheel parts (brake drums, shoes, flanges, brake housings, etc.), control wheels, control columns, landing gear frames, window, hatch and sitting brackets, as well as many other parts of aircraft and aviation engines.

Wrought magnesium-based alloys are supplied as bars, ribbons, profiles, and forgings. They have low ductility at low temperatures, so they are treated under pressure at high temperatures. In terms of the alloying elements present, they can be subdivided into the following systems:

- Mg+Mn system (e.g., MA8 alloy, which contains some amount of cerium) has high corrosion resistance and weldability;
- Mg+Al+Zn+Mn system (MA2-1 alloy) is well rolled and sheet stamped;
- Mg+Zn+Zr system (MA14, MA15, MA20 alloys) has high mechanical strength;
- Mg+ rare-earth elements (neodymium, zirconium) system (e.g., MA11, MA12 alloys) has high strength;
- Mg+Li system (ultralight alloys) (e.g., MA18, MA21) has low density and high strength.

The principal limitations that reduced the use of magnesium-based alloys in present-day aircraft and rocket manufacturing are low corrosion resistance, insufficient ductility at room temperature, low melting point, as well as fire hazard (Potemkin et al. 1982).

Titanium. Titanium density is approximately 4.5 g/cm^3, and in terms of this property, it takes an intermediate position between iron and aluminum, while its performance is higher than those of both elements. Its melting point is 1668 °C. The limitation of titanium is its low elastic modulus—112 GPa.

Titanium and Ti-based alloys have a very valuable set of properties—high strength in the range from cryogenic (−250 °C) to moderately high (300–600 °C) temperatures at much lower density than that of steel. They have high corrosion resistance and do not need protection against corrosion in atmospheric conditions, river and sea water, and in many other aggressive environments. High physico-mechanical properties of titanium and its alloys make them indispensable structural materials for manufacturing some parts and power units of current rocketry and aircraft. Titanium alloys are especially widely employed in aviation engines. In this field, titanium alloys ever more frequently substitute aluminum and stainless steel. On the other hand, titanium has an evident advantage of retaining its strength at temperatures up to 430 °C, which appear due to aerodynamic heating at high speeds (Taranenko 2014; Ostapchuk et al. 2015). In jet engines, titanium is used to manufacture compressor blades, turbine disks, and many other molded parts.

In military products, titanium is extensively used for producing parts of fuselage, airplane flaps, aerodynamic steering wheels, and supersonic missile bodies, which operate at elevated temperatures. From the viewpoint of production technology, titanium alloys are subdivided into

- wrought alloys,
- cast alloys, and
- powder alloys.

In terms of mechanical properties, they are subdivided into

- alloys with normal strength (structural alloys),
- high-strength alloys,
- heat-resistant alloys, and
- high-plasticity alloys.

As to their ability to be strengthened by heat treatment, titanium alloys are subdivided into hardenable and non-hardenable alloys.

Structural titanium-based alloys (of normal strength). VT20 alloy (its tensile strength σ_b = 932 MPa) is at present the most common titanium-based alloy. It is employed both in the structures of airframes, missile bodies, and in gas turbine engine parts. A large number of Su-35 airframe parts, as well as those of fuselage, wing, and fin welded units, are made of this alloy (Klimov and Kozlov 2017).

VT23 alloy ($UTS = 1080$ MPa) is a weldable alloy of wide application. It is used in monolithic, welded, and soldered structures of aircraft, missiles, and spacecraft. Technologies of semi-finished products (sheets, ribbons, stampings, and forgings) of the mass up to 3.5 tons have been developed. The alloy is widely employed in An-158, Su-29, Su-31M, etc. aircraft (Klimov and Kozlov 2017).

VT18ch alloy (UTS = 1000–1200 MPa) is a sheet structural alloy of high-temperature applications (up to 600 °C). It is intended for manufacturing aircraft skins, missile casings, and engine structures.

VT38 ($UTS = 970$ MPa) is a novel heat-resistant high-temperature low-oxidation alloy with operation temperature up to 650 °C. It is intended for manufacturing aircraft skins, missile casings, and engine structures.

High-strength titanium-based structural alloys. High-strength titanium alloys are used to manufacture critical parts and units: welded structures, turbines, stamped units, highly loaded parts, and structures. These alloys are capable of long-term operation at up to 400 °C and of short-term operation at up to 750 °C.

The specific feature of high-strength titanium-based alloys as structural materials is their increased sensitivity to stress concentrators.

VT22 alloy ($UTS = 1030$ MPa) is weldable and has high hardenability. It is widely used in aviation industry (Il-76, Il-86, Il-96, An-72, An-124, An-224, An-148, Yak-42, Be-200, Tu-204, MiG-29, T-50 aircraft). The alloy is intended for manufacturing large-sized parts of internal power set (beams, spars, frames, ribs, flaps, and slats), as well as for large-sized chassis parts and units of chassis, including the welded ones (traverse, main chassis beams, slot balls, brake levers).

Cast titanium-based alloys. Cast titanium alloys have low tendency to hot cracking and low linear (1%) and volume (3%) shrinkage. Shaped castings are produced by using cast iron, steel, and other special molds. Cast alloys have mechanical properties inferior to the respective wrought ones. Final heat treatment is not applied since that dramatically reduces their ductility.

VT1L ($UTS = 343$ MPa), VT5L ($UTS = 686$ MPa), and VT20L ($UTS = 932$ MPa) alloys are common in manufacturing highly loaded critical parts (cast bodies, turbines, impellers, etc.).

VT40L alloy ($UTS = 1030$ MPa) has an increased strength and high fatigue strength ($\sigma = 340$ MPa at 107 cycles and 50 Hz loading frequency) and is recommended for application instead of VT6L and VT20L alloys to manufacture various long-life parts of airframe assemblies and engines (Taranenko 2014; Ostapchuk et al. 2015).

The advantages of employing titanium-based alloys in manufacturing special-purpose structures (Vambol et al. 2011; Potemkin et al. 1982) are

- low density (4500 kg/m^3), which helps reduce structure weight;
- high mechanical strength: at elevated temperatures (250–500 °C) the characteristics of titanium-based alloys exceed those of high-strength aluminum- and magnesium-based ones;
- high corrosion resistance due to titanium capability to form on its surface a thin (5–15 μm) hard TiO_2 oxide film that is strongly connected to the bulk metal; and
- specific strength (strength-to-density ratio) of current titanium-based alloys is 30–35 and more, which is almost twice higher than that of alloyed steels.

Titanium alloy limitations are considered to be as follows:

- their prime cost (titanium is significantly more expensive than steels, aluminum, copper, and magnesium);
- high reactivity (especially in liquid state) with all atmospheric gases at elevated temperatures, which necessitates titanium melting and casting in vacuum or in an inert gas;
- poor anti-friction properties due to titanium sticking on numerous materials (titanium coupled with titanium cannot work in friction);
- titanium and its alloys are highly prone to hydrogen embrittlement and salt corrosion; and
- high reactivity, the tendency to grain growth at high temperatures and phase transformations during welding cycle cause problems in titanium welding.

Besides, machining of titanium and titanium-based alloys is difficult (cutting tool grip).

Refractory metals. Metals with melting points above 1800 °C are known as refractory ones. They include tungsten, molybdenum, chromium, tantalum, niobium, rhenium, hafnium, vanadium, zirconium, osmium, iridium, and ruthenium (Taranenko 2014; Ostapchuk and Semishov 2011).

The interest to refractory metals and the alloys based on them appeared in connection with the production of rockets, spaceships, nuclear reactors, and the development of power plants. The parts and units of those objects and facilities operate in the temperature range 1500–2000 °C.

The limitations of refractory metals are their cold brittleness (this is especially intrinsic to chromium, molybdenum, and tungsten), and their tendency to oxidizing. So, they can only be used with protective coatings.

Refractory metals are alloyed to diminish their tendency to brittle fracture at low temperatures and to improve their strength and heat resistance characteristics. The principal alloying elements for tungsten, molybdenum, and niobium are rhenium, carbon, zirconium, lanthanum, titanium, and boron.

All refractory metals and alloys have good ductility; niobium welds well. As of today, heat treatment of refractory alloys is not widely applied in industry. This is due to the lack of special equipment for quenching with fast cooling rate and for providing high temperature for alloy transition from ductile to brittle state.

Tungsten and its alloys. Tungsten density is 19.3 g/cm^3, its elastic modulus—350 GPa. Tungsten is produced by powder metallurgy methods.

Under normal conditions, pure tungsten is rather chemically resistant to perchloric, sulfuric, nitric, hydrofluoric acids, and aqua regia below 100 °C. When heated, it can dissolve in alkalis.

Tungsten powder is produced by reduction with hydrogen in tubular electric furnaces at 700...850 °C. Then it is molded at a pressure of 3…5 t/cm^2 and heated by electric current up to 3000 °C in a hydrogen environment. As a result, plastic tungsten rods are produced, which are treated by forging, drawing, or rolling.

The major alloying elements for tungsten are rhenium (up to 30%), niobium (up to 0.6–0.9%), zirconium (up to 0.12%), thorium dioxide ThO_2 (up to 2%), and tantalum carbide TaC (up to 0.4%). Hardening is achieved due to solid solutions of alloying elements and fine particles of refractory oxides and carbides (tungsten carbide accounts for nearly 65% of tungsten consumption).

To operate at high temperatures, tungsten alloys are protected with Al_2O_3, ZnO_2 coatings and (Hf+Ta) alloy, which are deposited by plasma spraying using plasmatrons. When tungsten is alloyed with non-metals, carbides (W_2C, WC), nitrides (WN, WN_2), borides (W_2B, WB, W_2B_5), and silicides (W_3Si_2, WSi_2) are formed. They have melting points 2300–2750 °C, high hardness, satisfactory cutting properties, and are resistant to oxidation (Taranenko 2014; Ostapchuk et al. 2015).

VV-2 alloy is a typical representative of tungsten-based alloys. It consists of tungsten and 0.3–2% niobium. Its tensile strength at 1000 °C is 240 MPa. Rupture strength after 500 h loading at 1500 °C is 50 MPa. Other designations of tungsten alloys are: American ones—AMS-7897, (90% W+6% Ni+4% Cu), ASTM-B-777-87: Grade 1; those of the United Kingdom—Elkanite 10W3 (72% W+Cu), Elkanite 20S (W+27% Ag), Mallory 1000 (90% W+Ni+Cu).

Currently, tungsten alloys are employed in electroslag, radio, and electrovacuum industries. In aviation engine units, tungsten alloys are employed in compressor parts, engine nozzle neck inserts. Inserts in nozzle necks of liquid-fuel rocket engines are made of tungsten alloyed with silver.

Special attention is given to tungsten alloys as the materials for manufacturing cutting and drilling tools (cutters, milling cutters, drills), fibers, wire, and rollers.

Non-aviation tungsten applications are those in damping weights for computer disk drive heads, balancing weights for ailerons in commercial aircraft, propellers of helicopters and guided missiles, kinetic energy penetrators for defeating heavy armor, in radiation protection, radioactive isotope containers, and gyroscope components.

Pure tungsten is used for hot filaments of electric lamps; cathodes and anodes of X-ray tubes; parts of electric amplifiers, rectifiers, welding electrodes; and to produce thermocouples that permit measuring temperatures in 1200–2000 °C range.

Molybdenum and its alloys. The most valuable property of molybdenum is its capability to retain mechanical properties at temperatures up to 1000 °C.

Molybdenum is produced by powder metallurgy methods. To improve its purity, molybdenum is remelted in electric arc furnaces. Remelted molybdenum is easily forged, stamped, and rolled. It is supplied as 0.1-mm-thick sheets, wire, tubes of 1.2 mm diameter and 0.1 mm wall thickness, sheets.

Molybdenum forms carbides, borides, and silicides. The principal alloying elements for it are: boron; silicon—to increase hardness; aluminum, chromium, cobalt, iron, nickel, tantalum, silicon—to increase ductility; and zirconium, titanium, niobium—to raise cold resistance. As a rule, the mass part of alloying element is 1% at most. In large amounts, molybdenum can only be alloyed with tungsten and tantalum—up to 20%. In alloys of VM-3 type, besides the abovementioned alloying elements, strengthening is achieved by introducing particles of titanium, zirconium, tantalum, and hafnium carbides. Industrial molybdenum alloys are subdivided into those strengthened by work hardening and the ones strengthened by heat treatment. To relax stresses in half-finished and ready-to-use products, 15-min annealing with polygonization at temperatures 800...1200 °C is employed.

Annually, more molybdenum than any other refractory metal is consumed.

Molybdenum has outstanding electrical properties and thermal conductivity, as well as relatively high tensile strength. Its heat conductivity is approximately 50% higher than that of steel, iron, or nickel alloys. For this reason, molybdenum finds wide application in radiator manufacturing. Its electrical conductivity is the highest of all refractory metals—about one-third that of copper, and it is higher than electric conductivity of nickel, platinum, or mercury. Molybdenum thermal expansion coefficient depends nearly linearly on temperature in a wide temperature range. This feature, in combination with high heat conductivity, accounts for its use in bimetallic thermocouples. Besides, engineering scientists developed methods of molybdenum powder alloying with potassium aluminosilicate to obtain a non-sagging microstructure comparable to that of tungsten.

As an alloying element, molybdenum is mainly employed as an alloying agent for alloyed and tool steels, stainless steels, and nickel- or cobalt-based superalloys to increase their high-temperature strength, durability, and corrosion resistance.

In electrotechnical and electronic industries, molybdenum is used in cathodes, cathode supports in radar devices, conductors for thorium cathodes, and magnetrons.

Molybdenum is important in rocket industry, where it is employed for manufacturing high-temperature structural parts, such as nozzles, leading edges of control surfaces, support blades, racks, entry cones, radiation screens, radiation radiators, turbine impellers, and pumps.

Mo-0.5Ti; Mo-50Rhe; TsM-2A3; VM-1, VM-2, VM-3, MR-47VP, TSM3 alloys have found the widest application in aircraft structures.

An example of BM-1 alloy is: Mo base, 0.4% Ti, 0.08 …0.25% Zr, 0.01% C, 0.6% Nb. It has tensile strength of 800 MPa at room temperature, and 100 MPa at 1800 °C.

In aircraft structures, molybdenum alloys are employed with protective coatings. Molybdenum can be used without protective coatings in disposable apparatuses and in vacuum. For protective coatings at temperatures below 1100 °C, one or several layers of metals (nickel, chromium, platinum) are used. Service life of such coatings is about 1000 h.

For temperatures 1100...1500 °C, coatings made of Zr_2O_3, Al_2O_3, Cr_2O_3 oxides with glass-like matrices are applied. At temperatures 1200–1300 °C, the operation life of such a coating is approximately several hundred hours. At 1500…1800 °C, $MoSi_2$-based coating lasts up to 30 h.

Molybdenum alloys are employed to manufacture aircraft engine nozzles, combustion chamber rings of supersonic aircraft. Molybdenum is used for manufacturing tail edges and nose cones of airplane streamlined bodies. It is considered as a promising material for fuel assemblies of nuclear reactors. Molybdenum alloys are also used in casting dies for aluminum, magnesium, zinc, and iron, in manufacturing heat shields for furnaces, structural parts, and heating elements.

Niobium and its alloys. Pure niobium is a low-strength but very ductile metal. It is easily forged, stamped, rolled, and drawn. It is supplied as sheets, wire, and pipes. Niobium is easily machined, is corrosion- and acid-resistant, and welds well. From the viewpoint of its importance and availability, it is close to molybdenum. Niobium limitations are rapid oxidation in the air and a high rate of new additives' diffusion when heated. Nearly all refractory metals raise the temperature of niobium recrystallization and melting. Niobium is alloyed with such elements as molybdenum, tungsten, vanadium, titanium, tantalum, aluminum, chromium, zinc, silicon, and boron. In terms of their structure, niobium-based alloys are subdivided into solid solutions and heterogeneous two-phase alloys. Niobium-based alloys are hardened and annealed at 1100...1300 °C for 1 h. The most common are alloys of Nb+Mo system (their example is VN-2 alloy, which contains 4.5% molybdenum, the rest being niobium) and of Nb+Mo+Zr+O system (VN-3, VN-4, VN-5 alloys). An example of VN-5A alloy composition is: molybdenum 6%, zinc 0.7%, carbon 0.1%, the rest is niobium.

Practically, niobium does not interact with uranium at temperatures up to 1100 °C, has high heat conductivity and high resistance to molten metals, it is also compatible with structural materials. These properties permit niobium to be applied in nuclear energy engineering. Heat exchanger tubes and fuel elements are made of that metal. It is considered as a promising material for nuclear engine parts in aircraft and onboard nuclear energy equipment.

Niobium alloys were employed for the "General Electric" and "Pratt & Whitney" companies widely employ niobium to manufacture turbine parts operating at 1100–1500 °C. Examples of niobium-based alloys are: VN-2, VN-3, RN-6, F-48, F-50, D-31, D-43, VN-2A. VN-2A alloy cold-worked at 20 °C has tensile strength 900 MPa, at 1200 °C–260 MPa, its rupture strength after 100 h loading at 1200 °C being 30 MPa.

Rhenium-based alloys. Of all elements, except tungsten and carbon, rhenium has the highest melting point. Its density is only exceeded by osmium, iridium, and platinum. There is no plastic-brittle transition temperature for pure rhenium. Rhenium is the only refractory metal that does not form transition temperature carbides.

Rhenium has high electrical resistivity in a wide temperature range; besides, it retains ductility and is not affected by the oxidation/reduction cycle, which happens to tungsten. One of the widest fields of rhenium application is that in mass spectrometer filaments, where it is available in commercial (99.99%) and zone-melting (99.999%) purity.

Rhenium is not affected by molten copper, silver, tin, or zinc. It readily dissolves in molten iron and nickel, but is stable in the presence of aluminum. Rhenium produces a significant strengthening effect on platinum. At elevated temperatures, rhenium resists the effect of hydrogen and inert atmosphere. It is resistant to corrosion caused by hydrochloric acid and sea water, as well as to the mechanical effects of electrical erosion.

Rhenium is a useful alloying element for other refractory metals; it considerably increases ductility and rupture strength of refractory metals and their alloys. Rhenium-based alloys are employed in nuclear reactors, semiconductors, vacuum tube components, thermocouples, gyroscopes, miniature rockets, electrical contacts, thermionic converters, and in other commercial and aerospace applications. Tungsten–rhenium alloys applied by vapor deposition are used to coat the surfaces of molybdenum targets in X-ray tube manufacturing. Other rhenium alloys (those with tungsten or molybdenum) are applied in hot filaments, mesh heaters, cup cathodes, ignition wires in photoflash lamps, and heat shields of re-entry space vehicles (with the measuring range up to 2600 °C).

Hafnium alloys. Hafnium (Hf) is very similar in its properties to those of zirconium. The best known of its compounds is hafnium dioxide. Hafnium density is 13.2 g/cm^3, its melting point is as high as 2227 °C. Hafnium has high corrosion resistance to hot water, steam, alkali, perchloric, and nitric acids, along with high radiation resistance. Pure hafnium has very high ductility and responds well to cold and hot forming, forging, rolling, and welding (Taranenko 2014).

Hafnium is supplied as powder, sponge, bars, castings, wire, sheets, tubes, foil, etc.

Commercially pure hafnium in the amounts of 1 …2% is used as an alloying element (e.g., in Zr, Ti, Ta, Ni, Co, W, Mo, Re, and low-weight alloys).

Hafnium-containing alloys are employed in manufacturing turbine blades, valves, nozzles, rocket bodies, fireproof crucibles and screens, steel-cutting tools, as protective coatings of outlet nozzles, for sitalls and glasses, in nuclear industry (fuel rods). Hafnium is a unique metal that can be used as a solder for welding low-carbon steels.

2.2 Peculiarities of Employing the Method in Investigations of Metals and Alloys

In manufacturing special-purpose structures, high-quality alloyed steels and alloys with high strength and heat resistance characteristics, which contain special alloying elements, are used. To examine the structure and chemical composition of the above-mentioned products, non-destructive research methods are commonly employed.

Scanning electron microscopy is one of the most universal methods of non-destructive studying and analyzing the microstructure characteristics of solid materials. Resolution of the serial instruments is about 10 nm, and if research is carried out with specialized laboratory devices, it reaches nearly 2.5 nm. Another important feature of the images acquired with a scanning electron microscope is that they are three-dimensional due to large focal depth of the device, which makes it possible to obtain more complete information about the specimen examined (Molyar et al.

1995; Ostapchuk and Semishov 2011; Protsenko et al. 2007; Beckert and Klemm 1984; Samsonova 1965).

Current scanning electron microscopes are provided with additional equipment allowing researchers to conduct X-ray microanalysis—measuring the concentrations of the chemical elements contained in the examined specimens with the localization within less than 1 μm due to using a focused electron probe. The specimen is not destroyed, and the relative error of the quantitative analysis is less than 1% at the sensitivity of up to 0.01% when a wavelength-dispersive spectrometer is employed. It should be noted that equipping electron microscope with X-ray microanalyzer makes it possible to investigate surface topography of a chosen specimen area, carry out qualitative analysis of the chemical elements present (revealing the presence of chemical elements without precise measurements of their concentrations), as well as study the spatial distribution of those elements, produce X-ray spatial-distribution maps during area scanning across the surface with an electron beam (Vambol et al. 2011; Protsenko and Sayenko 2002; Goldstein et al. 2017; Arkhipova and Osadchuk 2013).

When studying the structure of surface topography, the examined metal or alloy specimens require preparation. It consists in cutting a specimen of the necessary size (usually of 1–2 cm^3) to be placed in electron microscope chamber and cleaning their surfaces, especially in specimens of special-purpose structures that were exposed to high temperatures and mechanical loading. The specimen cleaning procedure commonly consists in solvent treatment in an ultrasonic bath or in surface wiping. In a number of cases, this operation can be preceded by an electrochemical treatment in special solutions or electrolytes to remove oxides, fat films, and other organic substances from the surface of the material. It should be emphasized that metal specimens, unlike dielectric ones, do not need conductive coatings to drain the electric charge, since they themselves have high electrical conductance (Vambol et al. 2011; Kay 1961).

The situation is quite different when preparing metal specimens for quantitative X-ray microanalysis (evaluating the concentrations of the chemical elements present in the specimen). In this case, it is necessary to get rid of the roughness and unevenness on the specimen surface. Actually, it is necessary to remove the structure of the surface that provides a topographic contrast in surface investigation by means of scanning electron microscope in the secondary electron mode. That is due to the following factor.

The relative error of the X-ray absorption correction in quantitative microanalysis is proportional (Potemkin et al. 1982) to

$$\coth \varphi \; \Delta\varphi,$$

where φ is the X-ray emission angle and $\Delta\varphi$ is the inclination of local surface areas. In the case of energy-dispersive spectrometer (EDS), the angle of emitted X-rays is determined as the angle between X-ray detector axis and the specimen surface. For wavelength-dispersive spectrometer (WDS), the X-ray emission angle is determined by the mutual positions, detector and specimen. Hence, the effect of the inclination of local surface areas $\Delta\varphi$ is to be minimized. Taking into account that X-ray analysis is, in fact, the analysis of the prepared surface layer of the specimen (the depth of electron beam penetration into the specimen is about 1 μm), it is necessary that this prepared surface should adequately represent the chemical composition of the specimen examined. The criteria of specimen preparation quality show that even small differences in the inclination of local areas affect the accuracy of measuring the intensities of characteristic X-ray radiation, which, in turn, influences the accuracy of measuring the concentrations of chemical elements. Significant errors in quantitative X-ray microanalysis appear even at differences in the heights of specimen surface relief of 2–4 μm and more.

To prepare specimens for quantitative X-ray microanalysis, at first, grinding is carried out on a grinding wheel, and then polishing, using the paste that is an organic substance with an abrasive filler. Several pastes with different sizes of polishing particles, from coarse to finer, are used in succession (Taranenko 2014; Ostapchuk et al. 2015; Pilyankevich and Klimovitsky 1976; Schimmel 1969).

It should be stressed that sample preparation criteria regarding the absence of roughness and a smooth surface are somewhat different when wavelength-dispersive or energy-dispersive spectrometers are employed.

Wavelength-dispersive spectrometers only register X-ray radiation whose wavelength corresponds to Bragg's law

$$n\lambda = 2d \sin\theta,$$

where n is an integer (1, 2, 3, ...), λ is the X-ray radiation wavelength, d is the interplanar spacing in the crystal, and θ is the angle of X-ray incidence on the crystal surface. In energy-dispersive spectrometers, the entire X-ray spectrum emitted by the specimen reaches the detector simultaneously, with further processing of the number of pulses which correspond to a certain wavelength of characteristic X-ray radiation. All X-ray radiation gets to the detector inside a certain solid angle, and there is no very strict dependence on the angle of X-ray radiation incidence on the detection system. Thus, a slight deviation from Bragg's condition in a wavelength-dispersive spectrometer leads to significant errors in measuring X-ray intensities and, as a rule, to errors in measuring the concentrations of chemical elements in quantitative X-ray microanalysis. Therefore, the preparation of specimens for a wavelength-dispersive spectrometer should be more careful (the absence of roughness and, ideally, an absolutely flat surface of the examined specimen) as compared to specimen preparation for an energy-dispersive spectrometer (Klimov and Kozlov 2017).

During specimen preparation for X-ray microanalysis, there is also an issue of infiltrating the polishing abrasive into the studied material. For example, a large number of silicon inclusions are observed in the specimen polishing with SiC-based sandpaper. After polishing with Al_2O_3 sandpaper, numerous Al inclusions are found in the specimen. Hence, it is recommended to use polishing materials that do not contain the chemical elements present in the specimen examined (Potemkin et al. 1982).

In the metals and alloys used for manufacturing special-purpose structures, light chemical elements ($Z < 10$) are usually present in low concentrations, but they significantly affect the properties of those structures. When that is the case, to determine concentrations of light elements with X-ray microanalysis methods is rather difficult since fluorescence yield of the elements with low atomic numbers is small. Besides, light elements emit in the long-wave range. Such emission is characterized with significant absorption; therefore, only a small part of X-ray intensity is registered.

This negative effect can be neutralized by using an optimal accelerating voltage, at which X-ray emission yield is maximum for the element investigated. The appearance of the maximum is due to the action of two opposite factors—an increase of X-ray intensity with a rise in accelerating voltage, on the one hand, and absorption enhancement on the other, because with an increase of electron energy in the beam, X-ray emission is generated deeper in the specimen and more of it is absorbed. Here, it should be taken into account that accelerating voltage should induce all characteristic X-ray radiation lines that correspond to all chemical elements present in the specimen (Potemkin et al. 1982).

The well-known problem in electron microscopy—carbon contamination on the studied specimen surface under the action of electron beam—should also be considered. The carbon deposit forms due to decomposition of hydrocarbons present in the microscope chamber. The growth rate of the carbon layer can reach several nanometers per second. With that, X-ray spectra can show the presence of carbon, on the objects which actually do not contain carbon or it can lead to an upward error in carbon. This issue is addressed by introducing into the electron microscope chamber a rod whose temperature is lower than the temperature of the examined specimen surface. In this case, molecules of organic substances would tend to concentrate on a colder surface rather than that of the specimen. Cooling rods effectively reduce the rate of impurity deposition when the rod is cooled with liquid nitrogen. Another way of dealing with the problem of carbon contamination is to employ ion pumps for forming higher vacuum in electron microscopes.

To investigate metals and alloys used for special-purpose structures by means of X-ray microanalysis methods, both energy-dispersive and wavelength-dispersive spectrometers are employed. Wavelength-dispersive spectrometers have energy resolution of less than 5 eV, whereas energy-dispersive ones—100 eV and more. Also, the sensitivity of wavelength-dispersive spectrometers is better than that of energy-dispersive devices. However, research with wavelength-dispersive spectrometers requires careful specimen preparation and the availability of reference samples whose chemical composition corresponds to the composition of the studied specimens, and more time is spent on the examination in comparison with energy-dispersive spectrometers. At the same time, energy-dispersive spectrometers make it possible to produce the results of qualitative and quantitative microanalysis literally in minutes (but with less accuracy compared to wavelength-dispersive spectrometers), as well as maps of spatial distribution of all present chemical elements over the surface of the examined specimen.

Thus, in summary, the stages of investigating metals and alloys for special-purpose structures look as follows:

1. Cutting specimens of the necessary sizes for placing them in the electron microscope chamber and cleaning their surfaces. Studying their topography by acquiring high-quality images of sample areas in secondary and backscattered electrons.
2. Analyzing the produced images of the surface structure of specimen areas to obtain information about the presumed high temperatures and the mechanical impacts to which special-purpose structures were exposed or otherwise, the information about the nature of the explosion and the properties of the explosive, as well as the temperatures acting on the structural elements during the explosion (if that did occur).
3. Preparing specimens for conducting X-ray microanalysis with energy-dispersive spectrometer (grinding, polishing). Obtaining the results of qualitative and quantitative microanalysis and producing distribution maps of all chemical elements present over the surface of the specimen.
4. Preparing specimens for conducting X-ray microanalysis by means of wavelength-dispersive spectrometer (better quality polishing in comparison with the specimen preparation for microanalysis with energy-dispersive spectrometer). With reliance on the results obtained with energy-dispersive spectrometer, the choice of research areas, selection of reference samples, and optimal accelerating voltages for microanalysis of all present elements, including light elements as well. Measuring the concentrations of the elements present with wavelength-dispersive spectrometer.
5. Using the results X-ray microanalysis of special-purpose structures manufactured both abroad and in Ukraine, which have been produced with wavelength-dispersive spectrometry, in implementing import substitution programs, as well as for controlling and determining the optimal compositions and concentrations of chemical elements in the implementation of manufacturing technologies for special-purpose structures with predetermined properties.

2.3 Results of Using Scanning Electron Microscopy in Examining Fragments of Special-Purpose Structures Made of Metals and Alloys

The development of novel rocket and artillery weaponry is of utmost importance in the war context since that can help the country ensure military superiority over its enemy and reduce losses in personnel and material resources. However, it should be noted that the development of any advanced systems, especially rocket and artillery weapons, is a difficult and expensive process which requires heavy investment in R&D, technology and production spheres.

Therefore, investigating the level achieved by the enemy country in this field, in particular, studying the materials used to manufacture various missile types can save substantial time and resources when developing analogous domestic systems.

Besides, that will help understand strengths and weaknesses of enemy's weaponry, find out which systems could be most effective in the confrontation, as well as assist in the development of more efficient anti-missile systems.

To this end, experts have conducted research into the chemical composition and structure of the materials of fragments of cruise, ballistic, air-launched ballistic and airborne missiles produced in the Russian Federation and North Korea.

The research was carried out with reliance on non-destructive methods of materials examination (the analysis of examination methods is given in (Taranenko 2014; Molyar et al. 1995)), i.e., scanning electron microscopy using a JXA-8200 X-ray microanalyzer, a JSM-6490LV scanning electron microscope, and an INKA Energy 350XT energy-dispersive spectrometer (JEOL Ltd, Japan).

2.3.1 Results of Examining Cruise Missile Fragments

2.3.1.1 Fragment of 3M-14 *Kalibr* Cruise Missile Engine

Figure 2.1 shows a fragment of 3M-14 *Kalibr* cruise missile engine.

Figure 2.2 presents an example of X-ray microanalysis results for an engine fragment of 3M-14 cruise missile of Kalibr complex. Chemical element concentrations are given in mass percentages.

Fig. 2.1 Fragment of 3M-14 *Kalibr* cruise missile engine

```
Unknown Specimen No. 4
 Group          : mil              Sample   : 3m14
 UNK No.        :     4            Comment  :
 Stage          :     X=    35.2355  Y=    43.4655  Z=    10.9855
 Acc. Voltage  :     15.0 (kV)     Probe Dia. : 0     Scan : Off
 Dated on Dec 12 19:49 2022
 WDS only           No. of accumulation : 1

Curr.(A)  : 1.967E-08
Element Peak(mm) Net(cps)    Bg-    Bg+  S.D.(%)    D.L.(ppm)
 1 Al     89.564     742.8   79.3   59.2     1.24         42
 2 Ti     87.472    3138.4   51.5   67.1     0.41        769
 3 V     174.201      41.0   17.8   26.8     5.05        161
 4 C     128.407      73.0   30.0   34.4     3.99         50
 5 Mn    146.182     202.2   28.3   28.3     2.45        137

ZAF Metal
Element  Mass(%)    Atom(%)  K(%)  K-raw(%)    ZAF       Z        A        F
 Al        1.719    2.9178   1.196    1.198  1.4374  0.9140  1.5726  1.0000
 Ti       96.207   94.8631  99.047  100.047  1.0016  1.0016  1.0009  0.9991
 V         0.449    0.4036   0.441    0.441  1.0186  1.0229  0.9973  0.9985
 C         0.092    0.3525   0.144   76.812  0.6400  1.0047  0.6371  1.0000
 Mn        1.755    1.4629   1.651    2.132  1.0629  0.9685  1.0974  1.0000
 ----------------------------------------------------------------
 Total   100.222 100.0000 102.479  180.630    Iteration = 3
```

Fig. 2.2 Results of X-ray microanalysis of a fragment of 3M-14 *Kalibr* cruise missile engine

The content of principal elements in the material of 3M-14 Kalibr cruise missile engine fragment, viz., titanium base predominantly alloyed with aluminum, shows that the material is a heat- and oxidation-resistant titanium alloy of OT4 (VT4) type, produced in accordance with GOST 19807-91 "Titanium and wrought titanium-based alloys" developed by the USSR Ministry of Aviation Industry.

The alloy of OT4 (VT4) type is a wrought Ti-based alloy used for manufacturing thin-walled complex-shaped parts that operate for a long time at the temperatures up to 300 °C.

2.3.2 Results of Examining Ballistic Missile Fragments

2.3.2.1 Fragment of the Outer Part of Hwasong-11 (KN-23) Missile Nozzle, North Korea

Figure 2.3 shows a view of a fragment of the outer part of the nozzle of the North Korean *Hwasong*-11 (KN-23) missile. Table 2.1 shows the contents of the main elements present in the material of the fragment of the outer part of the nozzle of the *Hwasong*-11 (KN-23) ballistic missile. The content of principal elements present in the material of an outer part fragment of *Hwasong*-11 (KN-23) ballistic missile nozzle shows that its composition corresponds to low-alloy temperature-resistant steel 15 KhM or 12 KhMF (Table 2.1).

Fig. 2.3 Fragment of the outer part of *Hwasong*-11 (KN-23) missile nozzle

Table 2.1 Chemical composition of fragment material in mass concentrations

Chemical elements	Mass concentrations, %
Cr, chromium	0.810
Mn, manganese	0.711
Ni, nickel	0.448
C, carbon	0.183
Fe, iron	97.675

15KhM and 12KhMF steels are used to manufacture bodies and inner elements for tubes, pump parts, pump plugs, and special fastenings of tubes that are to resist oxidation up to temperatures +500 °C.

```
Unknown Specimen No. 3
 Group          : mil               Sample   : x3
 UNK No.        :     3             Comment  :
 Stage          :     X=    79.2325  Y=    58.7680  Z=    10.9210
 Acc. Voltage :      15.0 (kV)      Probe Dia. : 6     Scan : Off
 Dated on Jan 15 18:55 2024
 WDS only          No. of accumulation : 1

Curr.(A)  : 1.009E-08
Element Peak(mm) Net(cps)    Bg-     Bg+  S.D.(%)    D.L.(ppm)
 1 Cr    159.309     61.3   14.6    12.6      3.43        195
 2 Si     77.143    186.2   22.6    18.6      2.56         72
 3 Mn    146.238     49.6   21.0    17.4      5.68        196
 4 Ni    115.224     31.6   37.2    32.2     10.06        356
 5 C     128.254     46.6    7.6     9.8      4.09         51
 6 Fe    134.762   7339.6   46.6    49.6      0.26        416

ZAF Metal
Element   Mass(%)    Atom(%)   K(%)   K-raw(%)    ZAF       Z        A        F
 Cr         0.810    0.8405   1.022    1.022  0.7926  1.0035  1.0063  0.7849
 Si         0.975    1.8732   0.661    0.661  1.4760  0.8918  1.6550  1.0000
 Mn         0.711    0.6987   0.695    0.695  1.0236  1.0213  1.0025  0.9997
 Ni         0.448    0.4123   0.428    0.428  1.0489  0.9859  1.0639  1.0000
 C          0.183    0.8232   0.180   95.646  1.0188  1.0015  1.0173  1.0000
 Fe        97.675   95.3522  98.498   98.684  1.0018  1.0014  1.0008  0.9996
 --------------------------------------------------------------------------
 Total    100.802 100.0000 101.483 197.134     Iteration = 3
```

Fig. 2.4 Results of fragment X-ray microanalysis

2.3.2.2 Fragment of Aluminum Power Structure Located Between Fuel Tank and Warhead of *Hwasong*-11 (KN-23) Missile, North Korea

The content of principal elements in the fragment of aluminum power structure located between the fuel tank and the warhead of *Hwasong*-11 (KN-23) ballistic missile shows the material is aluminum-based alloy of AMg6 type, which is characterized with low density (2.65 g/cm^3), high strength, resistance to corrosion and vibration loads (Tables 2.2 and 2.3).

This wrought aluminum–magnesium alloy finds wide application in special-purpose structures and corresponds to GOST 4784-97 "Aluminum and wrought aluminum alloys" (Fig. 2.5).

Table 2.2 Chemical composition of fragment material in mass concentrations

Chemical elements	Mass concentrations, %
Mg, magnesium	6.393
Ti, titanium	0.190
Al, aluminum	92.535
Mn, manganese	0.602

Table 2.3 Chemical composition of fragment material in mass concentrations

Chemical elements	Mass concentrations, %
Cr, chromium	1.262
Si, silicon	0.723
Mn, manganese	0.685
W, tungsten	0.218
C, carbon	0.193
Fe, iron	95.172
Ni, nickel	1.003

```
Unknown Specimen No. 1
 Group          : mil               Sample   : x4
 UNK No.        :    1              Comment  :
 Stage          :    X=    69.7265  Y=    58.7370  Z=    11.3590
 Acc. Voltage :    15.0 (kV)      Probe Dia. : 10     Scan : Off
 Dated on Jan 15 17:47 2024
 WDS only          No. of accumulation : 1

Curr.(A)  : 9.932E-09
Element  Peak(mm) Net(cps)   Bg-    Bg+  S.D.(%)    D.L.(ppm)
 1 Mg    107.516   1852.3   61.6   39.2     0.44        81
 2 Ti     87.472      2.7   13.1   10.3    43.07 ?     676
 3 Al     90.328  22266.2   62.0   33.2     0.21        83
 4 Mn    146.238     36.9   10.6    7.1     6.12       136

ZAF Metal
Element   Mass(%)    Atom(%)   K(%)   K-raw(%)    ZAF       Z         A         F
 Mg         6.393    7.1646   6.639    6.640   0.9630   0.9704   1.0540   0.9416
 Ti         0.190    0.1082   0.166    0.167   1.1484   1.1044   1.0404   0.9994
 Al        92.535   92.4288  84.000   84.848   1.0897   1.0013   1.0882   1.0000
 Mn         0.602    0.2984   0.525    0.525   1.1467   1.1343   1.0109   1.0000
 -------------------------------------------------------------------------------
 Total     99.720 100.0000  91.329   92.180     Iteration = 5
```

Fig. 2.5 Results of fragment X-ray microanalysis

2.3.2.3 Fragment of Fuel Tank Housing of *Hwasong*-11 (KN-23) Missile, North Korea

Figure 2.6 shows the results of EDX analysis of a fragment of the fuel tank body of the North Korean *Hwasong*-11 (KN-23) missile. Figure 2.7 shows a view of a fragment of the fuel tank body of the *Hwasong*-11 (KN-23) missile, North Korea. The content of principal elements in the material fragment of fuel tank housing of *Hwasong*-11 (KN-23) ballistic missile is made of shows that material composition corresponds to low-alloy high-temperature steel of 16 GNM type (Fig. 2.6).

16GNM steel is used to manufacture high-pressure boiler drums and parts of power equipment and pipelines with absolute pressure above 3.9 MPa for nuclear power plants, as well as other parts. Operating temperature of parts made of this steel amounts up to +425 °C.

Unknown Specimen No. 4
Group : mil Sample : x5
UNK No. : 4 Comment :
Stage : X= 54.5530 Y= 57.6665 Z= 11.1690
Acc. Voltage : 15.0 (kV) Probe Dia. : 10 Scan : Off
Dated on Jan 15 18:22 2024
WDS only No. of accumulation : 1

Curr.(A) : 1.006E-08

Element	Peak(mm)	Net(cps)	Bg-	Bg+	S.D.(%)	D.L.(ppm)
1 Cr	159.309	94.6	12.3	12.3	2.58	186
2 Si	77.058	137.7	27.0	16.6	3.09	74
3 Mn	146.238	47.7	22.0	16.0	5.85	197
4 W	223.600	2.3	3.1	3.8	23.41	360
5 C	128.254	49.3	9.2	10.2	4.02	54
6 Fe	134.713	7106.8	46.8	44.8	0.27	400
7 Ni	115.237	73.5	32.2	33.8	5.08	335

ZAF Metal

Element	Mass(%)	Atom(%)	K(%)	K-raw(%)	ZAF	Z	A	F
Cr	1.262	1.3607	1.582	1.582	0.7974	1.0029	1.0064	0.7900
Si	0.723	1.4438	0.490	0.490	1.4760	0.8914	1.6558	1.0000
Mn	0.685	0.6992	0.670	0.670	1.0227	1.0207	1.0026	0.9994
W	0.218	0.0665	0.140	0.140	1.5550	1.2583	1.2357	1.0000
C	0.193	0.9033	0.191	101.388	1.0148	1.0008	1.0140	1.0000
Fe	95.172	94.5685	94.019	94.019	1.0016	1.0013	1.0014	0.9990
Ni	1.003	0.9580	0.957	0.957	1.0477	0.9853	1.0635	0.9999
Total	99. 256	100.0000	98.049	199.246	Iteration = 3			

Fig. 2.6 Results of fragment X-ray microanalysis

Fig. 2.7 Specimen image

2.3.2.4 Fragment of Thrust Vector Control of *Hwasong*-11 (KN-23), North Korea

The research was carried out with a JXA-8200 X-ray microanalyzer produced by JEOL company (Japan).

Figure 2.8 presents the results of X-ray microanalysis (measurements of chemical element concentrations) obtained with X-ray microanalyzer.

As can be seen in Fig. 2.8, the specimen consists of tungsten. Currently, tungsten alloys are used in electroslag, radio-engineering, electrovacuum industries. In aircraft and missile engines, tungsten alloys are used to manufacture compressor parts, nozzle neck inserts, and inserts into nozzle necks of liquid- and solid-propellant missile engines (Ostapchuk et al. 2015) (Table 2.4).

2.3.2.5 Fragment of 9K720 *Iskander-M* Missile Body

Figures 2.9 and 2.11 show the view of fragments of the 9K720 *Iskander-M* missile body. The content of principal elements in body material of 9K720 *Iskander-M* hypersonic ballistic missile shows that the material composition corresponds to low-alloy heat-resistant martensitic steel of 15Kh5VF type (Figs. 2.10).

15Kh5VF steel is used for manufacturing bodies and internal elements of oil refineries and cracking pipes, pump and plug parts, special fastenings, pipes, plugs, and other parts that are to resist oxidation at temperatures up to +600 °C, of pipeline parts, etc.

```
Unknown Specimen No. 6
 Group          : mil              Sample   : rak-rul
 UNK No.        :    6             Comment  :
 Stage          :    X=    79.6520  Y=    62.3175  Z=    11.2695
 Acc. Voltage :      15.0 (kV)     Probe Dia. : 10     Scan : Off
 Dated on Feb 13 17:41 2024
 WDS only          No. of accumulation : 1

Curr.(A)  : 9.247E-09
Element Peak(mm) Net(cps)    Bg-    Bg+   S.D.(%)    D.L.(ppm)
 1 W    223.637    1500.6   16.5   10.3     0.37        771

ZAF Metal
Element   Mass(%)    Atom(%)   K(%)   K-raw(%)    ZAF       Z        A        F
 W          99.983 100.0000  99.983 100.083  1.0000  1.0000  1.0000  1.0000
 -------------------------------------------------------------------------
 Total      99.983 100.0000  99.983 100.083    Iteration = 2
```

Fig. 2.8 Results of specimen X-ray microanalysis

Table 2.4 Chemical composition of fragment material in mass concentrations

Chemical elements	Mass concentrations, %
W, tungsten	99.983

Fig. 2.9 Specimen image (two specimen sides are shown)

Unknown Specimen No. 3
Group : kin Sample : 9.6
UNK No. : 3 Comment :
Stage : X= 33.7065 Y= 60.3305 Z= 11.2140
Acc. Voltage : 15.0 (kV) Probe Dia. : 0 Scan : Off
Dated on May 23 18:56 2023
WDS only No. of accumulation : 1

Curr.(A) : 1.972E-08

Element		Peak(mm)	Net(cps)	Bg-	Bg+	S.D.(%)	D.L.(ppm)
1	Ni	115.254	124.7	79.2	72.6	4.22	269
2	C	128.254	41.0	19.8	19.4	5.45	39
3	Cr	159.334	343.7	28.1	24.5	1.30	132
4	Fe	134.582	13390.0	88.4	88.2	0.20	296
5	Al	90.304	1.8	28.5	21.3	112.62 ?	30
6	Mn	146.238	63.9	45.0	32.4	5.54	143
7	Ti	87.837	-3.4	198.1	167.7	100.00 ?	73
8	Si	77.034	115.0	67.8	36.8	4.07	59
9	W	102.398	20.2	109.7	109.4	15.53	564
10	V	174.201	16.5	21.7	19.1	10.26	153

ZAF Metal

Element	Mass(%)	Atom(%)	K(%)	K-raw(%)	ZAF	Z	A	F
Ni	0.900	0.8596	0.860	0.860	1.0462	0.9838	1.0636	0.9998
C	0.081	0.3785	0.081	43.006	1.0029	0.9993	1.0036	1.0000
Cr	2.249	2.4253	2.800	2.800	0.8032	1.0016	1.0065	0.7968
Fe	95.361	94.9139	94.414	94.414	1.0014	0.9998	1.0025	0.9990
Al	0.006	0.0132	0.003	0.003	1.8869	0.9148	2.0629	0.9999
Mn	0.467	0.4767	0.457	0.457	1.0216	1.0193	1.0029	0.9994
Ti	0.000	0.0000	0.000	0.000	0.0000	0.0000	0.0000	0.0000
Si	0.308	0.6142	0.209	0.209	1.4742	0.8902	1.6560	1.0000
W	0.469	0.1432	0.344	0.344	1.3665	1.3397	1.0200	1.0000
V	0.159	0.1754	0.177	0.177	0.8997	1.0205	1.0122	0.8711
Total	100.000	100.0000	99.346	142.272	Iteration = 3			

Fig. 2.10 Results of specimen microanalysis

Fig. 2.11 Specimen for examinations

2.3.2.6 9K720 *Iskander-M* Missile Body Plug

Figure 2.12 shows the backscattered electron image of specimen surface area. Specimen examinations have shown that the bulk of the examined material contains two phases that differ in color. As demonstrated by further studies of the chemical compositions of these phases, the lighter ones contain manganese in higher concentrations as compared to its concentrations in dark phases.

Brighter phase areas in the image correspond to the presence of higher concentrations of heavy chemical elements (those of higher atomic numbers) as compared to the dark areas' image. As demonstrated by further studies of the chemical compositions of these phases, the brighter ones contain manganese in higher concentrations as compared to its concentrations in dark phases.

X-ray microanalysis (measurement of chemical element concentrations) is performed both in dark and bright phase areas that can be seen in Fig. 2.12.

Figures 2.13 and 2.14 present the results of specimen X-ray microanalysis (measurements of chemical element concentrations) that have been obtained with microanalyzer in dark and bright phase areas, respectively.

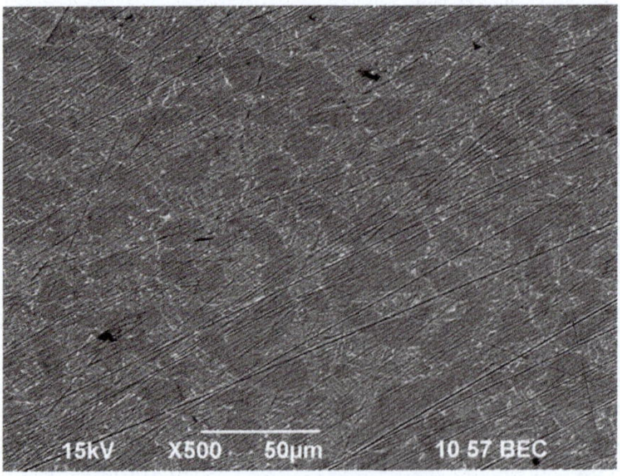

Fig. 2.12 Backscattered electron image of surface area of specimen cross section. Accelerating voltage is 15 kV. Magnification is 500, scale bar corresponds to 50 μm

```
Unknown Specimen No. 3
 Group          :                Sample   : 9,2
 UNK No.        :     3               Comment : dark
 Stage          :     X=    72.1190  Y=    27.6655  Z=    11.2385
 Acc. Voltage :       15.0 (kV)      Probe Dia. : 35     Scan : Off
 Dated on May 25 12:26 2023
 WDS only          No. of accumulation : 1

Curr.(A) : 1.966E-08
Element  Peak(mm)  Net(cps)   Bg-    Bg+   S.D.(%)    D.L.(ppm)
1 V     174.201      48.2   15.2   26.3    4.40        155
2 Al     90.299    1647.5   35.7   26.7    0.79         34
3 Mn    146.195     117.3   26.7   22.5    3.36        114
4 Ti     87.706   54535.5  249.6  243.7    0.10         84

ZAF Metal
Element   Mass(%)    Atom(%)   K(%)  K-raw(%)    ZAF       Z        A        F
 V          0.530    0.4882   0.518    0.518  1.0232  1.0260  0.9981  0.9991
 Al         4.474    7.7741   3.140    3.172  1.4248  0.9166  1.5544  1.0000
 Mn         0.946    0.8072   0.843    0.843  1.1225  1.0266  1.0934  1.0000
 Ti        94.040   90.9305  92.339   93.271  1.0061  1.0046  1.0020  0.9995
 ---------------------------------------------------------------------------
 Total    100.000 100.0000  96.840   97.804     Iteration = 4
```

Fig. 2.13 Results of specimen cross-sectional microanalysis (dark phase)

```
Unknown Specimen No. 4
 Group          :                Sample   : 9,2
 UNK No.        :     4               Comment : white
 Stage          :     X=    72.0855  Y=    27.6890  Z=    11.2385
 Acc. Voltage :       15.0 (kV)      Probe Dia. : 35     Scan : Off
 Dated on May 25 12:31 2023
 WDS only          No. of accumulation : 1

Curr.(A) : 1.963E-08
Element  Peak(mm)  Net(cps)   Bg-    Bg+   S.D.(%)    D.L.(ppm)
1 V     174.201      44.2   19.1   28.9    4.86        167
2 Al     90.304    1445.9   33.1   28.2    0.84         34
3 Mn    146.217     230.2   30.6   26.4    2.28        122
4 Ti     87.706   54354.6  256.5  234.5    0.10         84

ZAF Metal
Element   Mass(%)    Atom(%)   K(%)  K-raw(%)    ZAF       Z        A        F
V           0.486    0.4497   0.476    0.476  1.0215  1.0252  0.9980  0.9983
Al          3.941    6.8880   2.760    2.788  1.4295  0.9160  1.5607  1.0000
Mn          1.856    1.5916   1.656    1.656  1.1211  1.0258  1.0929  1.0000
Ti         93.717   91.0707  92.173   93.104  1.0047  1.0039  1.0019  0.9990
----------------------------------------------------------------------------
 Total    100.000 100.0000  97.065   98.024     Iteration = 4
```

Fig. 2.14 Results of specimen cross-sectional (bright phase) microanalysis

The content of principal elements of the material contained in the plug fragment of 9K720 *Iskander-M* missile body, viz., titanium base predominantly alloyed with aluminum, shows that the material is a heat- and corrosion-resistant titanium alloy of OT4 (VT4) type.

In terms of composition, the material belongs to near-α titanium alloys of Ti-Al-Mn system (OT4, VT4; in its improved version OT4U, VT4U manganese is substituted for vanadium), up to 0.9% Fe is allowed, which permits using ferroalloy master alloy for blending. Near-$\alpha\Sigma$ titanium alloys are structural ones and heat-resistant. Due to the presence of a small amount of β-phase, they have satisfactory cold and hot processing plasticity which is close to that of (α+β)-alloys. With the same plasticity, they have strength 10–20% higher than that of α-alloys, which is accounted for by grain refinement and intra-grain heterogenization when two-phase mixture of α and β phases is formed.

The presence of both dark and bright phases permits us to assume that this structure is a mixture of titanium α- and β-phases.

The globular and lamellar microstructure of α-phase with approximately the same proportion of these components, and a small amount of residual β-phase demonstrates that the alloy underwent standard rolling in the middle of the ($\alpha + \beta$) field at a temperature 40–50 °C than the β-transus (Ac_3). Respectively, the rolling temperature for the composition 4.5% Al,—1.5%, Mn with Ac_3 $\tilde{9}70$ °C (Taranenko 2014) should have been $\tilde{9}20$ °C. Rolling into sheets could have been carried out in several runs with intermediate heating. According to the grain size obtained, the total material reduction could have amounted to 70–80%. The final operation of part fabrication could have been cold sheet stamping.

2.3.3 *Results of Examining Fragments of Hypersonic Air-Launched Ballistic Missiles*

2.3.3.1 Fragment of Kh-47M2 *Kinzhal* (9-A-7660) Missile Body

Figure 2.15 shows a fragment of Kh-47M2 Kinzhal (9-A-7660) missile body. Figure 2.16 presents the backscattered electron image of specimen surface area. Bright areas (inclusions) on the image correspond to the presence of chemical elements heavier than those present in dark areas of the image.

X-ray microanalysis (measurements of chemical element concentrations) was carried out in both dark and bright areas (inclusions) that can be seen in Fig. 2.16.

Figures 2.17 and 2.18 present the results of specimen X-ray microanalysis (measurements of chemical element concentrations) obtained by means of X-ray microanalyzer in dark and bright areas (inclusions), respectively.

As can be seen in Figs. 2.17 and 2.18, the inclusions contain, apart from chemical elements present in the dark matrix, additional Fe (iron) and Mn (manganese). It is due to these elements that inclusions can be seen in the backscattered electron image (Fig. 2.16).

Fig. 2.15 Specimen for examinations (two sides of the sample are shown)

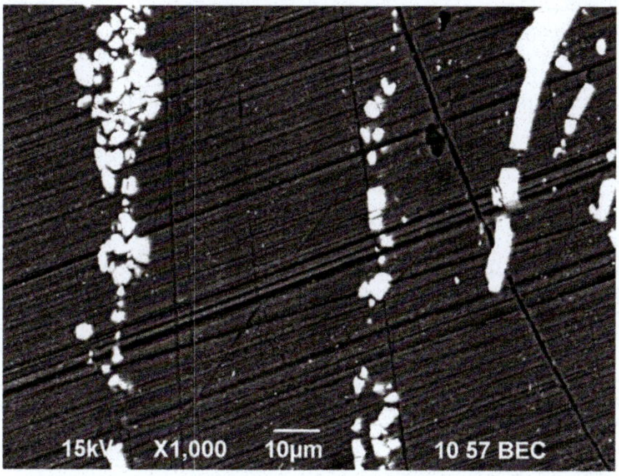

Fig. 2.16 Backscattered electron image of specimen surface area. Accelerating voltage is 15kV. Magnification is 1000, scale bar corresponds to 10 μm

The presence of iron and manganese in the inclusions is also confirmed by the 2*d* distributions of chemical elements on the surface that were obtained in X-ray characteristic radiation (Fig. 2.19).

The content of principal elements in the main (dark) area of Kh-47M2 Kinzhal (9-A-7660) hypersonic air-launched ballistic missile body fragment, which are shown in Figs. 2.17 and 2.18, shows that its material is aluminum alloy of AMg6 type, characterized with low density (2.65 g/cm^3), high strength, resistance to corrosion, and vibration loads.

This wrought aluminum-magnesium alloy is widely used in manufacturing special-purpose structures.

Inclusions (bright areas in Fig. 2.16) are secondary phases that formed due to the interaction of initial components. In the alloys of Al-Mg system, secondary phases

```
Unknown Specimen No. 1
 Group        :                Sample  : 8.1
 UNK No.      :    1             Comment :
 Stage        :    X=    79.8220  Y=    62.2600  Z=    11.0540
 Acc. Voltage :    15.0 (kV)     Probe Dia. : 0     Scan : Off
 Dated on May 22 15:10 2023
 WDS only        No. of accumulation : 1

Curr. (A) : 1.950E-08
Element Peak(mm) Net(cps)   Bg-    Bg+  S.D.(%)    D.L.(ppm)
1 Mg   107.462   4394.3  123.5   77.1    0.28        58
2 Al    90.284  41297.6  113.1   60.3    0.16        72
3 Si   228.274      1.6    2.0    1.6   45.07 ?      30

ZAF Metal
Element  Mass(%)   Atom(%)  K(%)  K-raw(%)    ZAF      Z        A        F
 Mg        7.745   8.5236   8.022    8.023  0.9546  0.9715  1.0426  0.9424
 Al       92.239  91.4606  99.571  188.119  0.9158  0.9640  0.9500  1.0000
 Si        0.016   0.0158   0.008    0.008  2.0550  0.9756  2.1063  1.0000
----------------------------------------------------------------------------
 Total   100.000 100.0000 107.601 196.149    Iteration = 5
```

Fig. 2.17 Results of bulk specimen microanalysis (dark areas in Fig. 2.16)

```
Unknown Specimen No. 2
 Group        :                Sample  : 8.1
 UNK No.      :    2             Comment : inclusion
 Stage        :    X=    79.7570  Y=    62.2185  Z=    11.0540
 Acc. Voltage :    15.0 (kV)     Probe Dia. : 0     Scan : Off
 Dated on May 22 15:03 2023
 WDS only        No. of accumulation : 1

Curr. (A) : 1.956E-08
Element Peak(mm) Net(cps)   Bg-    Bg+  S.D.(%)    D.L.(ppm)
1 Mg   107.474     84.6    65.7   48.2    3.28        44
2 Fe   134.608   1988.1   162.2   36.2    0.54       317
3 Al    90.294  27100.5    90.5   47.3    0.19        64
4 Mn   146.196   1716.9    23.8   22.1    0.77       125
5 Si   228.386    301.5     4.0    4.0    1.85        45

ZAF Metal
Element  Mass(%)   Atom(%)  K(%)  K-raw(%)    ZAF      Z        A        F
 Mg        0.214   0.2799   0.154    0.154  1.4159  0.9447  1.5335  0.9774
 Fe       15.035   8.5697  14.133   14.133  1.0852  1.0804  1.0044  1.0000
 Al       67.636  79.6781  65.141  123.069  1.0575  0.9374  1.1289  0.9994
 Mn       14.530   8.4184  14.101   18.207  1.0511  1.0400  1.0106  1.0000
 Si        2.585   3.0540   1.501    1.501  1.8318  0.9487  1.9310  1.0000
----------------------------------------------------------------------------

Total   100.000 100.0000  95.029 157.064
```

Fig. 2.18 Results of inclusion microanalysis (bright areas in Fig. 2.16)

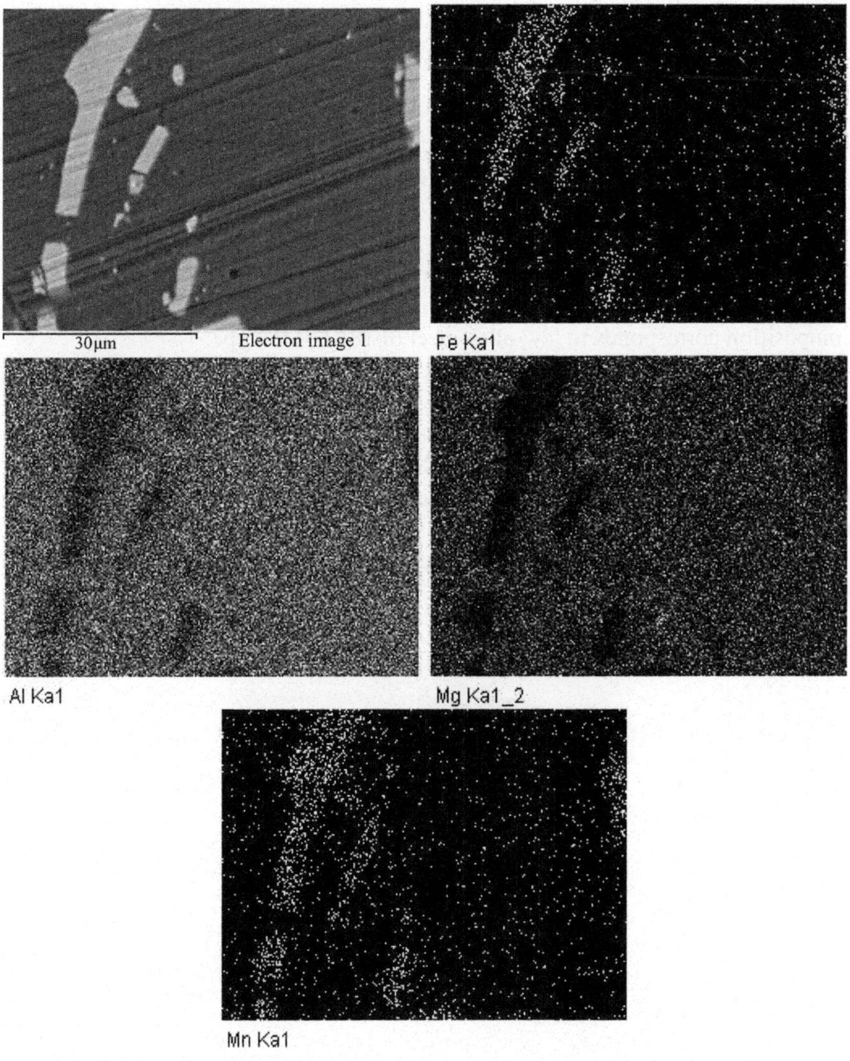

Fig. 2.19 Backscattered electron images of specimen surface area with 30 μm scale bar, along with X-ray characteristic radiation images of the same area ($2d$ distributions of chemical elements). White dots correspond to the presence of chemical element indicated under each image

of such compositions are, predominantly, Al6(Mn,Fe), α-Al(Fe,Mn)Si, Al3Fe, Alm(Mn,Fe), and Mg2Si intermetallic compounds [https://doi.org/10.1051/matecconf/202032602002].

2.3.3.2 Fragment of Tail Section of Kh-47M2 *Kinzhal* (9-A-7660) Missile Body

The content of principal elements in the material of the tail section fragment of Kh-47M2 *Kinzhal* (9-A-7660) hypersonic air-launched ballistic missile shows that its composition corresponds to low-alloy steel of 18Kh3MV type.

This steel is not brittle and has sufficient ductility and temperature resistance below 450–500 °C (Fig. 2.20).

Figure 2.21 presents the results of X-ray microanalysis of the specimen.

The content of principal elements in the material of the tail section fragment of Kh-47M2 *Kinzhal* (9-A-7660) hypersonic air-launched ballistic missile shows that its composition corresponds to low-alloy steel of 18Kh3MV type.

This steel is not brittle and has sufficient ductility and temperature resistance below 450–500 °C (Fig. 2.22).

Fig. 2.20 Specimen for examination

Unknown Specimen No. 2
Group : Sample : 8.2
UNK No. : 2 Comment :
Stage : X= 73.5770 Y= 39.9125 Z= 12.2195
Acc. Voltage : 15.0 (kV) Probe Dia. : 20 Scan : Off
Dated on May 25 12:14 2023
WDS only No. of accumulation : 1

Curr.(A) : 1.984E-08

Element		Peak(mm)	Net(cps)	Bg-	Bg+	S.D.(%)	D.L.(ppm)
1	Fe	134.578	13634.0	85.8	90.0	0.19	294
2	C	128.254	98.8	21.0	19.4	2.86	39
3	Cr	159.229	349.2	29.5	26.0	1.29	135
4	W	102.520	16.7	108.8	118.0	18.90	630
5	Si	228.274	32.9	8.6	5.2	6.57	58
6	Ni	115.302	112.8	75.8	62.6	4.44	250

ZAF Metal

Element	Mass(%)	Atom(%)	K(%)	K-raw(%)	ZAF	Z	A	F
Fe	96.081	95.3245	95.553	95.553	1.0017	1.0002	1.0024	0.9991
C	0.194	0.8998	0.194	103.088	1.0025	0.9997	1.0029	1.0000
Cr	2.266	2.4235	2.828	2.828	0.8013	1.0019	1.0064	0.7946
W	0.428	0.1296	0.313	0.314	1.3671	1.3402	1.0201	1.0000
Si	0.238	0.4717	0.161	0.161	1.4754	0.8906	1.6567	1.0000
Ni	0.793	0.7509	0.757	0.757	1.0467	0.9842	1.0637	0.9998
Total	100.000	100.0000	99.807	202.701	Iteration = 3			

Fig. 2.21 Results of specimen X-ray microanalysis

Fig. 2.22 Specimen for examination

2.3.3.3 Fragment of Inner Structure of Kh-47M2 *Kinzhal* (9-A-7660) Missile Body

Figure 2.23 presents the results of X-ray microanalysis of the specimen. Figure 2.24 shows a view of a fragment of the internal structure of the X-47M2 *Kinzhal* (9-A-7660) missile.

The specimen is not attracted by the magnet.

The content of principal elements present in the inner structure fragment of Kh-47M2 *Kinzhal* (9-A-7660) hypersonic air-launched ballistic missile shows that its material composition corresponds to heat-resistant steel of 12Kh25N16G7AR type.

In accordance with nomenclature, 12Kh25N16G7AR steel is used to produce sheets, wire, pipes, strips, and other parts operating below +950 °C under moderate loading; turbine parts (parts of gas pipeline systems, combustion chambers, membranes, sheet parts) operating at temperatures up to +1050 °C and under pressure up to 7760 kg/m^3; welding wire used in the welding of corrosion-resistant steels, including heat-strengthened ones (Fig. 2.24).

Unknown Specimen No. 2
Group : Sample : 8.3
UNK No. : 2 Comment :
Stage : X= 68.5830 Y= 59.0480 Z= 11.2880
Acc. Voltage : 15.0 (kV) Probe Dia. : 20 Scan : Off
Dated on May 22 16:14 2023
WDS only No. of accumulation : 1

Curr. (A) : 1.946E-08

Element	Peak (mm)	Net (cps)	Bg-	Bg+	S.D. (%)	D.L. (ppm)
1 Fe	134.618	7286.0	137.8	72.4	0.27	328
2 Cr	159.227	3350.5	28.0	27.0	0.39	137
3 Mn	146.156	832.1	63.1	34.2	0.82	192
4 Si	228.274	63.6	7.8	6.6	4.39	60
5 Ni	115.261	2132.8	73.6	70.4	0.71	260

ZAF Metal

Element	Mass (%)	Atom (%)	K (%)	K-raw (%)	ZAF	Z	A	F
Fe	52.156	51.5063	52.061	52.061	1.0112	1.0018	1.0240	0.9857
Cr	25.488	26.7831	27.663	27.663	0.9214	1.0034	1.0055	0.9132
Mn	6.764	6.7270	7.041	9.092	0.9606	0.9665	1.0042	0.9897
Si	0.466	0.9056	0.318	0.318	1.4628	0.8916	1.6408	1.0000
Ni	15.126	14.0781	14.598	14.598	1.0362	0.9860	1.0509	1.0000
Total	100.000	100.0000	101.681	103.732	Iteration = 4			

Fig. 2.23 Results of specimen X-ray microanalysis

Fig. 2.24 Specimen for examination

2.3.3.4 Fragment of Kh-47M2 *Kinzhal* (9-A-7660) Missile Power Structure

Figure 2.25 presents the results of X-ray microanalysis of the specimen. Figure 2.26 shows a view of a fragment of the Kh-47M2 *Kinzhal* (9-A-7660) missile.

The specimen is not attracted by the magnet.

In material composition, the power structure fragment of Kh-47M2 Kinzhal missile does not have considerable differences from the fragment of its body inner structure. This permits the conclusion that it is made of a heat-resistant steel with high chromium and nickel content.

According to element analysis, the composition close to it could be that of the Sv-30Kh25N16G7 wire which is used for welding high-alloy steels, low-alloy and carbon steels, for electric submerged arc welding, and for welding non-ferrous metals.

A more likely option is the steel of 12Kh25N16G7AR type, as well as the inner structure fragment of Kh-47M2 *Kinzhal* missile body, which was considered above. Both steel grades are used to produce sheets, wire, pipes, ribbon, and parts operating at temperatures up to +950 °C under moderate loading; parts of turbines (parts of gas pipelines, combustion chambers, membranes, sheet parts) operating at temperatures up to +1050 °C and under pressure up to 7760 kg/m^3.

```
Unknown Specimen No. 1
 Group          :              Sample  : 8.4
 UNK No.        :    1             Comment :
 Stage          :    X=   58.2525  Y=   62.7580  Z=   11.2415
 Acc. Voltage :     15.0 (kV)      Probe Dia. : 0     Scan : Off
 Dated on May 23 19:23 2023
 WDS only         No. of accumulation : 1

Curr.(A)  : 1.973E-08
Element  Peak(mm)  Net(cps)    Bg-     Bg+   S.D.(%)    D.L.(ppm)
1 Ni    115.198   2189.4     82.2    78.4     0.70       276
2 C     128.254     45.4     18.0    21.8     5.05        39
3 Cr    159.240   3289.7     31.7    34.0     0.39       148
4 Fe    134.577   7580.4    121.0    68.4     0.26       307
5 Si     77.043     65.9     50.2    33.8     5.87        53
6 Mn    146.217    764.6     59.9    39.2     1.20       161
ZAF Metal
Element   Mass(%)    Atom(%)  K(%)   K-raw(%)    ZAF       Z         A         F
 Ni        15.640  14.6442  15.094  15.094  1.0362  0.9861  1.0508  1.0000
 C          0.088   0.4042   0.089  47.603  0.9868  1.0015  0.9854  1.0000
 Cr        24.599  26.0048  26.789  26.789  0.9182  1.0035  1.0056  0.9099
Fe         54.008  53.1100  53.423  53.423  1.0100  1.0020  1.0233  0.9851
Si          0.175   0.3433   0.120   0.120  1.4676  0.8918  1.6458  1.0000
 Mn         5.490   5.4934   5.423   5.423  1.0125  1.0214  1.0020  0.9893
-----------------------------------------------------------------------------
 Total    100.000 100.0000 100.938 148.452     Iteration = 4
```

Fig. 2.25 Results of specimen X-ray microanalysis

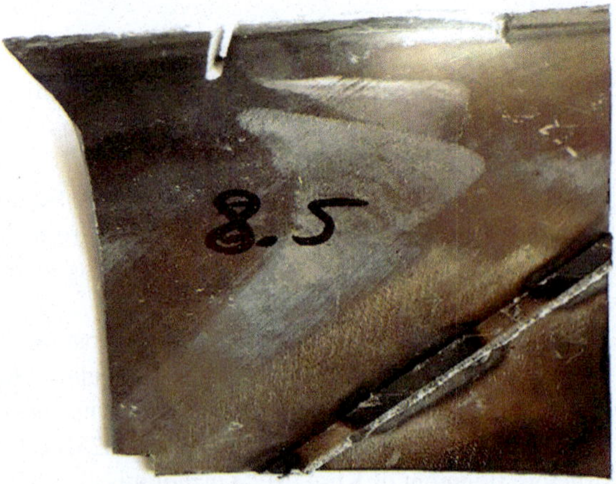

Fig. 2.26 Specimen for examination

2.3.3.5 Fragment of the Aerodynamic Flight Control of Kh-47M2 *Kinzhal* (9-A-7660) Missile

Specimen studies with scanning electron microscope have shown that in the bulk material examined there are inclusions containing chemical elements heavier than those present in the larger part of the bulk of the studied specimen bulk material (Fig. 2.26).

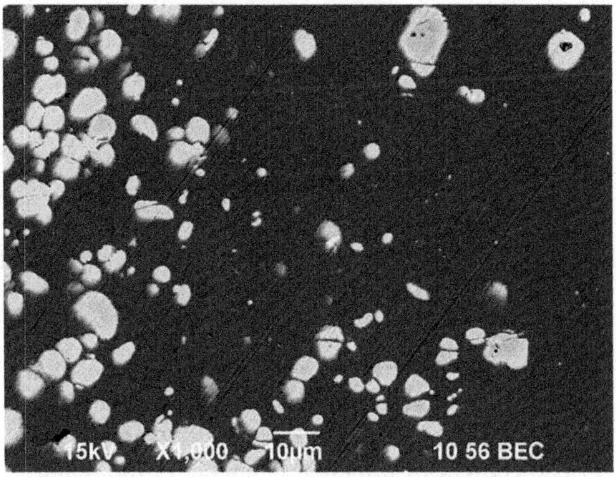

Fig. 2.27 Backscattered electron image of specimen surface. Accelerating voltage is 15 kV. Magnification is 1000, scale bar corresponds to 10 μm

Figure 2.27 presents the backscattered electron image of specimen surface area. Image bright areas (inclusions) correspond to the presence of chemical elements heavier than those present in the dark areas.

Unknown Specimen No. 4
Group : Sample : 8.5
UNK No. : 4 Comment :
Stage : X= 45.0800 Y= 61.4850 Z= 11.2655
Acc. Voltage : 15.0 (kV) Probe Dia. : 3 Scan : Off
Dated on May 25 14:33 2023
WDS only No. of accumulation : 1

Curr.(A) : 1.944E-08

Element	Peak(mm)	Net(cps)	Bg-	Bg+	S.D.(%)	D.L.(ppm)
1 Ti	87.709	901.7	216.3	189.5	0.90	78
2 Al	90.349	688.2	33.7	24.5	1.24	34
3 Cr	159.233	2586.7	32.6	34.1	0.45	151
4 Fe	134.613	224.0	49.2	50.6	1.93	226
5 Mo	173.338	339.7	20.9	19.9	1.28	133
6 Ni	115.221	8790.7	119.0	118.4	0.34	341
7 W	102.568	322.0	124.7	109.7	2.27	604

ZAF Metal

Element	Mass(%)	Atom(%)	K(%)	K-raw(%)	ZAF	Z	A	F
Ti	1.542	1.9169	1.544	1.560	0.9990	0.9977	1.0435	0.9595
Al	2.759	5.7643	1.361	1.364	1.9196	0.9139	2.1008	0.9999
Cr	21.102	23.9544	21.379	21.379	0.9787	0.9985	1.0218	0.9593
Fe	1.499	1.5976	1.602	1.602	0.9353	0.9957	1.0293	0.9127
Mo	4.118	2.5553	3.166	3.166	1.3009	1.0883	1.1962	0.9993
Ni	61.433	61.7676	61.563	61.563	0.9895	0.9783	1.0156	0.9959
W	7.547	2.4438	5.568	5.574	1.3555	1.3285	1.0203	1.0000
Total	100.000	100.0000	96.183	96.207	Iteration = 4			

Fig. 2.28 Results of microanalysis in specimen bulk (dark areas)

X-ray microanalysis (measurement of chemical element concentrations) has been carried out in both dark and bright areas (inclusions) that can be seen in Fig. 2.27.

Figures 2.28 and 2.29 present the results of specimen X-ray microanalysis (measurement of chemical element concentrations), which have been obtained by means of X-ray microanalyzer in dark and bright areas (inclusions), respectively.

As can be seen in Figs. 2.28 and 2.29, inclusions contain tungsten and molybdenum in high concentrations (45.605% W and 17.642% Mo) as compared to concentrations of these chemical elements in the specimen bulk. It is due to increased concentrations of these elements in the backscattered electron image (Fig. 2.27) that the inclusions are visible.

The increased W and Mo concentrations in the inclusions is also confirmed by the distribution of chemical elements obtained in X-ray characteristic radiation (Fig. 2.30).

The content of principal elements in the material of the fragment of Kh-47M2 Kinzhal (9-A-7660) hypersonic air-launched missile shows that its material composition corresponds to a nichrome alloy of KhN56VMTU type with high heat resistance and chemical stability.

Heat-resistant alloy of KhN56VMTU grade (another designation EP199) is used to manufacture highly loaded parts, connections, flanges, sheet-like parts for a limited service life at temperatures up to 950 °C (Fig. 2.31).

Unknown Specimen No. 3
Group : Sample : 8.5
UNK No. : 3 Comment : inclusive
Stage : X= 45.0960 Y= 61.4975 Z= 11.2655
Acc. Voltage : 15.0 (kV) Probe Dia. : 3 Scan : Off
Dated on May 25 14:29 2023
WDS only No. of accumulation : 1

Curr.(A) : 1.944E-08

Element	Peak(mm)	Net(cps)	Bg-	Bg+	S.D.(%)	D.L.(ppm)
1 Ti	87.732	795.2	281.4	249.6	1.03	89
2 Al	90.304	142.4	47.3	41.9	3.22	42
3 Cr	159.318	1665.5	48.0	42.5	0.56	176
4 Fe	134.672	85.5	83.0	62.6	4.56	273
5 Mo	173.359	1367.3	34.9	23.6	0.62	159
6 Ni	115.212	3231.4	141.8	188.8	0.58	403
7 W	102.531	2127.5	196.5	155.0	0.74	752

ZAF Metal

Element	Mass(%)	Atom(%)	K(%)	K-raw(%)	ZAF	Z	A	F
Ti	1.419	2.6863	1.362	1.375	0.9920	0.9025	1.1148	0.9860
Al	0.418	1.4064	0.282	0.282	1.4147	0.8300	1.7055	0.9994
Cr	13.657	23.8262	13.765	13.765	0.9449	0.9003	1.0630	0.9873
Fe	0.581	0.9433	0.611	0.611	0.9047	0.8938	1.0462	0.9675
Mo	17.642	16.6825	12.741	12.741	1.3187	0.9891	1.3336	0.9997
Ni	20.678	31.9514	22.630	22.630	0.8702	0.8730	1.0250	0.9725
W	45.605	22.5040	36.782	36.819	1.1808	1.1720	1.0075	1.0000
Total	100.000	100.0000	88.174	88.225	Iteration = 4			

Fig. 2.29 Results of microanalysis in bright areas (inclusions)

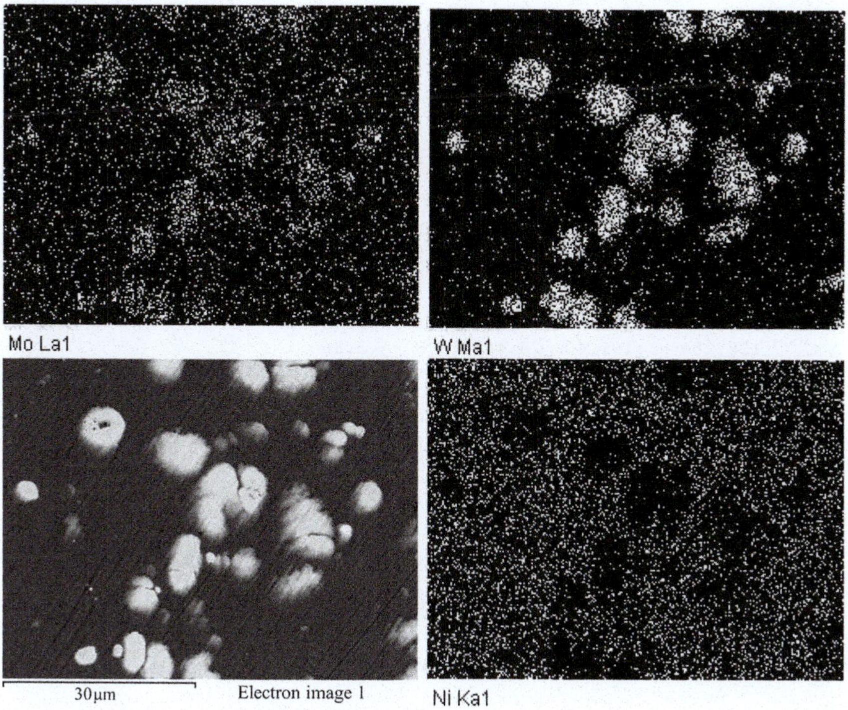

Fig. 2.30 Backscattered electron image of specimen surface area with 30 µm scale bar, and X-ray characteristic radiation images of the same area (2*d* distribution of chemical elements). White dots correspond to the presence of the element indicated under each image

Fig. 2.31 Specimen for examination

2.3.3.6 Plug of Kh-47M2 *Kinzhal* (9-A-7660) Missile Body

Figure 2.31 shows a view of a fragment of aerodynamic flight control of the X-47M *Kinzhal* (9-A-7660) missile. Figure 2.32 presents the results of X-ray microanalysis of the specimen.

The content of principal elements in the material of the plug in the body of Kh-47M2 *Kinzhal* (9-A-7660) hypersonic air-launched missile shows that the material composition corresponds to a heat- and corrosion-resistant steel of 08Kh16N13M2B type.

```
   Unknown Specimen No. 2
 Group        :                    Sample  : 8.6
 UNK No.      :     2          Comment :
 Stage        :     X=    77.3450  Y=    80.1565  Z=    11.4250
 Acc. Voltage :     15.0 (kV)      Probe Dia. : 20     Scan : Off
 Dated on May 23 19:28 2023
 WDS only          No. of accumulation : 1

Curr.(A) : 1.971E-08
Element Peak(mm) Net(cps)    Bg-    Bg+  S.D.(%)    D.L.(ppm)
1 Fe    134.586   7202.6  143.6   59.6     0.27       318
2 Si     77.417      4.9   49.2   43.4    63.72 ?      55
3 Mn    146.305    867.4   65.5   39.9     1.13       168
4 Cr    159.232   3233.8   32.7   29.7     0.40       151
5 C     128.254    101.4   24.4   19.8     2.86        42
6 Ni    115.222   2206.6   80.0   76.8     0.70       268
ZAF Metal
Element  Mass(%)    Atom(%)  K(%)   K-raw(%)     ZAF        Z        A        F
Fe      52.007  50.9925  50.812  50.812  1.0110  1.0018  1.0244  0.9851
 Si        0.014   0.0257   0.009   0.009  1.4654  0.8916  1.6436  1.0000
Mn         6.369   6.3478   6.215   6.215  1.0122  1.0212  1.0019  0.9893
Cr      25.774  27.1427  27.614  27.614  0.9220  1.0034  1.0054  0.9139
 C         0.198   0.9049   0.200 106.500  0.9791  1.0014  0.9777  1.0000
 Ni       15.638  14.5864  14.911  14.911  1.0359  0.9860  1.0507  1.0000
--------------------------------------------------------------------------
 Total   100.000 100.0000  99.761 206.061    Iteration = 4
```

Fig. 2.32 Results of specimen X-ray microanalysis

Fig. 2.33 Specimen for examination

Its main application is in forgings for disks and rotors, blades and bolts operating at temperatures up to 600 °C for a long time (Fig. 2.33).

2.3.3.7 Fragment of 3M22 *Zircon* Missile Body

Figure 2.33 shows a view of a fragment of the plug of the Kh-47M2 *Kinzhal* (9-A-7660) missile body. The results of examining specimen chemical composition with a JXA-8200 microanalyzer (JEOL company, Japan) are presented in Fig. 2.34 (Table 2.5).

The content of principal elements in the material of 3M22 Zircon hypersonic missile body shows that the material composition corresponds to a heat- and relaxation-resistant low-alloy steel of 30 KhMA type (Taranenko 2014).

30KhMA steel is used for producing bodies and inner elements of bodies, flanges, fastening parts that are to resist oxidation at temperatures up to +450 °C (Fig. 2.35).

```
Unknown Specimen No. 3
 Group          :                 Sample   : rak-10.2
 UNK No.        :    3               Comment :
 Stage          :    X=    66.7690   Y=    61.8735   Z=    10.9455
 Acc. Voltage :      15.0 (kV)       Probe Dia. : 10     Scan : Off
 Dated on Feb 13 18:11 2024
 WDS only          No. of accumulation : 1

Curr.(A)  : 1.001E-08
Element  Peak(mm)  Net(cps)    Bg-     Bg+    S.D.(%)     D.L.(ppm)
1 Mn      67.189      13.3    82.7    55.5    25.42        255
2 Cr     159.309      69.4    12.3    13.6     3.15        191
3 Si      77.098     202.9    20.6    20.8     2.44         73
4 Fe     134.692    7220.0    46.6    51.4     0.27        416
5 C      128.254      71.6    11.4    12.2     3.23         60

ZAF Metal
Element  Mass(%)    Atom(%)   K(%)   K-raw(%)     ZAF       Z        A        F
 Mn        0.127    0.1279    0.131    0.169   0.9716   0.9670   1.0047   1.0000
 Cr        0.925    0.9846    1.167    1.167   0.7926   1.0040   1.0062   0.7845
 Si        1.070    2.1081    0.725    0.725   1.4750   0.8922   1.6531   1.0000
 Fe       97.315   95.4696   95.994   95.994   1.0033   1.0024   1.0010   1.0000
 C         0.284    1.3097    0.278  148.036   1.0209   1.0020   1.0189   1.0000

---------------------------------------------------------------------------------
Total     99.721 100.0000   98.295 246.091     Iteration = 3
```

Fig. 2.34 Results of specimen X-ray microanalysis

Table 2.5 Chemical composition of the fragment material in mass concentrations

Chemical elements	Mass concentrations, %
Mn, manganese	0.127
Cr, chromium	0.925
Si, silicon	1.070
Fe, iron	97.315
C, carbon	0.284

Fig. 2.35 Specimen for examination

2.3.3.8 Fragment of 3M22 *Zircon* Missile Body Element

Figure 2.35 shows a view of the first fragment of the 3M22 *Zircon* missile body. The results of investigating the specimen chemical composition are presented in Fig. 2.36 and Table 2.6.

The content of principal elements of the fragment material of 3M22 Zircon supersonic missile shows that it is made of titanium alloy which consists of three main chemical elements (Ti, Al, V). This is a heat- and corrosion-resistant titanium alloy of OT4-0 type (Taranenko 2014) produced in accordance with GOST 19807-91 "Tita-

```
Unknown Specimen No. 5
 Group          :                 Sample   : rak-10.3
 UNK No.        :    5            Comment  :
 Stage          :    X=   57.5825  Y=    61.3670  Z=    10.8670
 Acc. Voltage  :    15.0 (kV)     Probe Dia. : 10     Scan : Off
 Dated on Feb 13 18:47 2024
 WDS only           No. of accumulation : 1

Curr.(A)  : 1.100E-08
Element Peak(mm) Net(cps)   Bg-    Bg+  S.D.(%)    D.L.(ppm)
1 Ti     87.720  31240.3  136.2  134.3    0.13       111
2 Al     90.328    360.4   15.2   13.1    1.71        51
3 V    174.057     20.6   10.3   18.8    7.65       204

ZAF Metal
Element  Mass(%)    Atom(%)  K(%)  K-raw(%)    ZAF      Z       A       F
 Ti       94.798  96.3489  94.539  95.494  1.0027  1.0020  1.0008  1.0000
 Al        1.835   3.3117   1.540   2.910  1.1914  0.8792  1.3551  1.0000
 V         0.355   0.3394   0.348   0.348  1.0204  1.0232  0.9972  1.0000
 -----------------------------------------------------------------------

 Total    96.988 100.0000  96.427  98.752    Iteration = 3
```

Fig. 2.36 Results of specimen X-ray microanalysis

Table 2.6 Chemical composition of fragment material in mass concentrations

Chemical elements	Mass concentrations, %
Ti, titanium	97.241
Al, aluminum	1.892
V, vanadium	0.366

nium and wrought titanium alloys" (developed by the USSR Ministry of Aviation Industry (Ostapchuk et al. 2015)).

Titanium alloy of OT4-0 type is a wrought titanium alloy used to manufacture thin-walled complex-shaped parts that operate at temperatures up to 350 °C for a long time and at 500–600 °C for a short term.

The content of principal elements of the material of the fragment of 3M22 *Zircon* supersonic missile shows that its chemical composition corresponds to the titanium alloy contained in R-31 air-to-air missile and in the body of R-800 *Oniks* supersonic anti-ship missile.

2.3.3.9 Fragment of Element of 3M22 *Zircon* Missile Body

Figures 2.37 and 2.39 show the second and third fragments of the 3M22 *Zircon* missile body. Investigations of 10.4 specimen chemical composition have been carried out with a JXA-8200 X-ray microanalyzer produced by JEOL company (Japan) (Fig. 2.37).

The results of investigating specimen chemical composition are presented in Fig. 2.38.

The content of principal elements of the material of 3M22 *Zircon* hypersonic missile body fragment shows that its composition corresponds to high-alloy heat-resistant steel of 08Kh16N13M2B (EI680) and 31Kh19N9MVBT (EI572) type (Taranenko 2014) (Table 2.7).

According to the classification, 08Kh16N13M2B (EI680) and 31X19H9MVBT (EI572) steels are used to manufacture disks and rotors, blades and fastening parts operating for a long time at temperatures up to +630 °C (Fig. 2.39).

Fig. 2.37 Specimen for examination

```
Unknown Specimen No. 3
 Group          :                  Sample  : rak-10.4
 UNK No.        :     3                  Comment :
 Stage          :     X=    46.4045  Y=    60.2560   Z=    10.9395
 Acc. Voltage :     15.0 (kV)      Probe Dia. : 10     Scan : Off
 Dated on Feb 13 19:15 2024
 WDS only          No. of accumulation : 1

Curr.(A) : 1.084E-08
Element Peak(mm) Net(cps)    Bg-     Bg+   S.D.(%)     D.L.(ppm)
1 Ti     87.837     58.0  113.9    87.1     6.30          98
2 Si     77.034    117.7   30.8    21.4     3.50          75
3 Fe    134.669   5687.2   43.4    37.0     0.30         348
4 Cr    159.209   1286.0   16.6    13.5     0.63         190
5 Ni    115.234    738.4   44.8    34.8     1.22         341

ZAF Metal
Element  Mass(%)     Atom(%)  K(%)   K-raw(%)    ZAF       Z        A         F
 Ti        0.168    0.1966   0.178    0.180  0.9439  1.0022  1.0195  0.9238
 Si        0.571    1.1383   0.389    0.389  1.4692  0.8922  1.6468  1.0000
 Fe       71.561   70.7523  69.825   69.825  1.0105  1.0024  1.0172  0.9911
 Cr       17.687   19.0474  19.967   19.967  0.8858  1.0040  1.0060  0.8770
 Ni        9.594    8.8653   8.922    8.922  1.0417  0.9865  1.0560 1.0000
 --------------------------------------------------------------------------

 Total    99.581 100.0000  99.281   99.283    Iteration = 4
```

Fig. 2.38 Results of specimen X-ray microanalysis

Fig. 2.39 Specimen for examination

Table 2.7 Chemical composition of fragment material in mass concentrations

Chemical elements	Mass concentrations, %
Ti, titanium	0.168
Si, silicon	0.571
Fe, iron	71.561
Cr, chromium	17.687
Ni, nickel	9.594

2.3.4 *Results of Examining Supersonic Anti-ship Missile Fragments*

2.3.4.1 Body Fragment of P-800 *Oniks* Supersonic Anti-ship Missile

Figure 2.40 gives an example of specimen X-ray microanalysis results. Chemical element concentrations are shown in mass percentages. Figure 2.41 shows a view of a fragment of the P-800 *Oniks* supersonic anti-ship missile.

The content of principal elements in the body fragment of P-800 Oniks supersonic anti-ship missile shows that it is produced from titanium alloy and consists of four principal chemical elements (Ti, Al, V, Mn). The material corresponds to heat- and corrosion-resistant titanium alloy of OT4-0 type, produced in accordance with GOST 19807-91 "Titanium and wrought titanium alloys" developed by USSR Ministry of Aviation Industry.

Titanium alloy of OT4-0 type is a wrought titanium alloy used to manufacture thin-walled complex-shaped parts that operate for a long time at temperatures below 350 °C and for short term—at temperatures up to 500–600 °C.

The content of principal elements in the material of the fragment of P-800 Oniks supersonic anti-ship missile body shows that its chemical composition corresponds to the titanium alloy from which the body of R-31 air-to-air missile is produced (Fig. 2.41).

```
Unknown Specimen No. 6
 Group         :                 Sample   : p800
 UNK No.       :     6              Comment :
 Stage         :     X=    23.0165  Y=    47.1455  Z=    11.2665
 Acc. Voltage  :     15.0 (kV)      Probe Dia. : 2     Scan : Off
 Dated on Dec 12 20:25 2022
 WDS only         No. of accumulation : 1

Curr.(A) : 1.966E-08
```

Element	Peak(mm)	Net(cps)	Bg-	Bg+	S.D.(%)	D.L.(ppm)
1 Al	89.488	392.3	826.6	65.1	2.66	109
2 Ti	87.472	3328.9	50.1	69.5	0.39	772
3 V	174.201	38.4	15.4	27.2	5.24	157
4 C	128.254	74.7	20.2	24.0	3.55	42
5 Mn	146.234	370.6	29.8	25.3	1.73	137

ZAF Metal

Element	Mass(%)	Atom(%)	K(%)	K-raw(%)	ZAF	Z	A	F
Al	0.912	1.4690	0.632	0.633	1.4440	0.9130	1.5816	1.0000
Ti	94.080	95.2963	105.115	106.177	0.9997	1.0005	1.0008	0.9984
V	0.420	0.3583	0.413	0.413	1.0162	1.0217	0.9972	0.9974
C	0.093	0.3361	0.148	78.693	0.6280	1.0034	0.6259	1.0000
Mn	3.213	2.5404	3.028	3.910	1.0611	0.9673	1.0969	1.0000
Total	100.718	100.0000	109.336	189.826	Iteration = 3			

Fig. 2.40 Results of specimen X-ray microanalysis

Fig. 2.41 Specimen for examination

2.3.5 Results of Examining Air-to-Air Missile Fragments

2.3.5.1 Fragment of R-37 Air-to-Air Missile

Figure 2.42 provides an example of specimen X-ray microanalysis. Chemical element concentrations are given in mass percentages. Figure 2.43 shows the R-37 air-to-air missile fragment. Figure 2.44 shows a view of the second fragment of the R-37 air-to-air missile.

```
Unknown Specimen No. 5
 Group           :                Sample   : r37
 UNK No.         :     5             Comment :
 Stage           :     X=    64.0680  Y=    34.9545  Z=    10.9385
 Acc. Voltage :     15.0 (kV)      Probe Dia. : 15     Scan : Off
 Dated on Dec 12 19:21 2022
 WDS only          No. of accumulation : 1

Curr.(A) : 1.965E-08
Element Peak(mm) Net(cps)    Bg-     Bg+  S.D.(%)    D.L.(ppm)
1 Al     89.569   1730.8    86.2    68.5     0.79        45
2 Ti     87.472   3021.8    49.3    69.0     0.41       768
3 V     174.201     41.7    15.8    25.4     4.88       155
4 Mn     67.189    197.3   112.1    81.3     2.96       154

ZAF Metal
Element  Mass(%)    Atom(%)  K(%)   K-raw(%)    ZAF       Z        A        F
Al         3.980   6.7711    2.789    2.795  1.4271  0.9160  1.5580  1.0000
Ti        94.954  91.9406   95.464   96.428  1.0051  1.0040  1.0017  0.9994
V          0.459   0.4137    0.449    0.449  1.0222  1.0253  0.9979  0.9991
Mn         1.047   0.8746    0.984    1.270  1.0644  0.9709  1.0963  1.0000
 ---------------------------------------------------------------

 Total    100.440 100.0000  99.686 100.942    Iteration = 4
```

Fig. 2.42 Results of specimen X-ray microanalysis

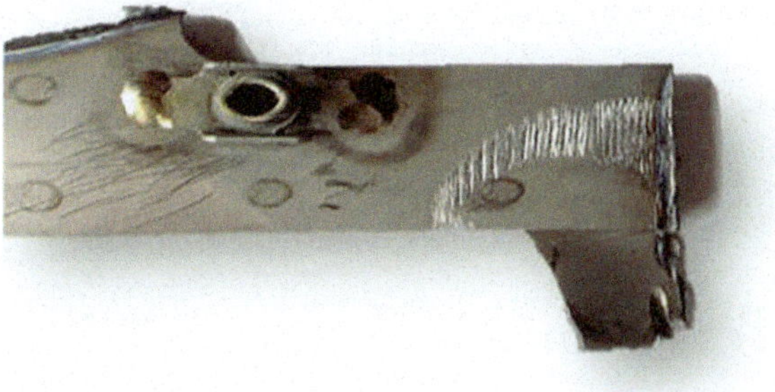

Fig. 2.43 Specimen for examination

Fig. 2.44 Specimen for examination

2.3.5.2 Fragment of R-31 Air-to-Air Missile Body

Figure 2.45 provides an example of the results of specimen fragment X-ray microanalysis. Chemical element concentrations are given in mass percentages.

The results of examining the body fragments of R-37 and R-31 air-to-air missiles show that they are made of titanium alloys and consist of four principal chemical elements (Ti, Al, V, Mn), but the concentrations of alloying elements vary. Vanadium (V) concentration is an exception—it is practically the same in all fragments and is approximately 0.4%.

The content of principal elements in the material of the fragment of R-37 air-to-air missile body corresponds to the material from which the plug of 9K720 *Iskander-M* missile body is produced—it is a heat- and corrosion-resistant titanium alloy of

```
Unknown Specimen No. 10
 Group          : mil             Sample    : x31
 UNK No.        :    10           Comment :
 Stage          :     X=    73.6620  Y=     29.7625   Z=     10.4145
 Acc. Voltage :     15.0 (kV)     Probe Dia. : 0      Scan : Off
 Dated on Dec 12 18:47 2022
 WDS only          No. of accumulation : 1

Curr.(A)  : 1.969E-08
Element Peak(mm) Net(cps)     Bg-     Bg+   S.D.(%)      D.L.(ppm)
1 Al      89.507     318.8   731.0    55.6      3.03          102
2 Mn      67.206     821.0   109.2    80.3      1.19          152
3 V      174.201      39.5    15.7    26.5      5.12          156
4 O      110.871     128.3    75.4    54.4      2.80           78
5 Ti      87.472    2967.4    47.2    71.9      0.42          769

ZAF Metal
Element  Mass(%)     Atom(%)   K(%)   K-raw(%)     ZAF        Z        A        F
 Al         0.744     1.2476   0.513     0.514  1.4519  0.9166   1.5840   1.0000
 Mn         4.340     3.5726   4.085     5.275  1.0624  0.9714   1.0937   1.0000
 V          0.433     0.3842   0.425     0.425  1.0190  1.0259   0.9972   0.9961
 O          1.194     6.2004   0.392     1.736  5.6022  0.9710   5.7694   1.0000
 Ti        93.831    88.5951  93.558    94.503  1.0029  1.0045   1.0007   0.9977

 --------------------------------------------------------------------------
 Total    100.542 100.0000   98.972 102.452     Iteration = 4
```

Fig. 2.45 Results of specimen X-ray microanalysis

OT4 (VT4) type. Its composition corresponds to the improved alloy variant—OT4U, VT4U.

The content of principal elements in the material of the fragment of R-31 air-to-air missile body, viz., titanium base predominantly alloyed with aluminum, shows that it is a heat- and corrosion-resistant titanium alloy of OT4-0 type produced in accordance with GOST 19807 "Titanium and wrought titanium alloys", developed by USSR ministry of Aviation Industry (Fig. 2.44).

Titanium alloy of OT4-0 type is a wrought titanium alloy used to manufacture thin-walled complex-shaped parts that operate for a long time at temperatures up to 350 °C, and for short term—at temperatures up to 500–600 °C.

2.3.6 *Results of Examining Loitering Munition Fragments*

2.3.6.1 Fragment of Warhead Body of *Geran-2* Loitering Munition

Figure 2.46 presents the results of specimen X-ray microanalysis.

The content of principal elements in the material of the body fragment of *Geran-2* loitering munition shows that the material composition corresponds to structural carbon steel VSt5ps of regular quality.

```
Unknown Specimen No. 1
 Group          : mil                Sample    : 31.10.22
UNK No.        :    1               Comment :
 Stage          :    X=    10.4100  Y=    45.9040  Z=    11.1160
 Acc. Voltage :    15.0 (kV)     Probe Dia. : 40     Scan : Off
 Dated on Oct 31 19:08 2022
 WDS only        No. of accumulation : 1
```

Curr.(A) : 2.092E-08

Element		Peak(mm)	Net(cps)	Bg-	Bg+	S.D.(%)	D.L.(ppm)
1	Mn	67.259	161.9	194.9	117.3	3.86	183
2	Si	77.342	85.3	50.8	45.4	4.99	53
3	Fe	134.642	15506.5	96.4	99.6	0.18	282
4	C	128.101	171.5	23.0	20.8	2.01	39

ZAF Metal

Element	Mass(%)	Atom(%)	K(%)	K-raw(%)	ZAF	Z	A	F
Mn	0.736	0.7373	0.758	0.979	0.9703	0.9658	1.0046	1.0000
Si	0.214	0.4196	0.145	0.145	1.4790	0.8912	1.6596	1.0000
Fe	98.750	97.3747	98.649	98.649	1.0010	1.0011	1.0000	1.0000
C	0.320	1.4683	0.319	169.754	1.0034	1.0006	1.0028	1.0000
Total	100.020	100.0000	99.871	269.527	Iteration = 3			

Fig. 2.46 Results of specimen X-ray microanalysis

VSt5ps steel is used to manufacture bodies and inner elements of tubes, sleeves, tubular gratings, and fastening elements of cross section up to 800 mm that operate at temperatures from 0 to +425 °C.

2.3.7 Results of Examining Solders in Fragments of Electronic Components of Means of Destruction

2.3.7.1 Solder of Fragments of Kh-101 Cruise Missile Components

Figure 2.47 presents an example of the results of X-ray microanalysis of solder in Kh-101 cruise missile electronic components (specimen 1). The microanalysis was carried out in the indicated rectangular area. The measured chemical element concentrations were averaged within this area which is shown in backscattered electron image. Chemical element concentrations are presented in mass percentages. Chemical element concentrations were calculated using K, L, M lines of characteristic X-ray radiation. The figure also shows the characteristic X-ray radiation spectrum for this surface area.

Examinations of specimen 2 showed the following chemical composition (wt.%): 0.2% N; 64.10% Sn; 35.70% Pb.

Element	Mass %
N K	0.29
Sn L	64.10
Pb M	35.61
Total	100.00

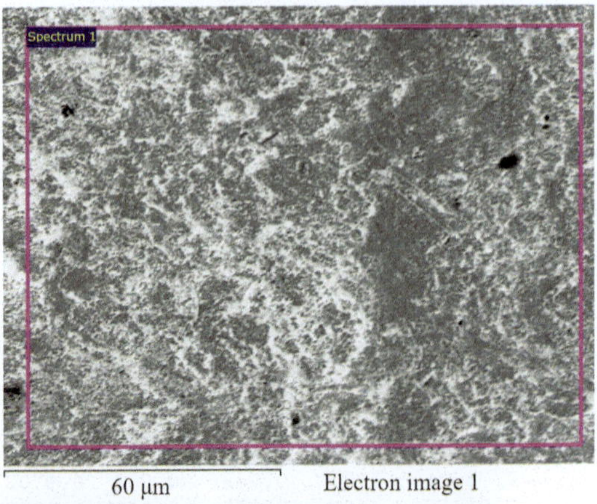

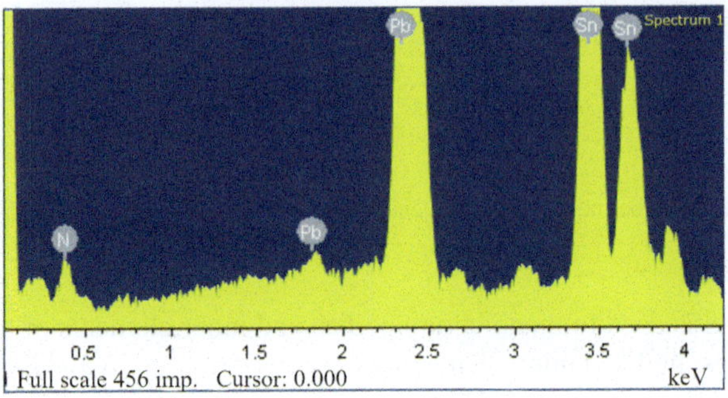

Fig. 2.47 Results of solder X-ray microanalysis in fragment of Kh-101 cruise missile (specimen 1)

2.3.7.2 Solder in Electronic Component Fragments of 3M-14 Cruise Missile of *Kalibr* Complex

Figure 2.48 presents an example of the results of X-ray microanalysis of solder in a fragment of electronic components of 3M-14 cruise missile of *Kalibr* complex (specimen 1). The microanalysis was performed in the indicated rectangular area. The measured chemical element concentrations were averaged within this area whose backscattered electron image is shown. Chemical element concentrations are presented in mass percentages. Chemical element concentrations were calculated using K, L, M lines of characteristic X-ray radiation. The figure also shows the spectrum of characteristic X-ray radiation for this measurement area.

Studies of specimens 2, 3, 4 have shown the following chemical compositions and chemical element concentrations (wt.%): Specimen 2: 1.43% N, 57.17% Sn, 41.40% Pb; Specimen 3: 1.42% N, 57.32% Sn, 41.26% Pb; Specimen 4: 1.60% N, 57.01% Sn, 41.39% Pb.

Element	Mass %
N K	1.56
Sn L	57.08
Pb M	41.36
Toltal	100.00

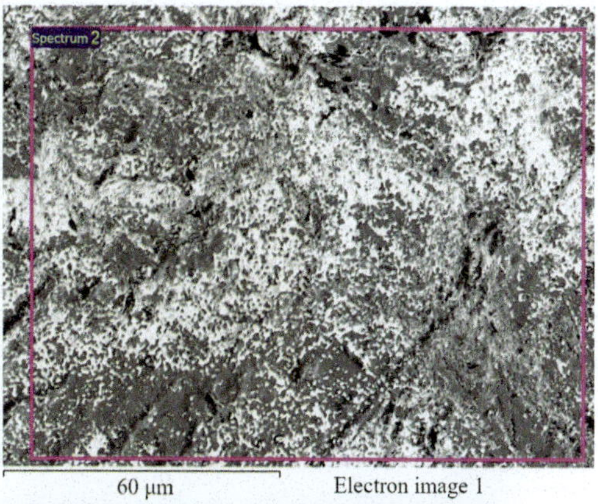

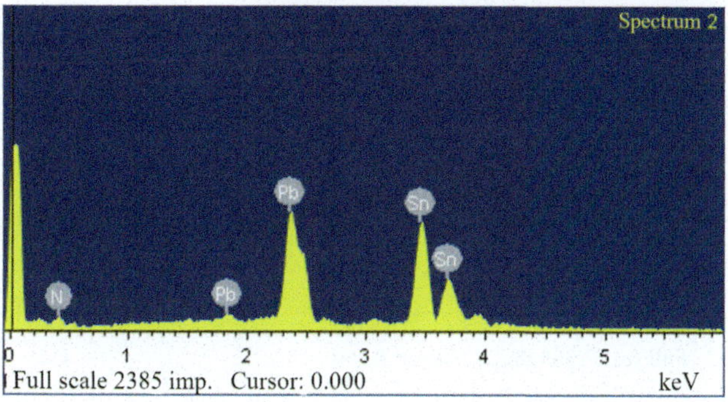

Fig. 2.48 Results of X-ray microanalysis of solder in a fragment of electronic components of 3M-14 cruise missile of *Kalibr* complex (specimen 1)

2.3.7.3 Solder of Electronic Component Fragment of S-300 Anti-aircraft Missile (1 Specimen)

Figure 2.49 presents an example of X-ray microanalysis results for a fragment of solder in the electronic components of complex S-300 anti-aircraft missile. Microanalysis has been carried out in the indicated rectangular area. The measured concentrations of chemical elements have been averaged in this area shown in the backscattered electron image. Chemical element concentrations are given in mass percentages. Chemical element concentrations have been calculated using K, L, M lines of characteristic X-ray radiation. The figure also shows the characteristic X-ray radiation spectrum for this measurement area.

Element	Mass %
N K	0.37
Sn L	57.87
Pb M	41.76
Total	100.00

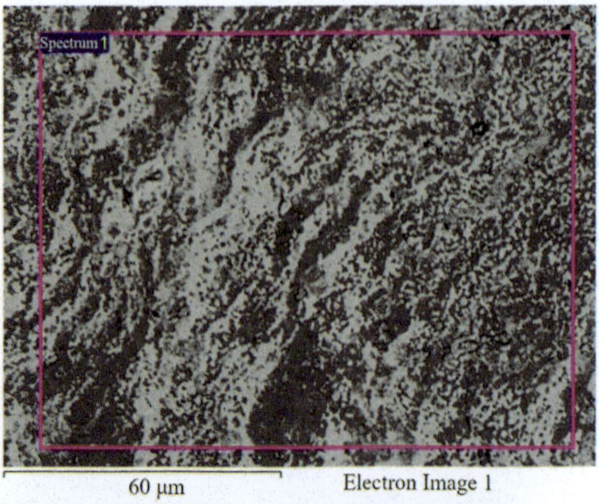

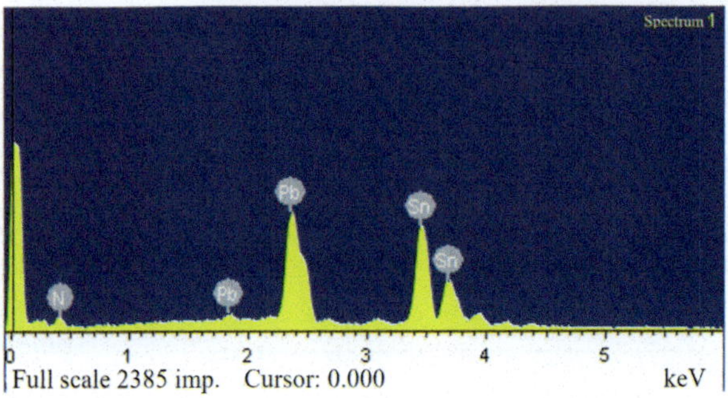

Fig. 2.49 Results of X-ray microanalysis of solder in electronic component fragment of complex S-300 anti-aircraft missile

2.3.7.4 Solder in Electronic Component Fragments of *Geran-2* Loitering Munition

Figure 2.50 presents an example of X-ray microanalysis results of solder in electronic components of *Geran-2* loitering munition (specimen 1). Microanalysis has been carried out in the indicated rectangular area. The measured chemical element concentrations have been averaged within this area shown in the backscattered electron image. Chemical element concentrations are given in mass percentages. Chemical element concentrations were calculated using K, L, M lines of characteristic X-ray radiation of this measurement area.

Examinations of specimens 2, 3, 4 demonstrated the following chemical compositions and concentrations of chemical elements: Specimen 2: 0.05% N, 67.81% Sn, 32.14% Pb; Specimen 3: N 0.06%, Sn 67.86%, Pb 32.08%; Specimen 4: 0.08% N, 67.73% Sn, 32.19% Pb.

The results of studying solders in fragments of electronic components of Kh-101 cruise missile, 3M-14 *Kalibr* complex cruise missile and anti-aircraft missile of S-300 complex have shown that the solders examined have practically identical

Element	Mass %
N K	0.07
Sn L	67.63
Pb M	32.30
Total	100.00

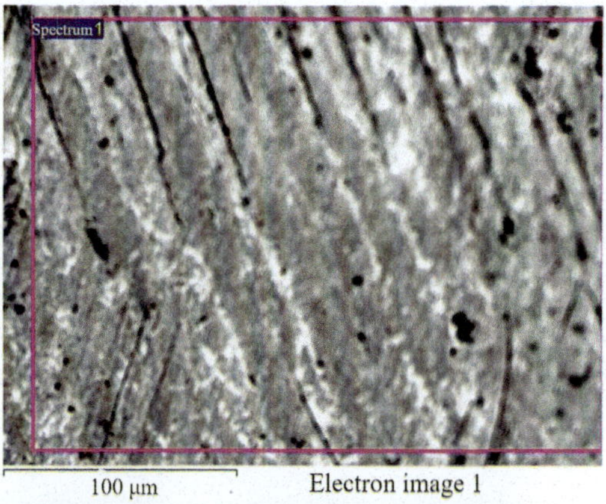

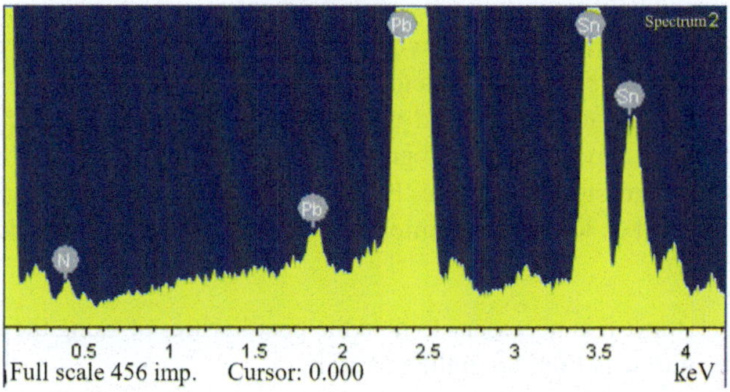

Fig. 2.50 Results of X-ray microanalysis of solder fragment of electronic components of *Geran-2* loitering munition (specimen 1)

chemical compositions that correspond to tin-lead solders of POS-63 and POS-60 grades, which are widely used in defense industry of the Russian Federation.

At the same time, examinations of solders in fragments of electronic components from *Geran-2* loitering munition have shown that the specimens examined have identical chemical compositions that correspond to tin-lead solders in compliance to GOST 21930 and GOST 21931-76, yet the maximum tin content in the specimens does not meet those standards.

Table 2.8 shows averaged values of chemical element concentrations for solders of all five fragments.

Conclusions

1. The chosen method of scanning electron microscopy and the developed mathematical model of signal generation during its application made it possible to determine with maximum accuracy the chemical composition and the structure of materials during examination of metal fragments of cruise, ballistic, air-launched ballistic and airborne missiles, loitering munitions made in the Russian Federation

Table 2.8 Averaged concentrations of chemical elements for all five fragments

Chemical elements	Mass concentrations of chemical elements in solders in %			
	Fragments of Kh-101 cruise missile electronic components	Fragments of *Kalibr* cruise missile electronic components	Fragment of electronic components of S-300 complex anti-aircraft missile	Fragment of *Geran-2* UAV electronic components
Sn, tin	64.10	57.14	57.87	67.76
Pb, lead	35.65	41.36	41.76	32.17
N, nitrogen	0.25	1.50	0.37	0.07

and North Korea, which, in turn, will save time and resources when developing domestic missile systems and systems of anti-missile defense.

2. Results of the research performed confirm that continuous upgrade of aircraft designs brought about a sustained growth in the specific strength of metals and alloys used in structures of cruise, ballistic, air-launched ballistic and airborne missiles, as well as loitering munitions produced in the Russian Federation and North Korea.

At the same time, it is necessary to note the peculiarities of using the abovementioned means of destruction. So, unlike aircraft:

(1) they are disposable and short-lived products; their operation life is from several minutes in ballistic, air-launched and hypersonic missiles to several hours in cruise missiles and loitering munitions;
(2) their structures are significantly less exposed to the effects of long-term asymmetrical cyclic loading.

Designers of ballistic, air-launched, and hypersonic missiles constantly work at increasing their maximum speeds at all stages of flight and enabling missiles to maneuver.

Designers of "disposable" means of destruction are more strictly limited by weight and size requirements and must provide the minimum weight of load-carrying structures.

Designers of means of destruction must rely on maximum application of technologically deployed materials of acceptable price to promote mass manufacturing of those weapons.

The specific features mentioned permitted using in the examined means of destruction many principal metal materials and alloys—heat- and corrosion-resistant steels, titanium, and aluminum alloys.

At the same time, research results show that Russian designers of cruise, ballistic, air-launched ballistic and airborne missiles used a limited number of steel and alloy grades developed in the 1970–1980s and implemented into mass production at the times of the USSR. This obviously resulted from the efforts to ensure technological feasibility and maximum productivity.

References

Arkhipova TF, Osadchuk AY (2013) Applied materials science. VNTU, Vinnytsia [in Ukrainian]
Batyrev VA (1982) X-ray spectral electron-probe microanalysis. Metallurgy, Moscow [in Russian]
Beckert M, Klemm H (1984) Handbuch der Metallographischen Ätzverfahren, 4th edn. VEB Deutscher Verlag für Grundstoffindustrie, Leipzig
Engel A, Klingele G (1986) Raster electron microscopy. Destruction: the handbook [in Russian]. Metallurgy, Moscow
Glauert AM (1977) Practical methods in electron microscopy. Elsevier
Goldstein JI, Newbury DE, Michael JR, Ritchie NWM, Scott JHJ, Joy DC (2017) Scanning electron microscopy and x-ray microanalysis. Springer, New York
Kay D (1961) Techniques for electron microscopy. Blackwell, Oxford
Klimov VN, Kozlov DM (2017) Modern aviation structural alloys. Publishing House of Samara University, Samara
Livshits BG, Kraposhin VS, Linetsky LY (1980) Physical properties of metals and alloys: a textbook for university. Metallurgy, Moscow [in Russian]
Molyar AG, Kotsyuba AA, Bychkov AS, Nechiporenko OY (1995) Structural materials in aircraft engineering. KVITS, Kyiv [in Russian]
Ostapchuk VV, Semishov NI (2011) Corrosion and methods of protecting aircraft alloys: textbook. Nat Aerospace Univ "KhAI", Kharkiv
Ostapchuk VV, Popova EG, Samoilov VY, Taranenko IM (2015) Use of non-metallic and composite materials in aerospace technology: tutorials. KhAI, Kharkiv [in Russian]
Pilyankevich AN (1961) Practice of electron microscopy. Mashgiz, Kyiv [in Russian]
Pilyankevich AN, Klimovitsky AM (1976) Electron microscopes. Technika, Kyiv [in Russian]
Potemkin AY, Sheidenman YI, Frolov YP et al (1982) Materials for aviation instrumentation and structures: a textbook for universities. Metallurgy, Moscow [in Russian]
Protsenko IY, Sayenko VA (2002) Thin metal films (technology and properties). Sumy State University, Sumy [in Ukrainian]
Protsenko IY, Chernous AM, Protsenko SI (2007) Tools and methods of film materials research. Sumy State University, Sumy [in Ukrainian]
Samsonova VG (1965) Physicochemical properties of elements. Handbook [in Russian]. Naukova Dumka, Kyiv
Schimmel G (1969) Elektronenmikroskopische Methodik. Springer, Berlin
Scott VD, Love G, Reed SJB (1995) Quantitative electron-probe microanalysis. Ellis Horwood, New York
Taranenko IM (2014) Material engineering and aviation structural: textbook part 2: non-metals composites. Nat Aerospace Univ "KhAI", Kharkiv
Vambol AA, Taranenko IM, Zhukovskiy NY (2011) Material engineering and aviation structural materials: workbook for laboratory course. Part 1: metals and alloys [in Ukrainian]. Nat Aerospace Univ "KhAI", Kharkiv
Weinberg F (1970) Tools and techniques in physical metallurgy in vol 2 (vol 1). Marcel Dekker. Inc., New York

Chapter 3
Application of Scanning Electron Microscopy for the Study of Fragments of Special-Purpose Structures Made of Composite Materials

This chapter provides a comprehensive consideration of various composite materials that are widely used in manufacturing current special-purpose structures (means of destruction), since those materials are characterized with high strength, heat resistance, low density, and thermal and electric conductivities along with other special properties.

Composite materials can be classified in terms of their structure, molding methods, matrix and filler materials, reinforcement techniques, their purpose, fields of application, etc. Professional literature provides a detailed analysis of the advantages and limitations of current composite materials from the viewpoint of their application for manufacturing elements of special-purpose structures.

The research into microstructural characteristics of the fragments of special-purpose structures (means of destruction), whose results are presented in this chapter, was carried out with the most versatile non-destructive scanning electron microscopy technique and the method of X-ray microanalysis.

To examine the structure, chemical composition, and elemental distribution on the surfaces and in the bulks of the fragments, as well as measure the dimensions of different layers of composite materials, sections of the structural elements of means of destruction used by Russian troops during hostilities in Ukraine were produced.

Physico-chemical composition, structure, and surface topography of the elements of a broad class of fragments of special-purpose means made of composite materials were determined.

The results obtained indicate that in manufacturing special-purpose structures (means of destruction) quite well-known composite materials were mostly used, namely, carbon fibers, fiberglass, organoplastics, boron-containing plastic composites, and Kevlar-based hybrid composite materials. The abovementioned materials are used both for manufacturing structural elements and for producing coatings. The main purpose of using the abovementioned composite materials in manufacturing special-purpose structures (means of destruction) is to increase their strength, heat resistance, reduce their weight, and radar signature (cross section).

V. L. Bogdanov et al., *Identification of Special-Purpose Structures by Their Fragments Based on Scanning Electron Microscopy*, Advanced Structured Materials 233,
https://doi.org/10.1007/978-3-031-88827-4_3

The use of scanning electron microscopy permitted a most accurate determination of the chemical composition, structure, and topology of the elements of special-purpose means (means of destruction) that were manufactured using composite materials.

3.1 Analysis of Current Composite Materials Used in Manufacturing Special-Purpose Structures

To develop special-purpose structures, especially those in the field of aircraft manufacturing and rocketry, there is a need for materials that have high strength, heat corrosion resistance and high-temperature strength, low density, heat- and electro-conductivity, as well as special properties.

Conventional metals and alloys do not fully meet the above requirements, and their physico-chemical characteristics are at the limit, i.e., the reserves of their further improvement are practically exhausted. For the practical application in this aspect, the materials that have the following properties are promising:

- they do not exist in nature;
- consist of two or more components of different chemical compositions or structures;
- have properties that differ from those of their components;
- are non-homogeneous in the microscale but homogeneous in macroscale; their composition, morphology, and distribution of components are designed in advance;
- the material is formed in the process of product fabrication.

While in the past non-metal products were mainly used as auxiliaries, now compositions intended for structural purposes are developed much more frequently; among them a special place is taken by composite materials.

It is generally accepted that to be considered as composite materials (CM, composites), substances are to consist of at least two components that are not dissolved in each other and do not absorb each other. For this reason, composite materials have properties that differ from the sum of properties of their component parts, and have specific characteristics inherent only in them. An important distinctive feature of composites is the possibility of producing materials with predetermined characteristics for dealing with the tasks posed in advance. In terms of structure, composites can be considered as an aggregate of reinforcing components (individual fibers, plates, or granules) and the matrix combining the reinforcement components into a single whole. It should be specially noted that the characteristics of composites are determined both by the ratio of their components, their structure (chemical, colloidal, supramolecular), and by the behavior of their ingredients and composites as a whole in the processes of fabrication and processing. The principal advantages of composite materials are their low density, high static and fatigue strength and stiffness, low coefficient of thermal expansion, and corrosion resistance. Due to using lighter materials (compared to aluminum and titanium alloys), a decrease in structural and technological joints, replacement of adhesive joints with mechanical fasteners, makes it possible to reduce the weight of the structure by 20–30%. Here, the greatest gain is achieved due to using composites in manufacturing fuselage elements. A great advantage of using composite materials in manufacturing means of destruction is the feasibility of fabricating high-quality complex aerodynamic surfaces and providing low radar visibility. Fibrous composite materials are superior to metals and alloys in terms of fatigue strength, heat resistance, vibration resistance, sound absorption, impact toughness, and other properties. Besides, owing to advanced means of mechanization used in producing composites, their cutting and laying out, it is possible to simultaneously reduce by 30–40% the labor intensity of fabricating structures as compared to that of aluminum alloys. Radically innovative composites based on polymers reinforced with high-strength carbon, boron, organic, and glass fibers are 2–3 times superior in specific strength and rigidity characteristics than traditional metals and alloys. Due to their properties, composite materials have found wide application in manufacturing various means of destruction, especially in rocketry, aircraft, and shipbuilding.

Strict and sometimes conflicting demands are made to modern planes and missiles regarding their resistance to overloads and high and low temperatures at the minimal mass and radar visibility. These requirements determine a continuous search for new light-weight and extremely rigid and high-strength composite materials capable of withstanding significant aerodynamic loads and retaining their strength at high temperatures.

The choice of matrix material is determined by the temperatures at which the composite is to operate. If the temperature does not exceed 100–200 °C, polymer matrices are used (polyester, epoxy, polyamide ones), which withstand heating to temperatures above 300 °C. At higher temperatures, metal matrices (Al, Mg, Ti) are employed. Their advantages are higher strength and heat resistance, while the disadvantage is the fact that the higher the density of such composite, the more difficult is its production and higher the probability of chemical reactions between the fibers and the matrix.

If the temperatures in the aircraft nose, leading edges of its wings, and engines exceed the melting point of metal matrix, ceramic matrices operating at the temperatures 1000–1700 °C are used, and at higher temperatures (up to 2500 °C) silicon–silicon composites are employed.

Composite materials can be classified in terms of structural elements, matrix materials, reinforcement method, structure, purpose, application areas, molding technique, etc. From the viewpoint of structural elements, composites are subdivided into fibrous, laminated, dispersion-hardened (matrix filled with inclusions), pseudo-alloys (matrix filled with low-melting material), and eutectic (produced by crystallization of eutectic alloys).

From the viewpoint of material's matrix, the composites are distinguished as those with metal, polymer, carbon, and ceramic matrices. Materials with polymer matrix are carbon fiber reinforced plastics, fiberglass materials, boron plastics, and metal plastics. Other composite materials have double names—the name of matrix and that of fibers, e.g., Cu-W, Al_2O_3-Al. Besides, there are hybrid composites in which reinforcing fibers of two or more kinds are used (e.g., carbon fiberglass ones). However, if reinforcing fibers have both different nature and different geometry, all of them are referred to as combined ones (e.g., titanium foil and boron fibers in aluminum matrix).

In terms of reinforcement methods, materials with oriented reinforcement and those with randomly arranged reinforcing fibers are distinguished. In the former case, composites have strong anisotropy of properties; in the latter, they are isotropic in the macrovolume. Isotropic composites include dispersion-hardened alloys, pseudo-alloys, and chaotically reinforced materials strengthened with fine needle-like particles (fiber segments or thread-like crystallites). Such composites are quasi-isotropic in microvolumes but isotropic in the bulk product.

Structurally, composite materials with frame, matrix, laminated, and combined structures are distinguished. Frame-structure composites include, e.g., pseudo-alloys produced by impregnation, matrix-structure ones—dispersion-strengthened and reinforced materials, laminar-structure materials—compositions produced of sets of alternating foil layers or sheets of materials of different nature or composition, and the materials with combined structures are a combination of the three groups.

From the viewpoint of their purpose, composites are subdivided into structural, corrosion-resistant, anti-friction, ablation, piezoelectric, heat-protective, etc. ones.

In terms of their applications, there are composites for aerospace, shipbuilding, automotive industries, electronics, medicine, etc.

In terms of molding techniques, composites can be subdivided into materials produced by layout, winding, pultrusion, pressing, etc.

From the viewpoint of mechanics, composite materials can be subdivided into strong materials (possessing high strength and stiffness) and specialized (those with certain physical properties, e.g., RF transparent or RF absorbent materials).

Advantages and Limitations of Composite Materials

The parts made of polymer composites have a wider range of frequencies of natural vibrations than metal parts. If for aluminum, titanium alloys, and steels the frequencies of natural vibrations are close to resonant ones, and the deviation from the resonance mode requires a change in geometric dimensions of the parts, for high-modulus polymer materials this is achieved mainly by a change of fiber orientation in certain layers. The use of composites, particularly carbon fiber reinforced plastics, whose specific feature is low coefficient of linear thermal expansion, allows reducing thermal stresses in structures. The advantages of composites are related to the available opportunity of broadly varying practically all material's properties, which is achieved by selecting the constituent components, their quantitative ratio, distribution, and orientation in the material bulk. This makes it possible to produce materials for multifunctional purposes, often those with reverse service properties.

In comparison with available structural materials, composites have the following *advantages*:

- low density (e.g., carbon plastic density is twice as low as the density of D16T alloy and 5.4 times as low as that of 30KhGSA steel);
- high specific strength and specific stiffness (fiberglass specific strength is 1.85 times higher than that of D16T alloy and by 1.33 times higher than in 30KhGSA steel);
- high chemical and corrosion resistance;
- manufacturability;
- high fatigue strength of fiber composite materials;
- a sharp reduction in the number of aircraft structural parts;
- special properties (RF transparency, dielectric permeability);
- possibility of power flow control in structures by providing the appropriate reinforcement direction.

The *limitations* of composite materials are

- a high price of the majority of composites as compared to metal alloys;
- low shear strength, crumpling of polymer composite materials;
- brittle fracture and low specific toughness;
- the impossibility of changing blank shapes of composites with thermoset matrices;
- the necessity of special safety precautions.

Using materials lighter in comparison with aluminum or titanium alloys, reducing the number of design and technological connectors, substitution of mechanical fastenings for adhesive connections allows a 20–30% reduction in structure mass. The greatest gain is achieved due to the employment of composites in fabricating fuselage structural elements. Of significant importance for using composite materials in manufacturing means of destruction is the feasibility of fabricating complex-shaped high-quality aerodynamic surfaces, and providing radar stealth property. Employing advanced automated equipment in producing semi-finished composites in their cutting and laying out also made possible a 30–40% reduction in labor intensity of structure manufacturing as compared to aluminum alloys. Due to their properties, composite materials have found wide application in manufacturing various means of destruction, especially in rocketry, aircraft-, and ship-building.

The most common composite materials used in manufacturing special-purpose structures are polymer composites, which, in turn, can be subdivided into the following widely spread types: carbon fiber reinforced plastics, organoplastics, fiberglass, boroplastics, and hybrid ones based on *Kevlar* (Ursache et al. 2024; Stanescu et al. 2014; Dhas and Arun 2022).

Carbon fiber plastics are composite materials in which carbon fibers serve as fillers. Carbon fibers are produced from man-made and natural fibers based on cellulose, acrylonitrile co-polymers, petroleum and coal tar pitches, etc. Composite carbon–graphite materials, which are the most heat-resistant composites able to withstand long-time temperatures up to 3000 °C in inert or reducing environments, are produced on the basis of carbon fibers and carbon matrix. Besides, these materials have very high mechanical properties of the fibers themselves. For example, in the case of an ideal crystal lattice structure without defects, when strained along the layer plane, their ultimate tensile strength is 10 GPa, and elastic modulus—1000 GPa. In real structures, the degree of orientation differs from the theoretical one; yet, ultra-high strength is achieved.

Industrially, carbon fibers (CFs) are produced by pyrolysis of original organic fibers in an inert gas environment. Polyacrylonitrile fibers (PAF), viscose, and pitches (products obtained through the distillation of coal, crude oil, natural asphalt, and some synthetic compounds) are most frequently used for pyrolysis.

Pitch-based production of CFs is more expensive. Besides, due to high porosity of pitch-based CFs, their mechanical properties are significantly inferior to the characteristics of PAF-based CFs.

Currently, CFs based on cellulose hydrate (CH) are considerably less frequently used. Depending on their mechanical properties, CH-based CFs are two to eight times more expensive as compared to CFs based on PAF; yet, this deficiency is compensated by high mechanical properties of the latter (e.g., ultimate tensile strength of PAF-based CFs can be 690–3445 MPa). However, it can be noted that lately the production cost of the abovementioned materials has been significantly reduced, and there are prospects of improving their properties, which permits an optimistic forecast of their further use.

Carbon products are industrially manufactured as textile structures (stapled, continuous threads, woven or nonwoven materials). The most common products are yarn, rovings, and mats. By yarn, we mean twisted threads consisting of cut fibers; roving consists of parallel or slightly twisted fibers; and mats (ribbons) are constructed from a large number of bundles or strands put next to each other or stitched into one piece.

To improve the CF bond to matrix and create better conditions during processing, fibers undergo additional treatment. This can be surface treatment and lubricant application. Polyvinyl alcohol, epoxy resins, polymides, and water are used for the treatment. Organic substances applied during lubrication can amount up to 7% mass fraction of fibers. It should be noted that surface treatment is able to improve shear strength by two to three times. The most widely used techniques of surface treatment are oxidation, deposition of organic or inorganic coating, and radiation treatment (Baker et al. 2004; Li et al. 2022).

Carbon fibers are also the basis for producing the most heat-resistant composite—carbon–carbon composite material (CCCM)—in which the adhesive matrix for carbon fibers is practically pure carbon. There are two methods of producing such material. Carbon fibers are impregnated with a special binder (e.g., phenol formaldehyde), which solidifies, and after that they are carbonated at high temperature (up to and above 2000 °C). Since the material becomes porous, it is once again impregnated with the binder and carbonated anew. This operation is repeated several times. The other technique is chemical vapor deposition of carbon at high temperatures and under high pressures. A combined method—impregnation with binder and coking followed by carbon deposition from vapor—is quite promising. The material produced can be used at temperatures up to 3000 °C if its surface is protected against oxidation.

The main task of improving such fibers is to create conditions that ensure the rise in ultimate tensile strength under the interlayer shear of polymer composite materials, which usually does not exceed 3.5 kg/mm^2. To this end, fibers are oxidized in a liquid or gaseous environment, which changes their surface structure considerably. It should be noted that potentially low cost of carbon fibers, combined with their capability to retain high values of strength and elasticity characteristics when heated, make these fibers hopeful strengtheners for composite materials with metal matrix. The main issues in developing such materials are high reactivity of carbon fibers in contact with the majority of metals and the difficulty of manipulating fibers due to their small sizes.

Carbon fiber reinforced plastics are mainly used for fabricating missile nose cones, and the parts subjected to maximum aerodynamic loads (missile engine nozzles etc.).

Organoplastics are composites whose fillers are organic man-made, less frequently—natural, fibers in the form of bundles, threads, fabrics, paper, etc. Organoplastics have high chemical resistance, low density, they are lighter than fiberglass and carbon plastics, have high tensile strength, high resistance to shock and dynamic loads but, at the same time, have low compressive and bending strength (Baker et al. 2004; Li et al. 2022). The progress of organofibers is necessitated by the development of light-weight structural materials that combine comparatively high specific strength and stiffness with the stability of properties under the action of alternating loads, abrupt change of temperatures and operating conditions. The earliest organofibers that, depending on their texture, were referred to as "fiberlites", "textolites", and "getinakses" contained either cut natural fiber randomly distributed in thermosetting binders, or fabrics of different weaves (cotton, linen) or cellulose paper layers. The scarcity of natural fibers, their low strength, stiffness, and heat and water resistance made researchers look for new organic fillers among man-made polymers.

In CF production, the undisputed leaders among organic fibers are aramid fibers known under *Kevlar* trademark. Aramid fibers are chemical fibers produced on the basis of linear fiber-forming polyamides in which at least 85% of amide groups are directly bound with two aromatic rings. Aromatic rings provide polyamide molecules having high rigidity, they also contribute to the predominant orientation of macromolecules along a single axis. Graphically, such system resembles a system of interconnected long rods. The presence of aromatic rings provides high chemical resistance and stability, while crystal structure determines thermal stability. *Kevlar* fibers, when heated to temperatures close to the decomposition temperature, do not have enthalpy zones. As a result, at relatively high temperatures, *Kevlar* has rather low ductility and rather low brittleness.

Current organofibers' production is based on using polyethylene terephthalate (e.g., lavsan) and polyamide (e.g., capron) fibers, which are rigid-chain and flexible-chain polymers. The most widely known high-strength and high-modulus organic fibers are aramid fibers based on liquid-crystal polyamides (Twaron, Terlon fibers) and their co-polymers (CBM, Armos, Technora fibers). As compared to fiberglass and carbon fibers, they are much more flexible; among the rest of organic fibers they have the highest ultimate tensile strength and elastic modulus.

Organoplastics are mainly used to manufacture parts that will be exposed to radiation, operate at ultra-high speeds, low temperatures (radar antenna radome, wing interceptor, sealing of the joint between the wing and the fuselage, engine pylon fairing).

In *boroplastic* composites, boron fiber fillers are inserted into a thermosetting polymer matrix, wherein fibers can be shaped as both mono threads and bundles braided with glass thread, or as ribbons in which boron threads are interwoven with other threads.

Among non-metal fibers that can be used as fillers, boron fibers have the highest values of specific strength and stiffness since their density is three times as low as that of steel. Beryllium fibers are an exception but their toxicity makes their use in the production of composite materials difficult (Baker et al. 2004; Li et al. 2022). The development of boron reinforcing fibers was caused by specific requirements to materials in aviation and missile industries. Fiberglass was efficiently used in casing and fairing fabrication; however, because of its low stiffness, CM based on them could not be used in manufacturing power elements, which necessitated their substitution for other composites in such structures.

The use of boroplastics is limited by a high cost of boron fiber production; therefore, they are mainly used in rocketry, in parts that are subjected to prolonged stresses (horizontal stabilizers, tail units, wing fastening, etc.).

Fiberglass composites are materials reinforced with glass fibers that are molded from molten inorganic glass. Both thermosetting synthetic resins and thermoplastic polymers are most frequently used as matrices.

Reinforcement with glass fiber is due to its unique properties: high tensile strength; heat, fire, moisture, and chemical resistance; and high thermal and electric characteristics. High tensile strength means high ultimate tensile strength and fiberglass specific strength—the ratio of ultimate tensile strength to material's density. By the latter characteristic, glass fibers are superior to steel wire. Fiberglass does not burn and has high melting point, which provides heat- and fire-resistance and permits it to be used at high temperatures. Humidity resistance is due to the fact that glass fibers do not absorb moisture, do not rot and retain high strength characteristics in humid environment. The chemical composition of glass affects fiber characteristics and, in the final analysis, the properties of composite materials.

The glass used for fiber fabrication is based on silicon, aluminum, magnesium, and calcium oxides. Besides, to provide special properties, small amounts of boron, sodium, zirconium, etc. are added to the glass. The choice of glass fiber is determined by fiberglass purpose and the technology of its processing. In producing general-purpose fiberglass based on polyester resins, reinforcing fillers are made of low-alkali E glass fibers. In producing glass-filled thermoplastics, G glass of 0.009–0.014 mm diameter and K glass of 0.013–0.014 mm diameter are used. For thermoplastics, cut fibers 3 or 6 mm long are more commonly used, but sometimes there are fibers up to 12 mm long. 6–12-mm-long fibers are considered as long, while those having the length of 3 mm and less are estimated to be short. In aviation and missile industries, S glass is used in fabricating high-strength fibers. As compared to E glass, it has approximately 40% higher strength characteristics. Besides, S glass has higher heat resistance and works well in alkaline and acid environments. The practice of using CM based on glass fibers has proved that maximum mechanical characteristics are realized for close-packed fibers with the minimal amount of binder. Current technologies permit binder fraction to be reduced to 20%. Nevertheless strength characteristics depend on the technology of fiber layout or winding as well. Therefore, in actual large-sized structures, binder content is within 30–32% interval.

Today, fiberglass products are least expensive to produce and are widely used. The absence of deformations at low and high temperatures, strength, and resistance to radiation and chemical reactions make this material the most suitable for fairing shells (Ursache et al. 2024).

Hybrid CMs—hybrid composite structures combining the advantages of their constituents also find application.

As is known, high-strength boron fibers easily withstand high bending loads, while carbon fibers work well under bending and torsion loads in arch structures. Boroplastics have higher fatigue characteristics. Boron-metal composite parts, laminated boroplastic overlays both reduce stresses and absorb vibration. Boron fibers are mainly used in products operating at medium and high temperatures.

In the second half of the twentieth century, alongside with boron fibers, researchers started to develop silicon-carbide fibers (SiC fibers), whose average ultimate tensile strength is as high as 3445 MPa. Elastic modulus of SiC fibers is equal to 448 GPa against 400 GPa of boron fibers (Baker et al. 2004; Li et al. 2022).

Besides, there is a current trend in structure engineering to develop hybrid *Kevlar*-based composites, when aramid fibers are combined with carbon or fiberglass ones. For example, aramid and carbon fibers work well in combination due to close values of their linear expansion coefficients. Thus, in *Kevlar*–carbon hybrids there are no limitations of carbon plastics but high price and high brittleness lead to complete fracture, especially under dynamic loads. But at the same time, deterioration of hybrid strength and stiffness as compared to pure carbon plastics occurs. Main limitations of fiberglass, e.g., its low flexibility, are absent in *Kevlar*–fiberglass plastic hybrids.

The main issue of attaining high characteristics of flexible-chain polymer fibers is to achieve high degrees of their orientation in the process of drawing and preventing macromolecule rupture. The higher the degree of orientation, the higher is tensile strength when the material is strained along fibers. Rigid-chain polymers even at high temperature themselves try to achieve unidirectional fiber orientation, so the stage of heat treatment is used in their production.

Metal–organoplastic composites are next-generation materials that consist of alternating layers—those of aluminum alloy and organoplastics—which reduces the load on metal and increases the durability of material as a whole. In comparison with conventional materials, they have high fracture toughness, which makes it possible to use them in fabricating fuselage structures.

The use of Alor (the material in which thin metal layers alternate with organic plastic layers) instead of traditional aluminum reduces structure mass by 10–20% when aramid filler is used, and halves it in the case of using a unidirectional filler, leading to a significant reduction of product mass. Alor has high damping characteristics, which permits it to be used in the structures exposed to high-frequency vibrations (missile and aircraft airframes, etc.).

3.2 Peculiarities of Employing the Method for Investigating Polymer Materials

Composite materials often have advantages over homogeneous materials in terms of such parameters as high electric resistance, low specific weight, good mechanical properties, chemical stability, and machinability. Electrodynamic properties of composition materials can vary widely with the changes in the concentration of inclusions, due to mechanical grinding of inclusions, etc. The utilization of composite materials makes it possible to separately consider their electrodynamic properties that mainly depend on the filler, and physico-mechanical properties that depend primarily and foremost on the composite material matrix. This permits the development of materials that combine high effectiveness in engineering applications and the necessary physical properties determining, e.g., material's resistance to various external impacts. For those reasons, magneto-dielectric composite materials are widely used in producing radar absorbent coatings.

The matrices of composite materials used for producing radar absorbent coatings are dielectrics, which immediately leads to certain difficulties in studying such coatings with electron microscopy and X-ray microanalysis techniques. When an electron beam hits a dielectric, absorbed electrons are accumulated on its surface due to the absence of charge flow to the ground. Such accumulation of electrons results in the emergence of charged areas on the specimen surface, which deflect the incident electron beam in an irregular way during further scanning. This causes severe image distortions and errors in quantitative microanalysis.

There are three techniques that can help prevent the abovementioned charging effect (Baker et al. 2004; Li et al. 2022; Ursache et al. 2024): (1) applying a conductive coating; (2) operating at low acceleration voltage; and (3) using the technique of one-time exposure. Of these procedures, conductive coating application (Newbury and Yakowitz 1975; Dikin et al. 2006; Ahmad et al. 2021) is the most common. When employing it, a thin film of carbon or other conductive material (aluminum, platinum, gold) is applied by thermal evaporation in vacuum or by cathode sputtering. The conductive layer is made as thin as possible, but it is necessary that it should be continuous for providing good ground connection. Usually, layers 10–100 mm thick are suitable, but it is desirable that the layer should be as thin as possible so that it does not impede observation of fine details on the specimen surface. During electron microscope operation in the secondary electron mode, the criterion of choosing the material for sputtering is to provide the maximum number of secondary electrons emitted from the specimen. By considering the physical aspects of secondary electron emission, it was shown (Woodruff and Delchar 1994) that a gold coating 8–10 nm thick can provide optimal secondary electron emission. In practice, platinum is also used for those purposes. However, in X-ray microanalysis, it can be advised to apply the coatings consisting of elements with low atomic numbers.

As mentioned above, to achieve a good ground connection, the conductive coating should be continuous. However, in a number of cases this is not possible to achieve. For example, in investigating composite coatings that absorb electromagnetic waves, there often is the issue of high roughness of the studied surface, which makes the application of a continuous conductive film impossible. In such cases, two other methods of removing the effect of surface charging are employed in the investigations of dielectric materials.

According to Oatley (1972), the reduction of surface charging effects is caused by the peculiarities of electron emission from the solid material surface. The dependence of electron emission factor, which takes into account both backscattered primary electrons and secondary electrons has a pronounced maximum. For dielectrics, there is a range in which the number of emitted electrons exceeds the number of incident electrons, i.e., $\delta + \eta > 1$, where δ is the backscattering coefficient of electrons. This range is limited by two values of incident electron energies E_0^1 and E_0^2, for which $\delta + \eta = 1$ and which are referred to as the first and the second critical points, respectively. E_0^1 has the value of the order of hundreds of electron volts, while E_0^2 varies from 1 to 10 keV. If the energy of the incident beam electrons is lower than E_0^1, $\delta + \eta < 1$ and the specimen emits less electrons than hit it, a negative charge is accumulated on the specimen. This charge reduces the effective energy of incident electrons, contributing to a further decrease of $\delta + \eta$ value. Such situation persists till the specimen is charged to the level sufficient for the total beam deflection. If the energy of beam electrons is between E_0^1 and E_0^2, more electrons leave the specimen than get onto it, i.e., $\delta + \eta > 1$. This results in a positive specimen charge, which reduces the effective value of δ, since low-energy secondary electrons are attracted to the specimen. Due to these processes, the effective value of $\delta + \eta$ becomes equal to 1, and at a small positive charge of specimen surface, dynamic equilibrium is established, at which the emitted current becomes equal to that of the beam. Operation in the $E > E_0^2$ mode again causes a negative surface charge, which reduces the effective E value and increases $\delta + \eta$ until equilibrium is established with an effective value of E equal to E_0^2. This equilibrium is adverse since E_0^2 can considerably vary on the specimen surface, causing large variations of the final surface potential from point to point. A weak charge drain due to low surface conduction can change this situation significantly, leading to a complicated image behavior in time. Thus, it is desirable to work with the energy values that are within $E_0^1 < E < E_0^2$, which is usually achieved at the values of energy E close to 5 keV. In such conditions, dielectrics can be investigated without deposition of conductive coatings. However, at such low accelerating voltages, the parameters of scanning electron microscope deteriorate significantly, since in that case the electron source brightness becomes much lower. The way out from this situation is to use cathodes (electron sources) with a higher brightness, e.g., auto-emission ones.

Yet another technique to be taken into account when it is necessary to investigate dielectrics without depositing a coating on them is one-time scanning of the examined specimen. This method is employed in measuring chemical element concentrations along the scan line by means of X-ray microanalysis. However, here it should be noted that such a technique does not provide enough time for registering a sufficient numbers of X-ray quanta for reliable measurements of chemical elements' concentrations. With that, it should be mentioned that measuring concentrations or values proportional to concentrations along the scan line (not necessarily single-time measurements) is very effective in studying composite specimens with an electroconductive film deposited. This is especially useful in investigating specimen sections of radar absorbent coatings since such examinations make it possible to study changes in chemical elements' concentrations at the transitions from one layer to another.

Lately, besides the abovementioned methods of working with non-conductive specimens, current microscopes utilize the low-vacuum mode providing charge drain from the specimen surface directly through a low-vacuum environment.

In studying composite radar absorbent coatings, there appears a task of identifying the inclusions, such as electroconductive particles of various shapes, electroconductive fibers and man-made tissues, electroconductive and dielectric thin layers, etc. There, it is necessary to produce high-quality images of such inclusions for determining their geometric dimensions and location in the composite matrix. Besides, it is also essential to determine their chemical composition by means of X-ray microanalysis.

Since electrical properties of a composite, in our case, generally depend in many respects on its dielectric matrix, high-quality secondary electron images with sufficient resolution are achieved with the techniques considered above (in many cases, by film deposition).

X-ray microanalysis of fine inclusions contained in a composite material has some peculiarities resulting from the analysis of inclusions whose sizes are smaller than the region of generating X-ray emission induced by the incident electron beam. X-ray microanalysis of such kind of objects is usually somewhat difficult since most often it is not easy to find out whether the registered characteristic X-ray radiation is generated by the elements of the region studied or by those of the surrounding areas. This issue is addressed by increasing the resolution in X-ray emission regime by reducing the accelerating voltage and the electron beam current. With that, the intensity of characteristic X-ray emission is sacrificed in the hope of achieving the situation when the volume of direct excitation is inside the examined specimen region. Thus, in examining fine inclusions in composites, the employment of energy-dispersive spectrometers is more effective as compared to studies with wavelength-dispersive ones, since an energy-dispersive spectrometer needs the electron beam current which is by an order of magnitude lower as compared to the beam current needed for a wavelength-dispersive spectrometer. Besides, the requirements to specimen preparation for investigations by means of an energy-dispersive spectrometer are much less strict than those to preparing specimens for studies with a wavelength-dispersive spectrometer.

When examining composite materials with dielectric matrices, it is often necessary to deposit on the specimen a thin electroconductive film that is not related to the removal of surface charges. In this case, the purpose of coating the specimen with a conductive film to reduce its damage caused by the heat released during specimen surface bombardment with an electron beam, since the damage mentioned makes it impossible to carry out X-ray microanalysis and produce high-quality images of the specimen surface.

The value of local heating caused by electron beam can be evaluated by the following formula (Goldstein and Yakowitz 1975), which is very useful in practical work:

$$\Delta T = 4.8E_0 i / k d_b,$$

where ΔT is the local rise of specimen temperature at the point of beam incidence (in °C); E_0 is accelerating voltage (in kV); i is the electron beam current (in μA); k is the thermal conductivity of the composite matrix (in W·°C/cm); and d_b is the beam diameter (in μm).

Metals often have thermal conductivity $k = 1$, and in that case the temperature rise is of the order of 10 °C; for plastics, $k \approx 2 \cdot 10^{-3}$, hence, the temperature rise is of the order of 500–1000 °C. Such temperature change at the point of beam incidence will naturally make studies with X-ray microanalysis practically impossible. Here we would note that the presence of aluminum coating just 20 nm thick reduces ΔT by a factor of 3.

An effective mode of scanning electron microscope functioning in the investigations of composite materials is to produce X-ray images of specimen surface areas. The method of producing X-ray images by specimen scanning across its area is, in essence, scanning X-ray spectroscopy. In this mode, an amplified signal from the detector (an energy-dispersive spectrometer or a wavelength-dispersive spectrometer), which is proportional to the intensity of characteristic X-ray radiation, is employed to modulate the brightness of the microscope cathode-ray tube. The scan of cathode-ray tube is synchronized with the microscope electron beam. Thus, the image on the screen of the cathode-ray tube is formed due to varying the intensity of X-ray radiation from the specimen surface. When such images are produced, a scan system with image magnification is used, which is coupled with an electron microscope. X-ray image is the map of chemical elements' distribution over the specimen surface. The image in area scanned can contain shades from black to white. The places of high concentrations of the chemical element within the scan region will be almost white, the places where element concentration is low will be grey, and those where the element is absent—black. Besides, in currently employed microscopes, it is possible to produce images that show the distribution of several chemical elements over the surface in various colors.

When using a detector with energy dispersion, the time necessary for producing a single satisfactory X-ray image can be quite long (up to 15 min). This is because the necessary number of X-ray photons per 1 image should be at least 20,000. Since a scanning electron microscope often operates with the value of electron beam current 10–10 A, the number of photons generated per second is rather small (about 104 photons per second). So, if the specimen is not damaged by the beam, it is advisable to apply higher currents (in excess of 1nA) both in producing maps of chemical elements' distribution and at all stages of X-ray microanalysis. Resolution deterioration caused by an increase in beam current is of no special importance since the region of X-ray emission generation is determined mainly by electron diffusion and not by the diameter of electron beam. Since it is difficult to reduce the effective region of

X-ray emission generation to less than 1 μm, the permissible maximum X-ray image enlargement is approximately 3000 times.

Because of the requirement for the X-ray emission source in a wavelength-dispersive spectrometer to be located strictly on the Rowland focusing circle, scanning across a large surface area when obtaining X-ray images leads to a reduction in X-ray emission intensity, which is registered on the edges of the scan region. Such problem is absent when energy-dispersive spectrometers are employed.

In the studies of composite radar absorbent coatings, of special interest is the investigation of the layers the composite consists of, since the information on their dimensions, chemical composition, and location in the entire structure makes it possible to develop a physical model of the process of electromagnetic radiation absorption in the coating, as well as calculate certain radio physical parameters of the coating. In the investigations of composite layers, a sample section (slice) is produced, to be followed by electroconductive film deposition, which would permit researchers to achieve high-quality secondary and backscattered electron images.

As to performing X-ray microanalysis, there is an issue of conducting measurements on layer boundaries and near them. Here, the following three main aspects are to be taken into account (Herman and Lemasters 2012):

1. *Resolution which is directly related to the size of X-ray generation region.* If there are changes in the concentration of the analyzed element within the region of X-ray generation, the measured concentration would, evidently, be an average concentration within this region. Therefore, it is difficult to carry out accurate measurements on the boundaries between composite layers in the case when the region of X-ray emission covers both layers. It should be noted that this issue is analogous to that emerging in studying fine composite inclusions.
2. *Absorption of X-ray radiation.* In quantitative X-ray microanalysis, absorption adjustment can be taken into account only for homogeneous specimens. When the region generating X-ray emission is different in composition from the analyzed area that can introduce uncertainty.
3. *X-ray fluorescence.* In measurements near layer boundaries, besides characteristic and continuous X-ray radiation from the principal region of measurements, which appears as a result of electron beam bombardment, secondary fluorescence of the neighboring layer elements can emerge. This secondary fluorescence can also affect the results of measuring X-ray radiation intensity.

As noted above, the region of generating primary X-ray emission is larger than the diameter of the focused electron beam. Hence, correct measurement of the chemical composition on the composite layer boundaries in the region whose dimensions are smaller than those of the region of X-ray emission is not possible. At large concentration gradients near layer boundaries, proper studies also cannot be carried out. Due to this, the size of X-ray generation region should be made as small as possible.

The dimensions of the region of X-ray emission depend on the energy of the incident electron beam E_0, the average atomic number of the specimen, and the characteristic line of X-ray line. The size of the region of X-ray emission can be reduced to 1 μm by decreasing E_0 and choosing an appropriate X-ray line for measurement. Usually, during conducting X-ray microanalysis, the intensities of characteristic X-ray radiation are measured using K-line. However, with reducing E_0, it is necessary to use L-line. That way, the size of the X-ray emission region, which is smaller than 0.5 μm, can be determined in analyzing L line of such elements as Ti, Cr, Mn, Fe, Ni, Co, and Cu at the accelerating voltage of the microscope higher than 10 kV.

When an electroconductive coating is present on the sections of composite radar absorbent specimens, the problem of surface roughness of the studied specimen persists. In this situation, the measurements of chemical elements' concentrations are conducted by previously choosing the areas with smooth surface and normalizing to 100% the sum of the measured concentrations of chemical elements.

3.3 Research Results of Special-Purpose Structure Fragments Made of Composite Materials

3.3.1 Results of Examining Cruise Missile Fragments

To investigate the chemical composition, structure, and surface topography of special-purpose composite materials, fragments of Kh-101, 9M727 *Iskander-K* and 3M-14 *Kalibr* cruise missiles, produced by the Russian Federation; *Hwasong-11* (KN-23) (North Korea); 3M22 *Zirkon* aeroballistic missile (RF); *Shakhed-136*, upgraded Shakhed-136 (Iran) and *Lancet* (RF) loitering munitions (LM); *Orlan* unmanned aerial complex (UAC) (RF); and 9M544 *Tornado-C* guided missile (RF) were examined. The studies were carried out with reliance on the methods of non-destructive material examination, viz., scanning electron microscopy. Determination of the chemical composition, structure, and surface topography of the specimen surface materials was carried out with a JXA-8200 X-ray microanalyzer produced by JEOL Ltd. (Japan), and a JSM-6490LV scanning electron microscope produced by JEOL Ltd. (Japan) equipped with an INKA Energy 350XT energy-dispersive spectrometer. The electron microscope functioned in the mode of producing the secondary and backscattered electron images of surface areas of the specimens examined, and had the following operation characteristics: accelerating voltage was up to 30 keV, magnification up to ×300000, and resolution—6 nm. Microanalysis was performed for elements starting from $Z = 6$.

3.3.1.1 Fragment of Kh-101 Missile Wing

Figure 3.1 presents secondary electron images (SEI) of the surface of grey coating of Kh-101 cruise missile wing.

Fig. 3.1 Secondary electron image (SEI) of grey coating fragment of Kh-101 cruise missile wing. Accelerating voltage is 15 kV. Magnification is 2500, scale bar corresponds to 10 μm

Fig. 3.2 Secondary electron image (SEI) of the next-to-grey layer in coating of Kh-101 cruise missile wing. Accelerating voltage is 15 kV. Magnification is 100, scale bar corresponds to 100 μm

Figures 3.2 and 3.3 present the secondary electron images (SEI) of the next-to-grey layer in coating of Kh-101 cruise missile at different magnifications. Figure 3.3 shows the fiber sizes of artificial tissue seen in Fig. 3.2.

Figure 3.4 shows the secondary electron image (SEI) of a fragment of dark-spotted grey surface of the coating of Kh-101 cruise missile wing.

To study the structure and measure the concentration of chemical elements in the bulk of the fragment and coating, as well as determine the thickness of different coating layers, a section of Kh-101 cruise missile wing fragment was prepared (Fig. 3.5).

Figure 3.6 presents a secondary electron image of Kh-101 cruise missile wing section in the bulk of the principal material the wing is produced of.

Figures 3.7 and 3.8 show the secondary electron images (SEI) at different magnifications of Kh-101 cruise missile wing section on different areas of arc-shaped inserts in the wing bulk (principal wing material).

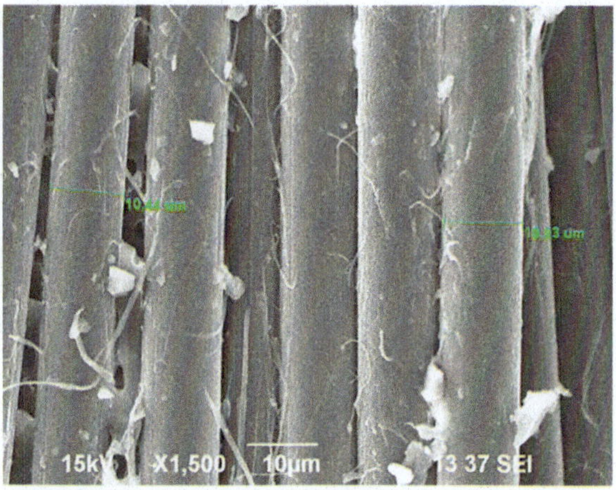

Fig. 3.3 Secondary electron image (SEI) of the next-to-grey layer in coating of Kh-101 cruise missile wing fragment. Accelerating voltage is 15 kV. Magnification is 1500, scalebar corresponds to 10 μm. The size of fibers (thickness) is shown

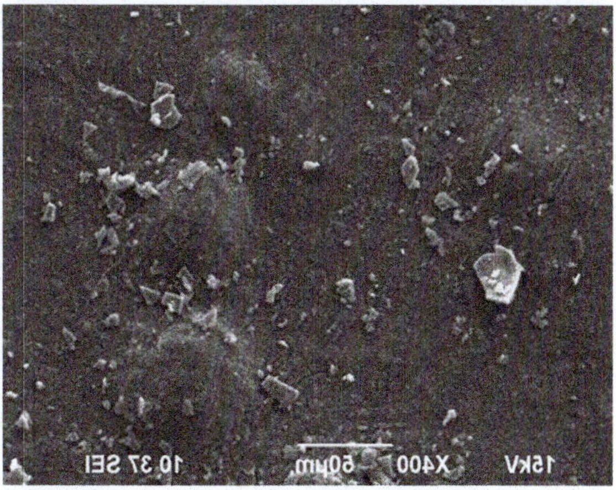

Fig. 3.4 Secondary electron image (SEI) of a fragment of dark-spotted grey surface of the coating of Kh-101 cruise missile wing. Accelerating voltage is 15 kV. Magnification is 400, bar corresponds to 50 μm

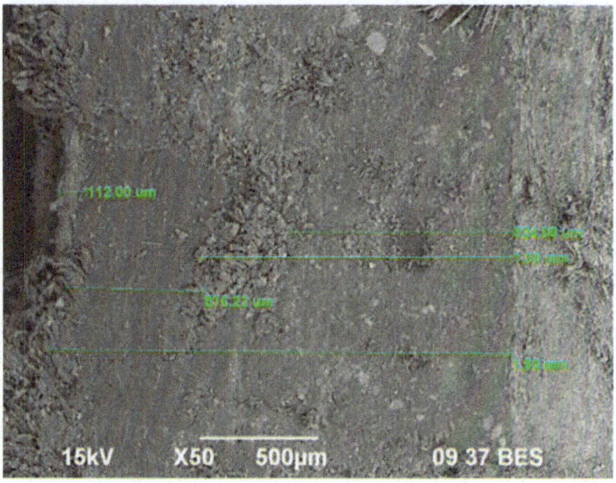

Fig. 3.5 Backscattered electron image of Kh-101 cruise missile wing fragment in the zone of the layer reinforced with a man-made tissue (this layer follows the grey one). Accelerating voltage is 15 kV. Magnification is 50, scale bar corresponds to 500 μm

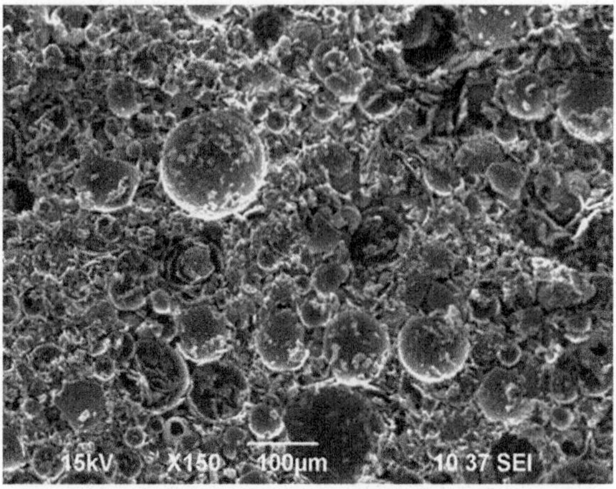

Fig. 3.6 Secondary electron image (SEI) of Kh-101 cruise missile wing section in the bulk of wing principal material. Accelerating voltage is 15 kV. Magnification is 150, scale bar corresponds to 100 µm

Fig. 3.7 Secondary electron image of Kh-101 cruise missile wing section in the domain of arc-shaped inserts in the wing bulk—the principal wing material. Accelerating voltage is 15 kV. Magnification is 3000, scalebar corresponds to 5 µm

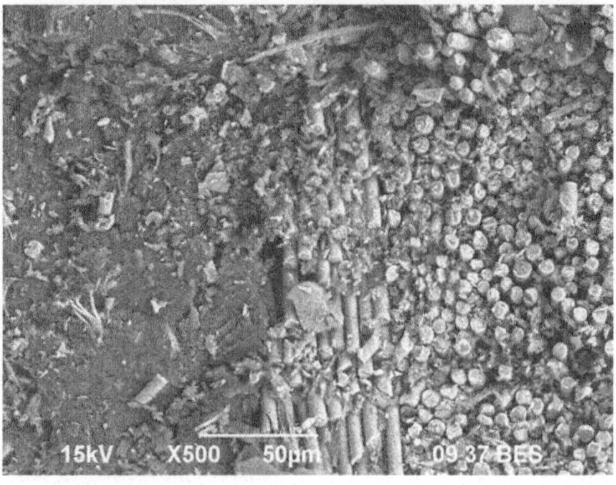

Fig. 3.8 Secondary electron image of Kh-101 cruise missile wing section in the area of arc-shaped inserts in the wing bulk (principal wing material). Accelerating voltage is 15 kV. Magnification is 500, scale bar corresponds to 50 µm. The coating layer can be seen in the left-hand part of the image

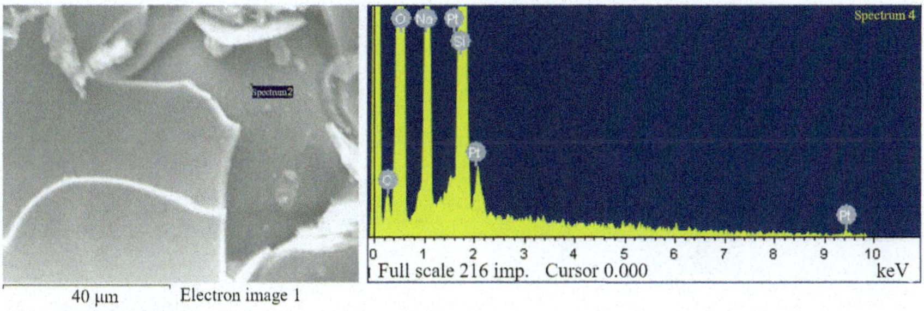

Element	Weight %
C K	8.44
O K	49.85
Na K	9.63
Si K	32.08
Total	100.00

Fig. 3.9 X-ray microanalysis results at a point in the zone of bulk of principal material of Kh-101 cruise missile wing

The comparison of the secondary electron image of Kh-101 cruise missile wing section in the zone of arc-shaped inserts in the wing bulk shows that these areas have different structures. The wing material in the area of wing bulk section has the foamed structure. The wing material in the zone of arc-shaped inserts in the wing bulk contains layers of fibers about 5 µm thick which are oriented in perpendicular directions.

Figure 3.9 presents an example of X-ray microanalysis in the domain of the bulk of the principal material of Kh-101 cruise missile wing. The microanalysis was performed at the points that are shown in secondary electron images. Chemical element concentrations are presented in mass percentages. Concentrations of all chemical elements were measured on K lines of the characteristic X-ray emission. Figure 3.9 also presents characteristic X-ray spectrum at measurement point.

Figure 3.10 presents an example of X-ray microanalysis results in the zone of the layer (next to the grey one) reinforced with artificial tissue. Microanalysis was performed in the rectangular area shown in secondary electron image. Chemical element concentrations are presented in mass percentages. Concentrations of all chemical elements were measured on K lines of characteristic X-ray emission. The figure also shows the spectrum of characteristic X-ray emission from measurement area.

Figure 3.11 presents an example of X-ray microanalysis in the zone of the layer reinforced with artificial tissue (next to the grey one). The microanalysis was performed at the point shown in the secondary electron image. Chemical element concentrations were measured on K lines of characteristic X-ray emission and are shown in mass percentages. The figure also presents the spectrum of characteristic X-ray emission at measurement point.

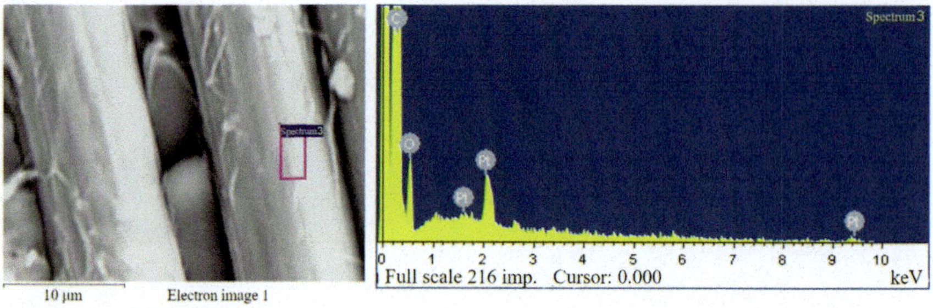

Element	Weight %
C K	87.69
O K	12.31
Total	100.00

Fig. 3.10 Results of X-ray microanalysis in the zone of layer reinforced with man-made tissue (the layer following the grey one). Microanalysis (chemical composition) of artificial tissue fibers

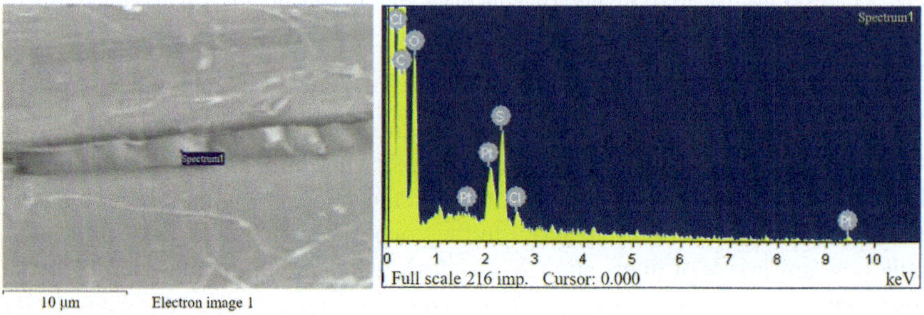

Element	Weight %
C K	79.19
O K	17.94
S K	2.43
Cl K	0.44
Total	100.00

Fig. 3.11 Results of X-ray microanalysis in the zone of layer reinforced with man-made tissue (the layer following the grey one). Microanalysis (chemical composition) of the matrix (binder)

Figure 3.12 presents an example of the results of X-ray microanalysis in the zone of arc-shaped inserts in the wing bulk (the principal wing material). Microanalysis was performed at the point shown in the secondary electron image. Chemical element concentrations are shown in mass percentages. Concentrations of all chemical elements were measured on K lines of characteristic X-ray emission. The figure also presents the spectrum of characteristic X-ray emission at measurement point.

Figure 3.13 presents an example of the results of X-ray microanalysis on the surface of the grey layer of wing coating. Microanalysis was performed in a rectangular area shown in the secondary electron image. Chemical element concentrations

Element

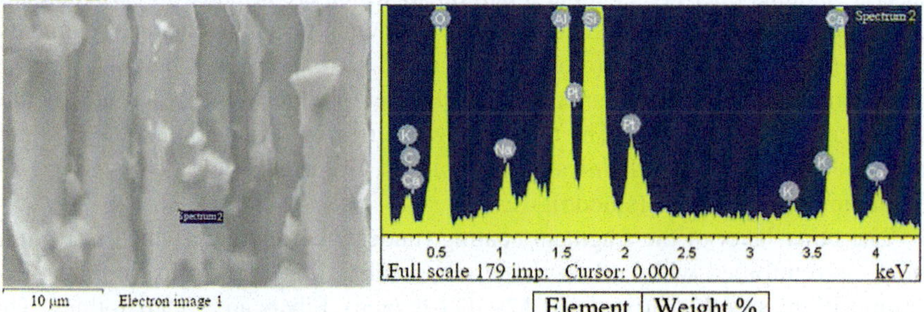

Element	Weight %
C K	6.14
O K	44.21
Na K	1.27
Al K	7.11
Si K	27.41
K K	0.66
Ca K	13.21
Total	100.00

Fig. 3.12 Results of X-ray microanalysis in the zone of arc-shaped inserts in the wing bulk. Microanalysis (chemical composition of fibers)

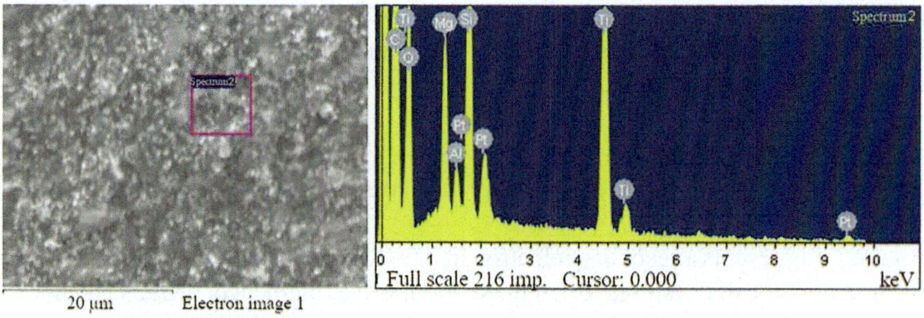

Element	Weight %
C K	52.55
O K	25.82
Mg K	2.99
Al K	0.60
Si K	4.19
Ti K	13.85
Total	100.00

Fig. 3.13 Results of X-ray microanalysis in the domain of arc-shaped inserts in the wing bulk. Microanalysis (chemical composition of fibers)

are presented in mass percentages. Concentrations of all chemical elements were measured on K lines of characteristic X-ray emission. Figure 3.13 also presents the spectrum of characteristic X-ray emission from measurement area.

Studies of the chemical composition and structure of a wing fragment of Kh-101 cruise missile showed that the main bulk material of the wing has foamed structure (most likely—foamed glass) of the following chemical composition: 31% Si (silicon), 51% O (oxygen), 7.5% C (carbon), and 10.5% Na (sodium). There are arc-shaped inserts in the main bulk of the wing. Wing material in the zone of arc-shaped inserts contains fiber layers about 5 μm thick, which are oriented in perpendicular directions. The insert material contains the following principal chemical elements: 30% Si (silicon) and 45% O (oxygen).

The fragment has two-layer coating, each layer of which has a different structure and chemical composition. The top coating layer (grey layer) about 100 μm thick consists of the following chemical elements: 52% C (carbon), 26% O (oxygen), 14% Ti (titanium), and 4% Si (silicon).

The other layer of the fragment coating (under the grey one), which is 1.9 mm thick, is reinforced with carbon tissue whose fibers have the following chemical composition: 88% C (carbon) and 12% O (oxygen). Fibers are 10.5 μm thick. The matrix (binder) of this composite has practically the same chemical composition as its fibers: 80% C (carbon) and 17% O (oxygen). This fact suggests the assumption that the other reinforced coating layer could belong to the reinforced carbon–carbon material. It should be emphasized that carbon-based composite materials have high strength and stiffness, and low mass; they often have higher strength than steel, but are much lighter.

Hence, the coating layer reinforced with carbon-based tissue, the principal wing material of apparently foamed glass, and arc-shaped reinforcing inserts into the main wing material, which have a high silicon content, result in high wing strength. Besides, carbon-based fibers (tubes) in the coating layer make it a non-magnetic dielectric RF absorbent material.

3.3.1.2 Fragment of 9M727 *Iskander-K* Cruise Missile Coating

Figure 3.14 presents a secondary electron image (SEI) of a dark surface (the surface of the middle—the thickest—layer) of the specimen of a fragment of 9M727 Iskander-K cruise missile body coating (hereinafter—the specimen).

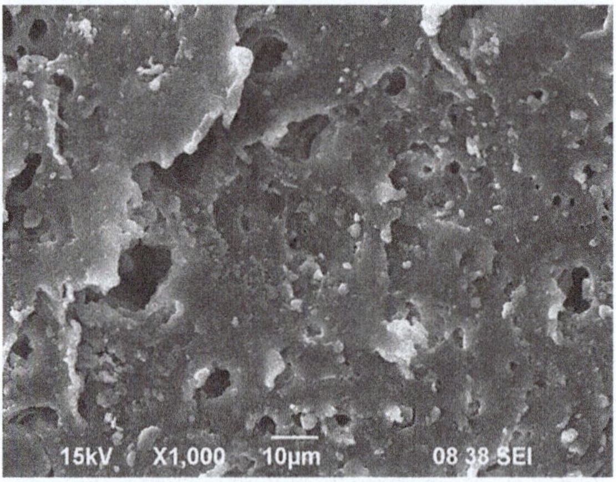

Fig. 3.14 Secondary electron image of dark surface (the surface of the middle—the thickest—layer) of the specimen. Accelerating voltage is 15 kV. Magnification is 1000, scale bar corresponds to 10 μm

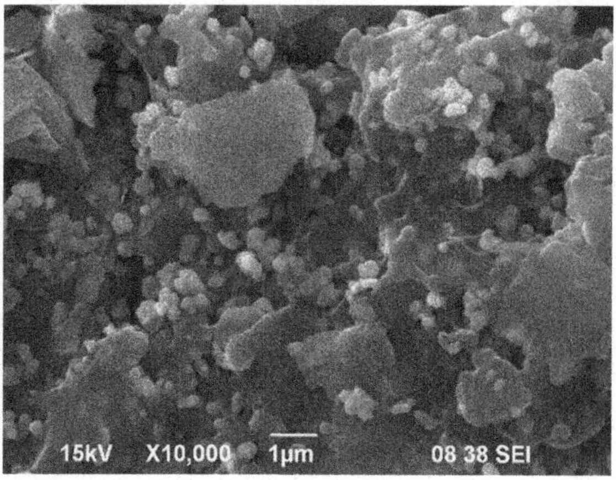

Fig. 3.15 Secondary electron image of specimen grey surface. Accelerating voltage is 15 kV. Magnification is 10,000, scale bar corresponds to 1 μm

Figure 3.15 presents the secondary electron image (SEI) of specimen grey surface.

Figure 3.16 presents the secondary electron image (SEI) of the specimen light-colored surface at various magnifications.

To study the structure and measure the concentrations of chemical elements in the coating bulk, as well as measure coating thickness, a specimen section was prepared. The backscattered electron image produced with a JSM-6490LV scanning electron microscope is shown in Fig. 3.17.

It can be seen in this figure that the specimen is a coating which contains three structurally different layers. The thickness of the whole coating is 1930 μm. The thickness of the grey layer is about 150 μm, and that of light-colored layer—from 30 to 50 μm. The thickness of the dark—middle and the thickest—layer of the specimen is about 1700 μm, respectively.

It can be seen that in the bulk (of the section) of coating dark layer (the thickest specimen layer) there are high-density tube-shaped inclusions, which are about 7 μm thick (Fig. 3.18).

Figure 3.19 gives an example of the results of X-ray microanalysis in the bulk of dark specimen coating (the thickest layer of the specimen). Microanalysis was performed at the point on tube surface that is shown in the secondary electron image. Chemical element concentrations are presented in mass percentages. Concentrations of all chemical elements were measured on K lines of characteristic X-ray emission. Figure 3.19 also presents the spectrum of characteristic X-ray emission of the measurement area. It can be seen from Fig. 3.19 that tube-shaped inclusions which are present in the bulk of dark layer of the coating contain C (carbon) in high concentrations (over 90%). The high carbon content in the tube-shaped inclusions is also confirmed by the spatial distribution of chemical elements.

Figure 3.20 presents an example of the results of X-ray microanalysis in the bulk of dark specimen coating. Microanalysis was performed in the indicated rectangular area located beyond tubes in the binder zone. The measured concentrations of chemical elements were averaged within the area shown in the secondary electron

Fig. 3.16 Secondary electron image of the specimen light surface. Accelerating voltage is 15 kV. Magnification is 5000, scale bar corresponds to 5 μm

Fig. 3.17 Backscattered electron image of the whole specimen section. Accelerating voltage is 15 kV. Magnification is ×40, scale bar corresponds to 500 μm

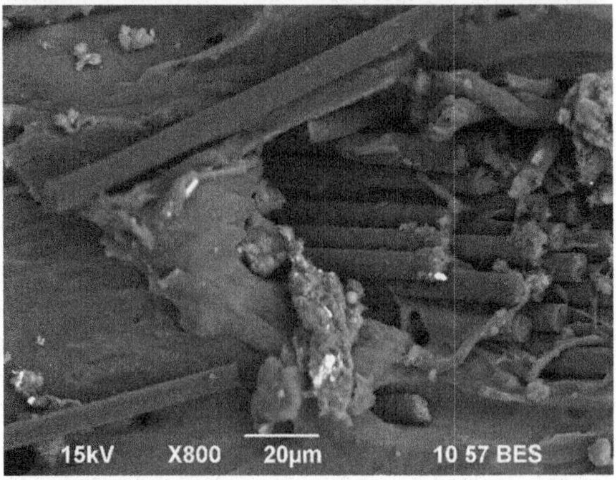

Fig. 3.18 Backscattered electron image of specimen section in the zone of dark layer (the thickest layer of the specimen). Accelerating voltage is 15 kV. Magnification is ×800, scale bar corresponds to 20 μm

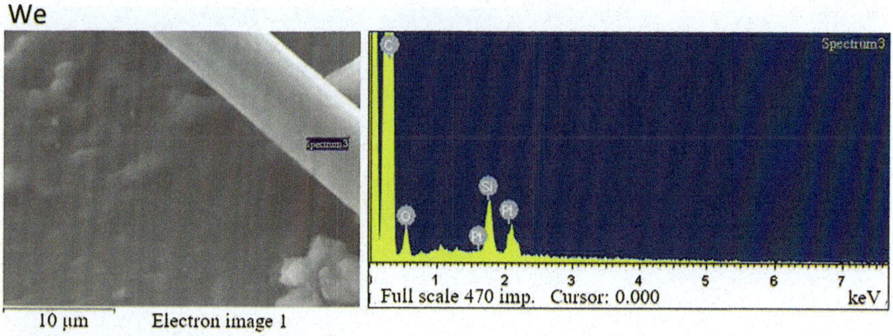

Element	Weight %
C K	93.04
O K	5.78
Si K	1.17
Total	100.00

Fig. 3.19 Results of X-ray microanalysis in the bulk of specimen dark layer. Microanalysis was performed at a point on tube surface

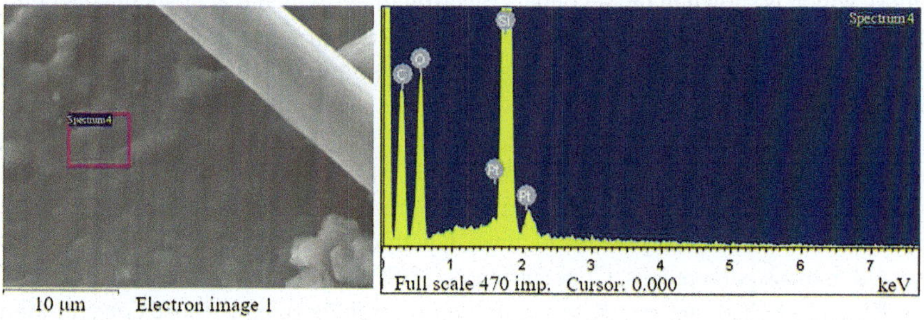

Element	Weight %
C K	49.67
O K	24.27
Si K	26.06
Total	100.00

Fig. 3.20 Results of X-ray microanalysis in the bulk of dark coating layer of the specimen. Microanalysis was performed in the indicated rectangular area located beyond tubes in the binder zone

image. Chemical element concentrations are presented in mass percentages. Concentrations of all chemical elements were measured on K lines of characteristic X-ray emission. This figure also shows the spectrum of characteristic X-ray emission of the measurement area.

Figure 3.21 shows an example of the results of X-ray microanalysis of the surface of specimen grey coating. Microanalysis was performed in the indicated rectangular area (the measured concentrations of chemical elements were averaged within this

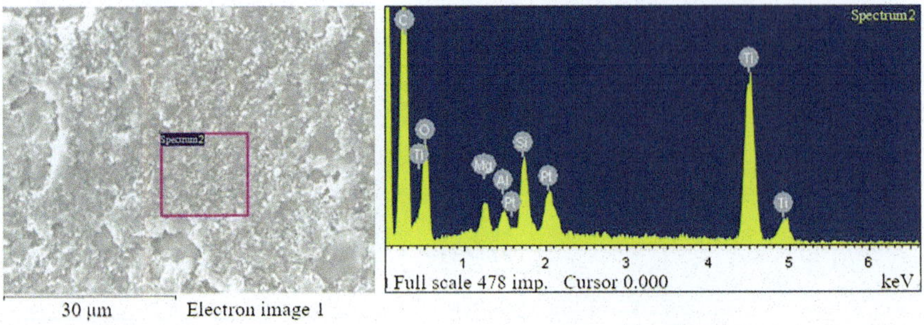

Element	Weight %
C K	48.21
O K	26.04
Mg K	1.32
Al K	0.67
Si K	2.29
Ti K	21.48
Total	100.00

Fig. 3.21 Results of X-ray microanalysis on the specimen grey surface

area), which is shown in the secondary electron image. Chemical element concentrations are given in mass percentages. Concentrations of all chemical elements were measured on K lines of characteristic X-ray emission. This figure also presents the spectrum of characteristic X-ray emission of this measurement area.

Figure 3.22 gives an example of the results of X-ray microanalysis of the surface of specimen light-colored coating. Microanalysis was performed in the indicated rectangular area (the measured concentrations were measured within this area), which is shown in the secondary electron image. Chemical element concentrations are presented in mass percentages. Concentrations of all chemical elements were measured on K lines of characteristic X-ray emission (except Zn (zinc), whose concentration was measured on L line). The figure also presents the spectrum of characteristic X-ray emission of this measurement area.

The analysis of examining the structure and chemical composition of the specimen has shown that it is a 1930-μm-thick coating that consists of three layers of different structures and chemical compositions. Grey layer is approximately 150-μm-thick, light-colored layer—30 μm, and dark layer has a thickness of 1700 μm and is a composite material in whose bulk high-density tube-shaped inclusions nearly 7 μm thick are observed. These inclusions contain C (carbon) in high concentration (over 90%, Fig. 3.19). The presence of carbon in tube-shaped inclusions is also confirmed by the spatial distribution of chemical elements The matrix—binder of this polymer—is a polymer and consists of the following principal chemical elements: 26% Si (silicon), 24% O (oxygen), and 50% C (carbon) (see Fig. 3.20).

Currently, to produce RF absorbent materials, carbon-containing materials, including carbon tubes, are widely used as fillers of polymer matrices. So, the presence of carbon tubes with carbon concentrations above 90% in the main coating layer gives reason to consider the fragment of 9M727 cruise missile of *Iskander-K* system (the specimen) as a non-magnetic dielectric RF absorbent material.

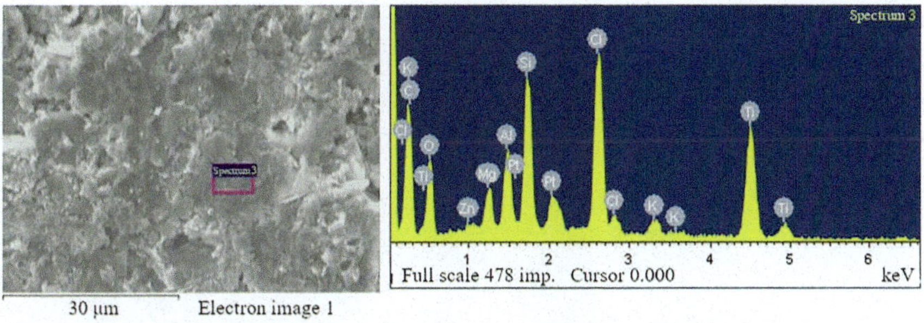

Element	Weight %
C K	35.51
O K	23.13
Mg K	1.92
Al K	2.95
Si K	6.09
Cl K	11.77
K K	1.59
Ti K	16.80
Zn L	0.24
Total	100.00

Fig. 3.22 Results of X-ray microanalysis on the specimen grey surface

3.3.1.3 Fragment of the Leading Edge of the Wing from 9M727 Cruise Missile of *Iskander-K* Complex

Figure 3.23 presents a secondary electron image (SEI) of brown coating of wing leading edge.

Figure 3.24 presents the secondary electron image (SEI) of the dark surface of the coating of wing leading edge.

To examine the structure and measure the concentrations of chemical elements in coating bulk, as well as determine the thicknesses of different coating layers, a section of a fragment of the coating of wing leading edge was prepared. Secondary electron image was produced with a JSM-6490LV electron microscope and is shown in Fig. 3.25.

Figure 3.27 presents an example of the results of X-ray microanalysis in the bulk (section) of the brown layer of wing leading edge. Microanalysis was performed in the indicated rectangular area. The measured concentrations of chemical elements were averaged within this area, which is shown in the secondary electron image. Chemical element concentrations are presented in mass percentages. Concentrations of all chemical elements were measured on K lines of characteristic X-ray emission. The figure also presents the spectrum of characteristic X-ray emission of this measurement area (Fig. 3.26).

Figure 3.28 presents an example of the results of X-ray microanalysis in the bulk (section) of brown layer of the coating of wing leading edge. Microanalysis was performed in the indicated rectangular area located in matrix zone. The measured concentrations of chemical elements were averaged within the area shown in the secondary electron image. Chemical element concentrations are presented in mass percentages. Concentrations of all chemical elements were measured on K lines of

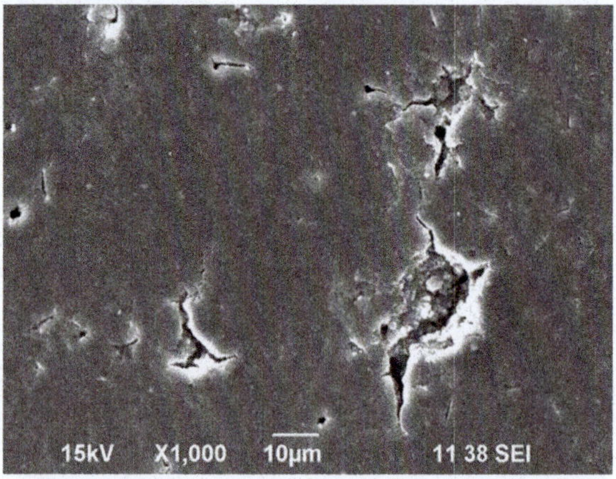

Fig. 3.23 Secondary electron image (SEI) of brown coating of wing leading edge. Accelerating voltage is 15 kV. Magnification is 1000, scale bar corresponds to 10 μm

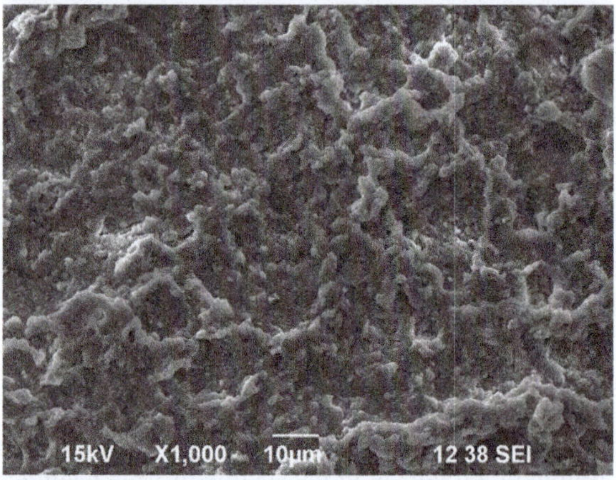

Fig. 3.24 Secondary electron image (SEI) of dark coating of wing leading edge. Accelerating voltage is 15 kV. Magnification is 1000, scale bar corresponds to 10 μm

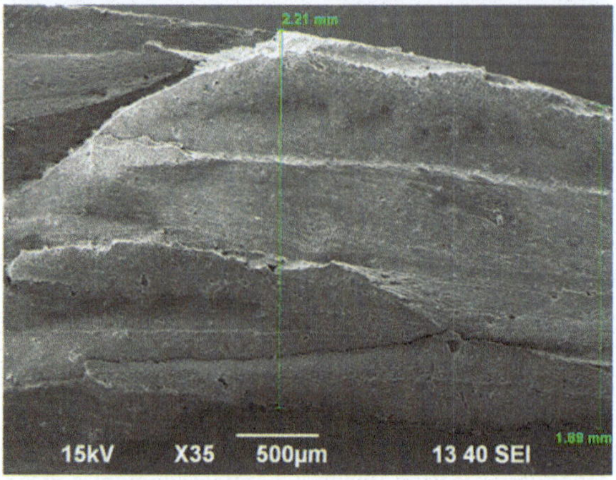

Fig. 3.25 Secondary electron image of a section of the coating of wing leading edge. Accelerating voltage is 15 kV. Magnification is ×35, scale bar corresponds to 500 μm. Brown coating layer size is shown

Fig. 3.26 Backscattered electron image of the coating of wing leading edge. Accelerating voltage is 15 kV. Magnification is ×100, scale bar corresponds to 100 µm

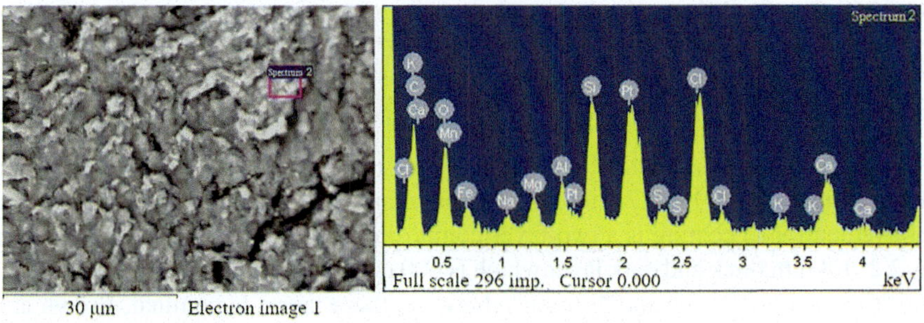

Element	Weight %
C K	29.82
O K	19.79
Na K	0.81
Mg K	1.33
Al K	1.52
Si K	4.43

Element	Weight %
S K	0.74
Cl K	7.38
K K	1.12
Ca K	4.10
Mn K	2.48
Fe K	26.47
Total	100.00

Fig. 3.27 Results of X-ray microanalysis in the bulk (section) of brown coating layer of wing leading edge. Microanalysis was performed in the region of increased iron concentration (the chemical composition of separated iron particles was determined)

characteristic X-ray emission (except Zn (zinc), whose concentration was measured on L line). Figure 3.28 also presents the spectrum of characteristic X-ray emission of this measurement area.

Chemical composition of dark layer matrix is identical to that of brown layer. Chemical element concentrations differ by several percentage points.

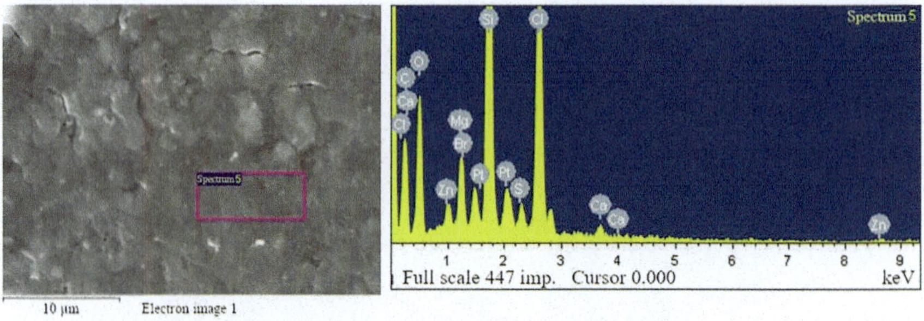

Element	Weight %
C K	40.26
O K	26.47
Mg K	0.83
Al K	0.35
Si K	12.57

Element	Weight %
S K	1.15
Cl K	16.09
K K	0.38
Ca K	0.58
Zn L	2.39
Total	100.00

Fig. 3.28 Results of X-ray microanalysis in the bulk (section) of brown layer of the coating of wing leading edge. Microanalysis was performed in the indicated rectangular area located in matrix zone

The studies of the chemical composition and structure of a fragment of coating on the leading edge of 9M727 cruise missile of Iskander-K system have shown that the abovementioned coating is a two-layer composite material with brown layer which is 2.2–1.8 mm thick and dark layer which is 600 μm thick. In terms of chemical composition and properties, the difference between these layers is nominal. Brown and dark layers have identical chemical compositions of matrices. The principal chemical elements of the matrices are 13 % Si (silicon), 26% O (oxygen), 40% C (carbon), and 16% Cl (chlorine). Concentrations of chemical elements in matrices differ by small percentage points. In both layers, there are zones containing considerable Fe (iron) concentrations (up to 30%), yet the dark layer shows a larger number of regions with increased iron concentration.

Taking into account the fact that iron is a ferromagnetic, the fragment of the coating of wing leading edge of 9M727 cruise missile of *Iskander-K* system can be attributed to the class of magnetic RF absorbent materials.

3.3.1.4 Fragments of Nose Cone Coating of 9M727 *Iskander* Cruise Missile

Specimen 1

Figure 3.29 presents the backscattered electron image (BEC) of brown surface of specimen 1.

Fig. 3.29 Backscattered electron image (BEC) of *specimen 1* brown surface. Magnification is 500, scale bar corresponds to 50 μm

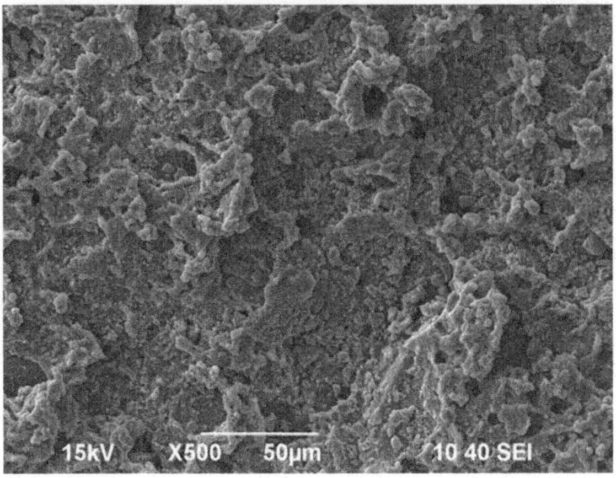

Fig. 3.30 Secondary electron image (SEI) of specimen 1 dark surface. Accelerating voltage is 15 kV. Magnification is 500, scale bar corresponds to 50 μm

Figure 3.30 shows the secondary electron image (SEI) of *specimen 1* dark surface.

To study the structure and measure chemical element concentrations in the coating bulk, as well as measure coating thickness, a section of *specimen 1* was prepared. The secondary electron image of the section was obtained with a JSM-6490LV electron microscope and is presented in Figs. 3.31, 3.32, and 3.33.

It can be seen in Figs. 3.31, 3.32, and 3.33 that specimen 1 is a coating that consists of three layers of different structures. The thickness of the whole coating is 3120 μm. The thickness of coating top layer is approximately 120 μm and the thickness of the bottom one is approximately 170 μm. The thickness of the middle layer is, respectively, 2830 μm.

As seen in Fig. 3.34, there are high-density inclusions in the form of particles of different sizes from 0.5 to 10 μm in the bulk of coating middle interlayer.

As seen in Figs. 3.35 and 3.36, similar particles, but less densely distributed over the bulk, are present in the bulk of top and bottom layers.

Fig. 3.31 Secondary electron image of specimen 1. Accelerating voltage is 15 kV. Magnification is ×25, scale bar corresponds to 1 mm

Fig. 3.32 Backscattered electron image of specimen 1. Accelerating voltage is 15 kV. Magnification is 200, scale bar corresponds to 100 μm. The thickness of the top layer is shown

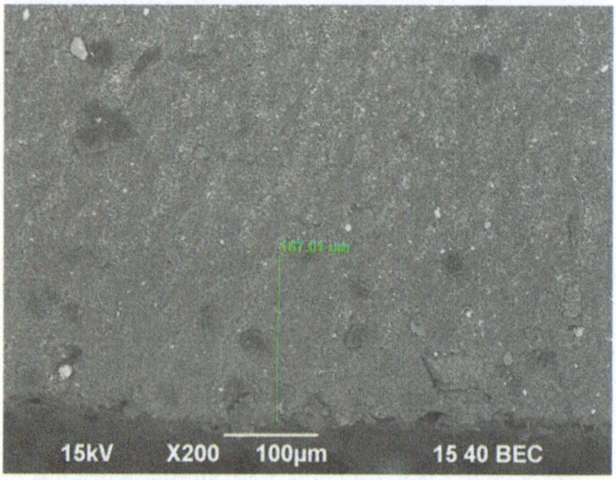

Fig. 3.33 Backscattered electron image of *specimen 1* section. Accelerating voltage is 15kV. Magnification is ×200, scale bar corresponds to 100 μm. The thickness of bottom layer is shown

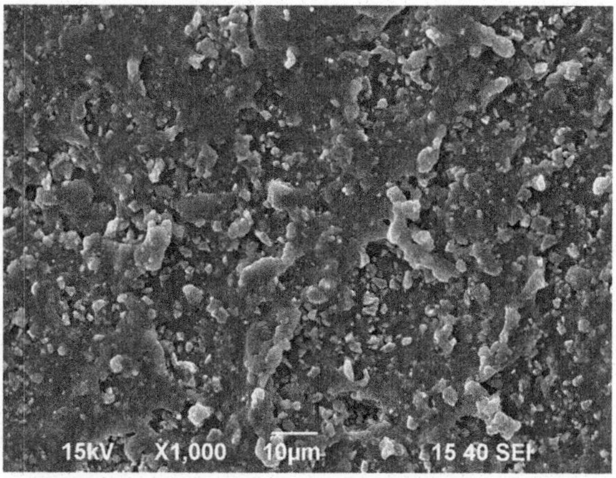

Fig. 3.34 Secondary electron image of *specimen 1* section in the zone of coating middle layer. Accelerating voltage is 15kV. Magnification is ×1000, scale bar corresponds to 10 μm

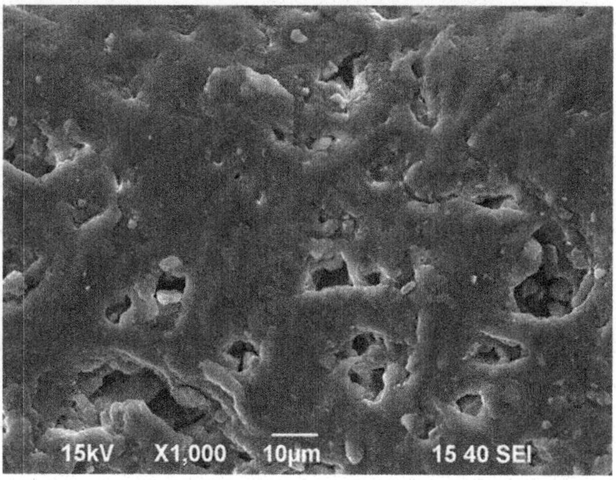

Fig. 3.35 Secondary electron image of *specimen 1* section in the zone of coating top layer. Accelerating voltage is 15 kV. Magnification is ×1000, scale bar corresponds to 15 μm

As can be seen in Figs. 3.37 and 3.38, similar particles but of lower density are present on the surfaces of the bottom and top layers.

Figure 3.39 presents an example of the results of X-ray microanalysis in the bulk of middle (the thickest) layer of *specimen 1* coating. Microanalysis was performed at the point on particle surface shown in the secondary electron image. Chemical element concentrations are given in mass percentages. Concentrations of all chemical elements were measured on K lines of characteristic X-ray emission (except Ba (barium), whose concentration was measured on line L). Figure 3.39 also shows the spectrum of characteristic X-ray emission of this measurement point.

Figure 3.40 shows an example of the results of X-ray microanalysis on brown surface of *specimen 1* coating. Microanalysis was performed in the indicated rectangular area (the measured concentrations of chemical elements were averaged within this area), which is shown in the secondary electron image. Chemical element concentrations are presented in mass percentages. Concentrations of all chemical elements

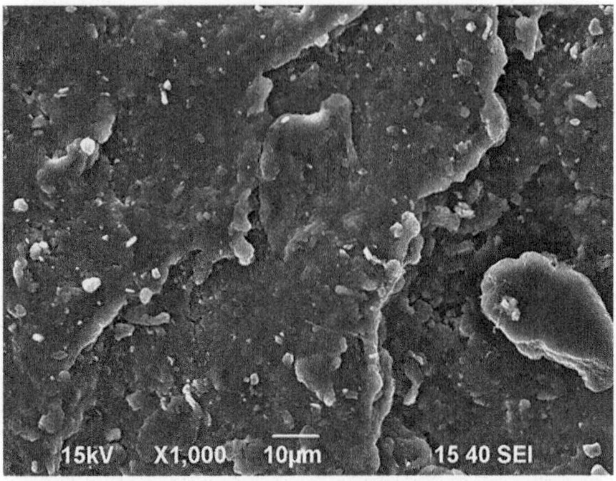

Fig. 3.36 Secondary electron image of *specimen 1* in the zone of coating bottom layer. Accelerating voltage is 15 kV. Magnification is ×1000, scale bar corresponds to 10 μm

Fig. 3.37 Backscattered electron image of the surface of specimen 1 top layer. Accelerating voltage is 15 kV. Magnification is ×2000, scale bar corresponds to 10 μm

Fig. 3.38 Backscattered electron image of the surface of specimen 1 bottom layer. Accelerating voltage is 15 kV. Magnification is ×2000, scale bar corresponds to 10 μm

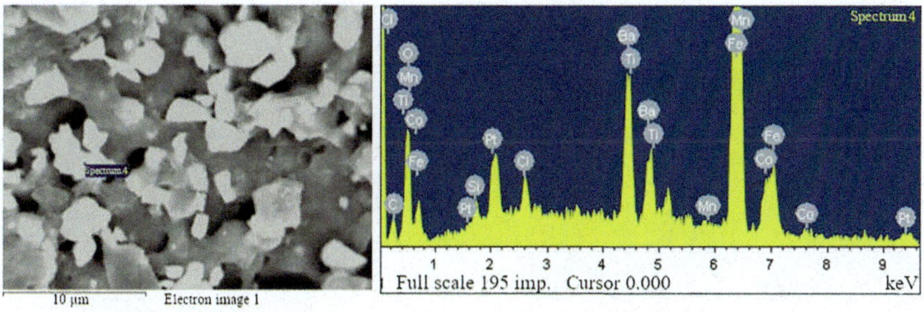

Element	Weight %
C K	5.55
O K	7.29
Si K	0.35
Cl K	1.06
Ti K	0.36

Element	Weight %
Mn K	0.37
Fe K	53.68
Co K	7.43
Ba L	23.91
Total	100.00

Fig. 3.39 Results of X-ray microanalysis in the bulk of middle layer of *specimen 1* coating. Microanalysis was performed at the point located on particle surface

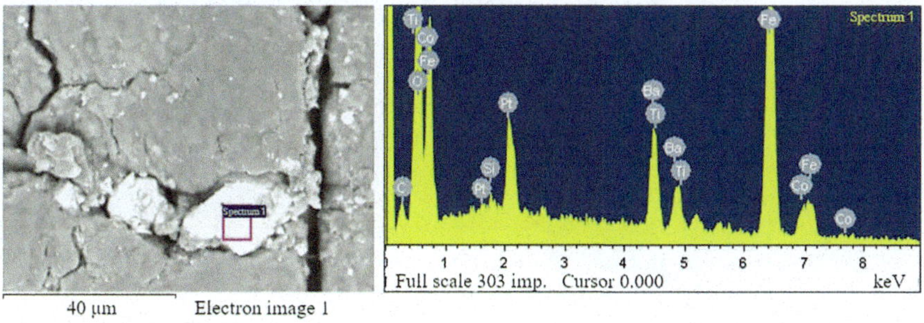

Element	Weight %
C K	7.12
O K	28.09
Si K	0.13
Ti K	0.69

Element	Weight %
Fe K	41.90
Co K	4.97
Ba L	17.11
Total	100.00

Fig. 3.40 Results of X-ray microanalysis on brown surface of *specimen 1* coating performed at indicated rectangular area shown in the secondary electron image (the measured concentrations of chemical elements were averaged within it)

were measured on K lines of characteristic X-ray emission (except Ba (barium), whose concentration was measured on L line). The figure also presents the spectrum of characteristic X-ray emission of this measurement area.

Figure 3.41 presents an example of the results of X-ray microanalysis on the surface of *specimen 1* dark coating. Microanalysis was performed at the indicated rectangular area (the measured concentrations of chemical elements were averaged within this area), which is shown in the secondary electron image. Chemical element

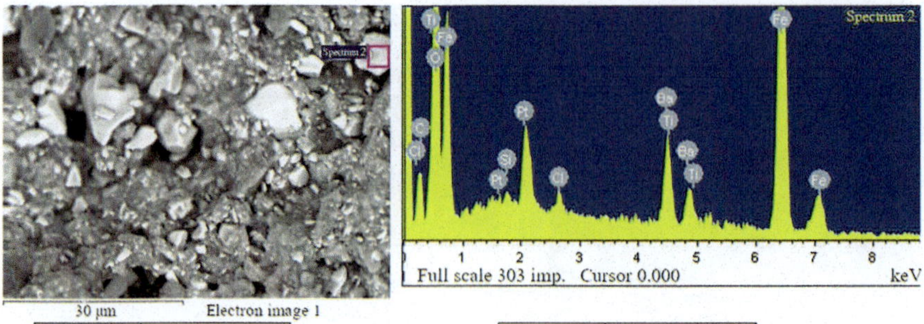

Element	Weight %
C K	12.07
O K	32.20
Si K	0.31
Cl K	0.47
Ti K	1.10
Fe K	42.61
Ba L	11.24
Total	100.00

Fig. 3.41 Results of X-ray microanalysis on the surface of black *specimen 1* coating in the indicated rectangular area (the measured concentrations of chemical elements were averaged within this area), which is shown in the secondary electron image

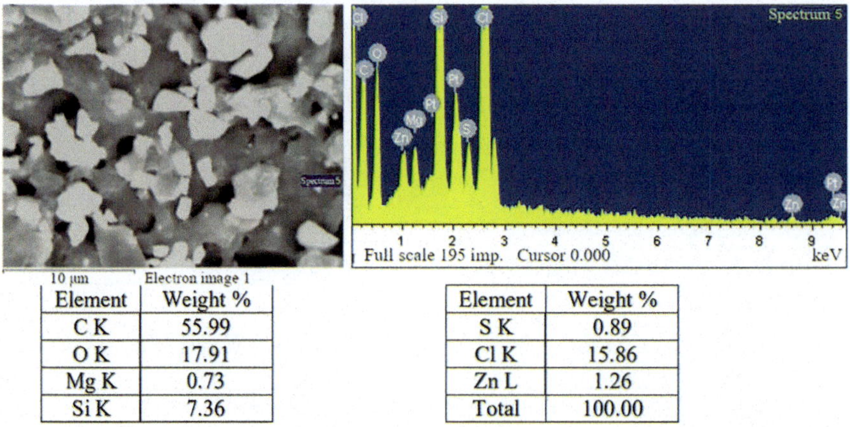

Element	Weight %
C K	55.99
O K	17.91
Mg K	0.73
Si K	7.36
S K	0.89
Cl K	15.86
Zn L	1.26
Total	100.00

Fig. 3.42 Results of X-ray microanalysis in the bulk of the middle layer of *specimen 1* coating. Microanalysis was performed at the point located beyond particles in the binder zone

concentrations are presented in mass percentages. The concentrations of all chemical elements were measured on K lines of characteristic X-ray emission (except Ba (barium), whose concentration was measured on L line). Figure 3.41 also presents the spectrum of characteristic X-ray emission of this measurement area.

Figure 3.42 shows an example of the results of microanalysis in the bulk of the middle layer of specimen 1 coating. Microanalysis was performed at the point located beyond particles and situated in the zone of binder. Chemical element concentrations were measured at the point shown in the secondary electron image. Chemical element concentrations are given in mass percentages. Concentrations of all chemical elements were measured on K lines of characteristic X-ray emission (except Zn (zinc), whose concentration was measured on L line). The figure also presents the spectrum of characteristic X-ray emission at the measurement point.

It can be seen in Fig. 3.39 that the inclusions in the form of particles, which are present in the bulk of the middle layer of specimen 1 coating, contain Fe (iron) and Ba (barium) in high concentrations (over 50% Fe and over 20% Ba). The high concentrations of iron and barium in the inclusions are also confirmed by the spatial distribution of chemical elements, see Fig. 3.43.

Figure 3.44 presents an example of the results of X-ray microanalysis of the brown surface of specimen 1 coating. Microanalysis was performed in the indicated rectangular area (the measured chemical element concentrations were averaged within it), which is shown in the secondary electron image. Chemical element concentrations are listed in mass percentages. Concentrations of all chemical elements (except Zn (zinc)) were measured on K lines of characteristic X-ray emission. The figure also presents the spectrum of characteristic X-ray emission of this measurement area.

Figure 3.45 shows an example of X-ray microanalysis of the surface of specimen 1 dark coating. Microanalysis was performed in the indicated rectangular area (the measured chemical element concentrations were averaged within this area), which is shown in the secondary electron image. Chemical element concentrations are presented in mass percentages. Concentrations of all chemical elements (except Zn (zinc) and Ba (barium), whose concentrations were measured on L line) were measured on K lines of characteristic X-ray emission. Figure 3.45 also presents the spectrum of characteristic X-ray emission of this measurement area.

The analysis of studying the structure and chemical composition of specimen 1 has shown that this specimen is a composite 3120-μm-thick coating, which consists of three layers which are different in structure and chemical composition. The top layer is about 120 μm thick, the middle one is 2830 μm, and the bottom layer is 170 μm thick. All three layers are composite materials which contain inclusions in the form of particles of various sizes from 0.5 to 10 μm. The highest density of particles presence in the bulk is observed in the middle coating layer. It can be seen in Fig. 3.39 that inclusions in the form of particles that are present in the bulk of the middle layer of specimen 1 coating contain Fe (iron) and Ba (barium) in high concentrations (above 50% Fe, above 20% Ba). The presence of high concentrations of iron and barium in the particles is also confirmed by the spatial distribution of chemical elements, see Fig. 3.43. Matrix—the binder of this composite—consists of the following principal chemical elements: 8.8% Si (silicon), 33.4% O (oxygen), 40% C (carbon), and 13% Cl (chlorine) (see Fig. 3.44).

The presence of Fe-rich particles (iron is a ferromagnetic) in the bulk of the fragment of brown nose cone coating of 9M727 cruise missile of *Iskander-K* system (*specimen 1*) allows this specimen to be attributed to the class of magnetic RF absorbent materials.

Specimen 2

Figure 3.46 presents the backscattered electron image (BES) of lustrous black surface of *specimen 2*. Particles of different sizes from 5 to 50 μm are observed on this surface.

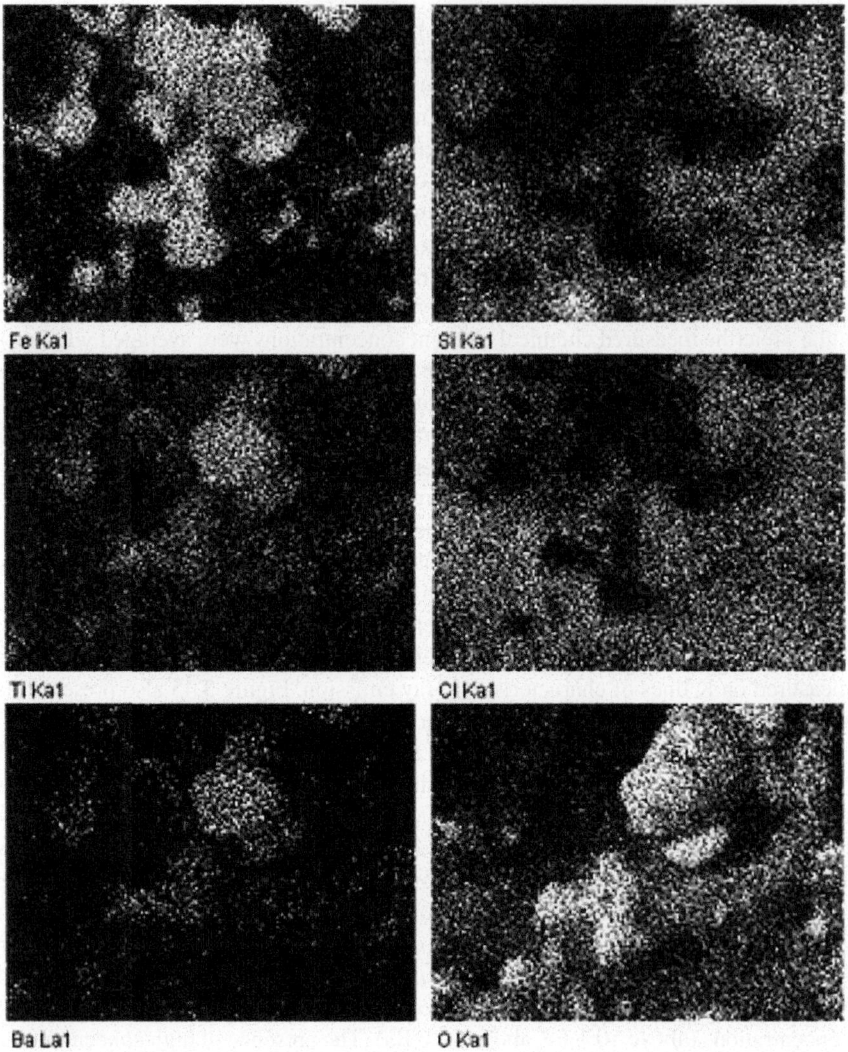

Fig. 3.43 X-ray characteristic emission images of the same area (spatial distribution of chemical elements over the studied specimen section). White dots correspond to the presence of the chemical element indicated under each image

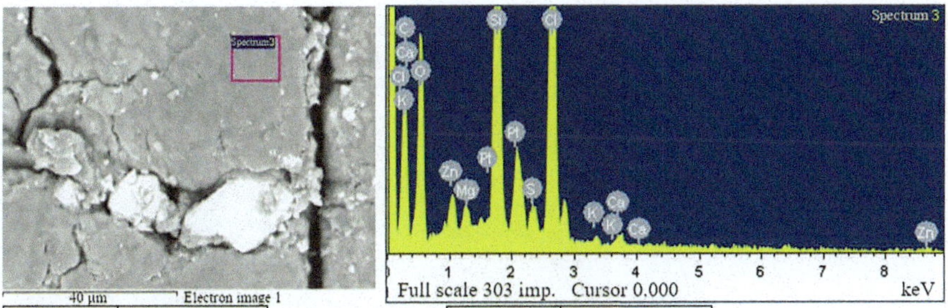

Element	Weight %
C K	39.86
O K	33.45
Mg K	0.59
Si K	8.79
S K	0.74
Cl K	13.33
K K	0.36
Ca K	0.55
Zn L	2.34
Total	100.00

Fig. 3.44 Results of X-ray microanalysis on brown surface of *specimen 1*

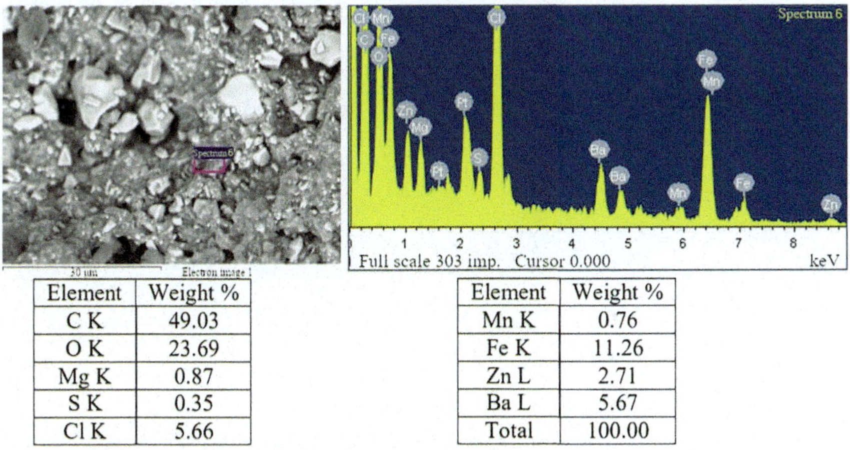

Element	Weight %
C K	49.03
O K	23.69
Mg K	0.87
S K	0.35
Cl K	5.66
Mn K	0.76
Fe K	11.26
Zn L	2.71
Ba L	5.67
Total	100.00

Fig. 3.45 Results of X-ray microanalysis on dark surface of *specimen 1*

Figures 3.47 and 3.48 present secondary electron images (SEI) of matt black surface of *specimen 2* at different magnifications.

Particles are not observed in the images of *specimen 2* matt black surface. To study the structure and measure chemical element concentrations in the coating bulk, as well as measure coating thickness, a section of specimen 2 was prepared. The section images obtained with a JSM-6490LV electron microscope are shown in Figs. 3.49, 3.50, and 3.51.

It is seen in Fig. 3.49 that *specimen 2* is a coating consisting of two layers of different structures. The whole coating is 3230 µm thick. The matt black coating layer is approximately 175 µm thick, the lustrous black coating layer is approximately 3055 µm thick.

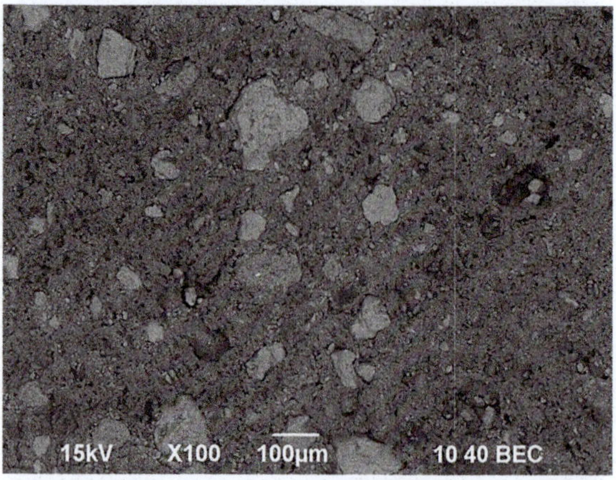

Fig. 3.46 Backscattered electron image of lustrous black surface of *specimen 2*. Accelerating voltage is 15 kV. Magnification is ×100, scale bar corresponds to 100 µm

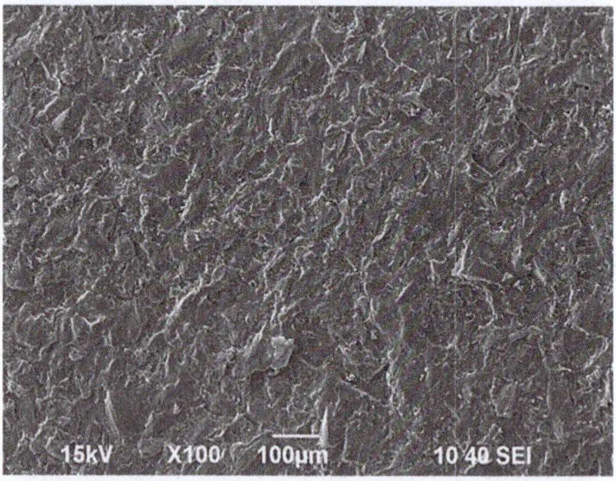

Fig. 3.47 Secondary electron image of matt black surface of *specimen 2*. Accelerating voltage is 15 kV. Magnification is ×100, scale bar corresponds to 100 µm

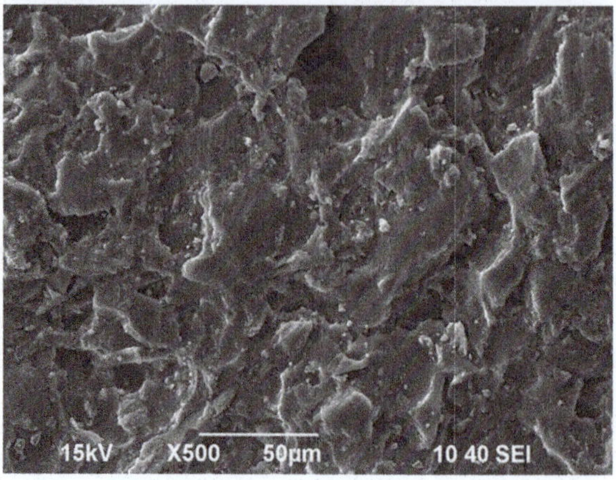

Fig. 3.48 Secondary electron image of matt black surface of specimen 2. Accelerating voltage is 15 kV. Magnification is ×500, scale bar corresponds to 50 µm

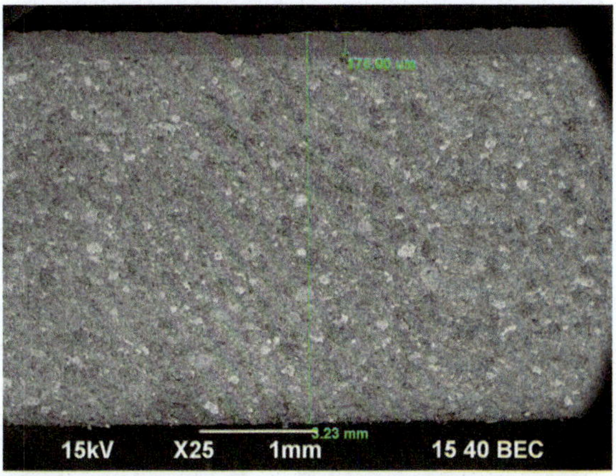

Fig. 3.49 Backscattered electron image of specimen 2 section. Accelerating voltage is 15 kV. Magnification is ×25, scale bar corresponds to 1 mm

Fig. 3.50 Backscattered electron image of *specimen 2* section in the zone of lustrous black layer. Accelerating voltage is 15 kV. Magnification is ×500, scale bar corresponds to 50 μm

Fig. 3.51 Secondary electron image of *specimen 2* section in the zone of matt black layer. Accelerating voltage is 15 kV. Magnification is ×350, scale bar corresponds to 50 μm

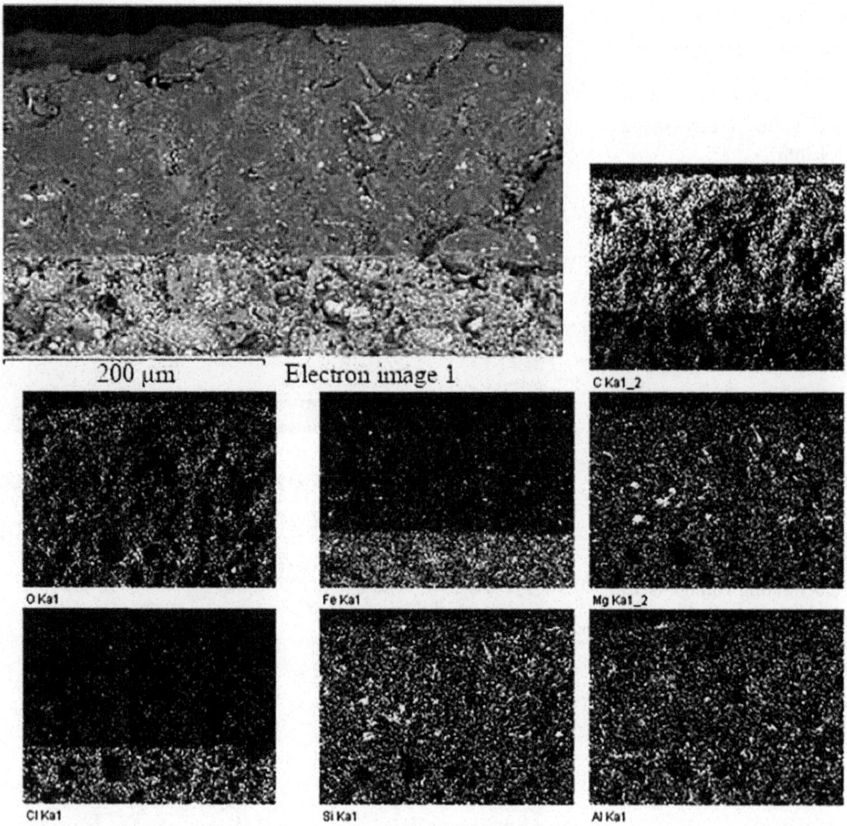

Fig. 3.52 Secondary electron image of specimen 2 section in the zone of the boundary between matt black layer (top layer) and lustrous black layer (bottom layer) with scale bar corresponding 200 μm, and X-ray characteristic emission images of the same area (spatial distribution of chemical elements over the section of the specimen). White dots correspond to the presence of the chemical element indicated under each image

Figures 3.49 and 3.50 show that over the bulk of lustrous black layer there are high-density inclusions: particles of different sizes from 5 to 50 μm.

It can be seen in Fig. 3.51 that no particles are observed in the zone of matt black layer.

It is seen in Fig. 3.52 that Fe (iron) content rises dramatically in the lustrous black layer as compared to matt black layer of *specimen 2*.

Figure 3.53 presents an example of the results of X-ray microanalysis on the particle surface in the bulk of lustrous black layer of *specimen 2*. Microanalysis was performed in the indicated rectangular area (measured concentrations of chemical elements were averaged within this area), which is shown in the secondary electron image. Chemical element concentrations are given in mass percentages. Concen-

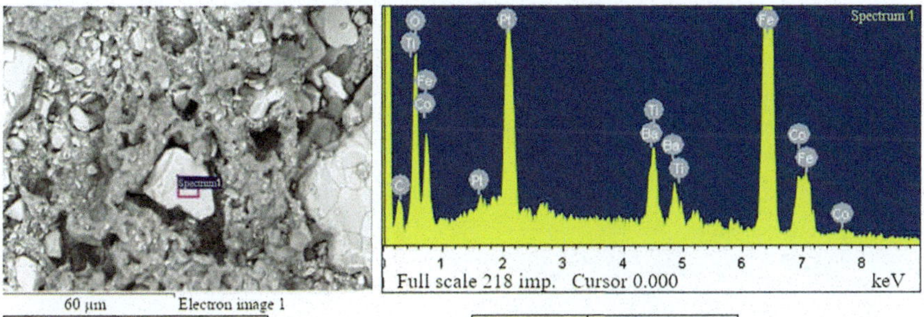

Element	Weight %
C K	10.29
O K	12.01
Ti K	1.41

Element	Weight %
Fe K	57.33
Ba L	9.81
CoK	9.17
Total	100.00

Fig. 3.53 Results of X-ray microanalysis on particle surface in the bulk of lustrous black layer of specimen 2

trations of all chemical elements were measured on K lines of characteristic X-ray emission of this measurement area.

Figures 3.54 and 3.55 present the results of determining chemical element concentrations along the scan line in the bulk (of section) of lustrous black layer in *specimen 2*.

It can be seen in Fig. 3.53 that the inclusions (particles) present in the bulk of lustrous black layer of the coating contain a large amount (up to 60%) of Fe (iron). The presence of iron in the inclusions is also confirmed by the spatial distribution of chemical elements (Fig. 3.52) and the Fe concentration profile along the scan line (Fig. 3.55).

Figure 3.56 presents an example of X-ray microanalysis of the surface of *specimen 2* lustrous black coating. Microanalysis was performed in the indicated rectangular area (the measured concentrations of chemical elements were averaged within this area), which is shown in the secondary electron image. Chemical element concentrations are given in mass percentages. Concentrations of all chemical elements were measured on K lines of characteristic X-ray emission of this measurement area.

Figure 3.57 presents an example of the results of X-ray microanalysis in the bulk of matt black layer of *specimen 2* coating. Microanalysis was performed in the indicated rectangular area (the measured concentrations of chemical elements were averaged within this area), which is shown in the secondary electron image. Chemical element concentrations are presented in mass percentages. Concentrations of all chemical elements were measured on K lines of characteristic X-ray emission (except Zn (zinc), whose concentration was measured on L line). Figure 3.57 also presents the spectrum of characteristic X-ray emission of this measurement area.

The analysis of the results of studying the structure and chemical composition of specimen 2 has shown that this specimen is a 3230-µm-thick composite coating

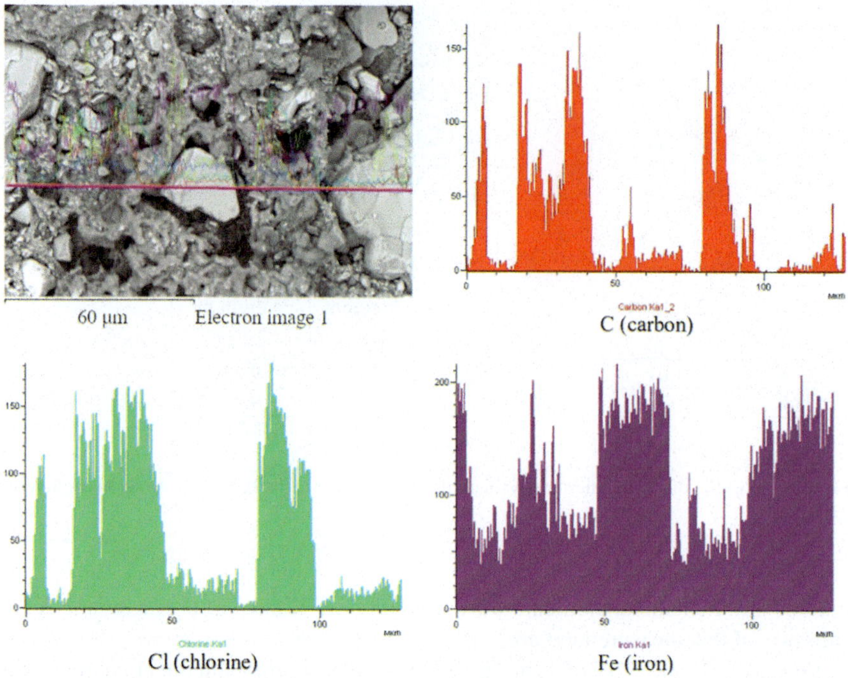

Fig. 3.54 Concentration profiles of C, Cl, Fe (arb. units) along the scan line shown in secondary electron image (horizontal line)

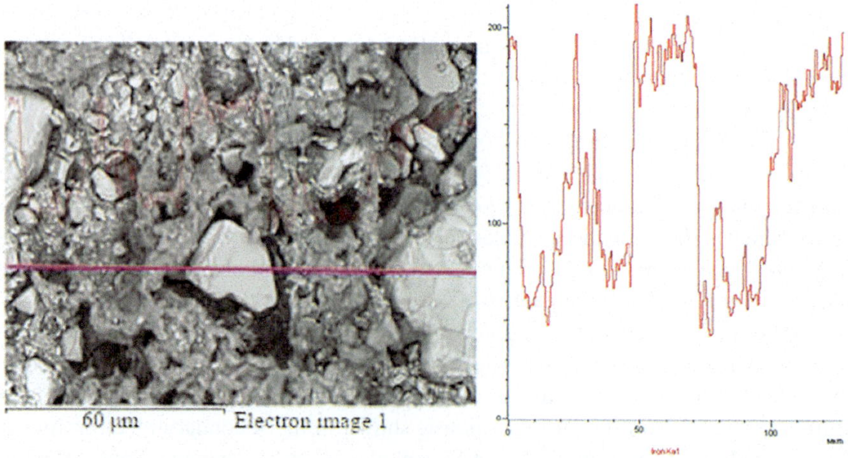

Fig. 3.55 Fe (iron) concentration profile (arb. units) along the scan line shown in secondary electron image (horizontal line)

Element	Weight %
C K	49,13
O K	18.00
Mg K	0.41
Si K	4.91
S K	0.71
Cl K	12.56
Ca K	0.36
Ti K	0.91
Mn K	1.29
Fe K	11.71
Total	100.00

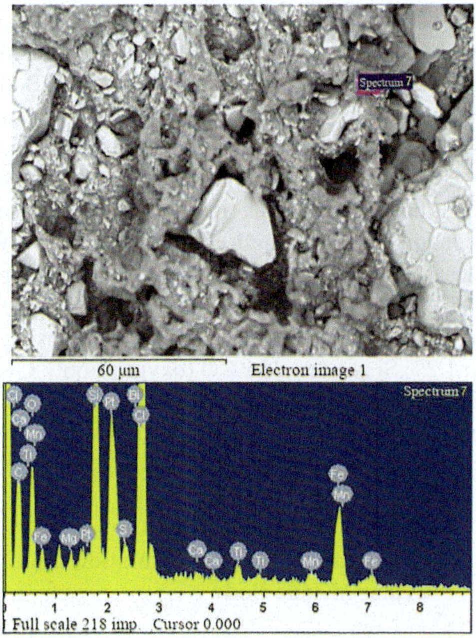

Fig. 3.56 Results of X-ray microanalysis on the surface of lustrous black coating of *specimen 2*

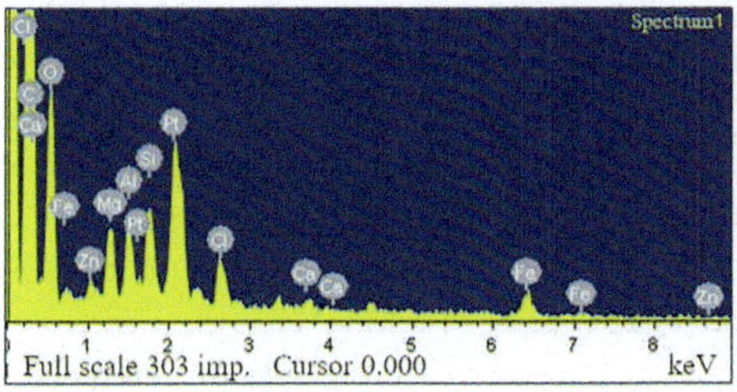

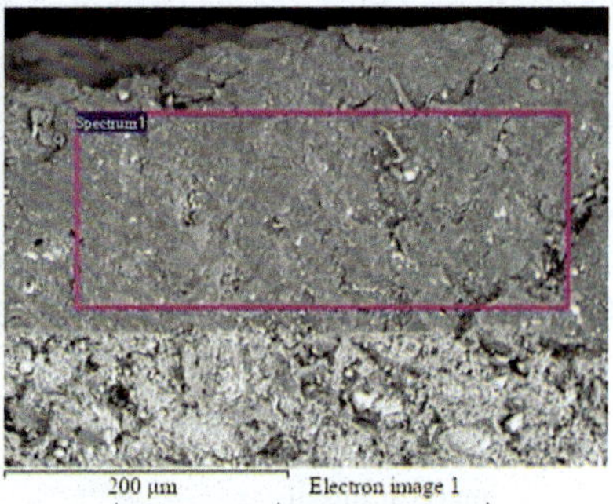

Element	Weight %
C K	65.54
O K	26.49
Mg K	1.17
Al K	0.97
Si K	1.20
Cl K	1.10
Ca K	0.37
Fe K	2.36
Zn L	0.79
Total	100.00

Fig. 3.57 Results of X-ray microanalysis in the bulk of matt black layer of *specimen 2* coating

which consists of two layers of different structures and chemical compositions. The thickness of matt black coating layer is approximately 175 μm, and that of lustrous black coating layer is approximately 3055 μm. Both layers are composite materials with inclusions (particles) of different sizes (from 5 to 50 μm). The presence of inclusions in the form of high-density particles in the bulk of lustrous black coating layer (as compared to matt black coating layer) is noticed. It can be seen in Fig. 3.53 that the inclusions (particles) present in the bulk of lustrous black coating layer contain Fe (iron) in high concentrations (up to 60%). The presence of iron in the particles is also confirmed by the spatial distribution of chemical elements (Fig. 3.52) and the Fe concentration profiles along the scan line (Fig. 3.55). The matrix (binder) of matt black material of the coating layer consists of such principal chemical elements: 1.2% Si (silicon), 26% O (oxygen), 66% C (carbon), and 1.1% Cl (chlorine), see Fig. 3.57.

Iron is a ferromagnetic, so *specimen 2* under study belongs to the class of magnetic RF absorbent materials.

3.3.1.5 Fragment of Thermal Insulation Coating of 9M723 *Iskander-M* Cruise Missile Body

The examined fragment of thermal insulation coating of 9M7 23 *Iskander-M* missile is a composite material which consists of several layers: the dark-colored topmost one, followed by a porous layer, under it is the layer reinforced with artificial longitudinal and transverse glass fibers, then there is the thickest foam glass layer, and a green layer is located at the bottom.

Figures 3.58, 3.59, and 3.60 present the secondary electron image (SEI) of dark-colored (further—dark) surface of the coating at different magnifications.

Figures 3.61 and 3.62 show secondary electron images of the green surface of missile body coating at different magnifications.

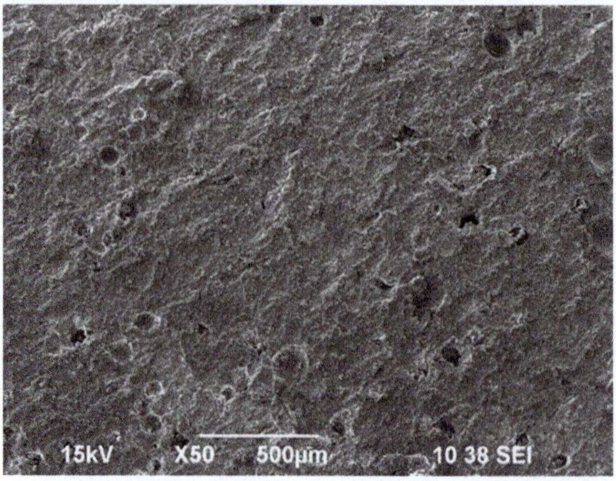

Fig. 3.58 Secondary electron image of missile body dark coating. Accelerating voltage is 15 kV, magnification ×50, scale bar corresponds to 500 μm

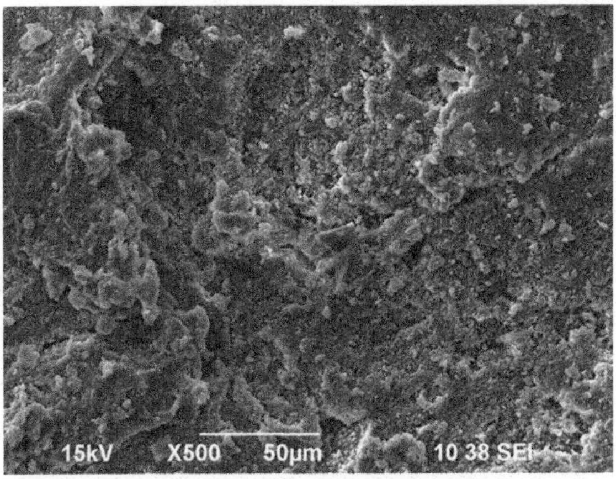

Fig. 3.59 Secondary electron image of missile body dark coating. Accelerating voltage is 15 kV, magnification ×500, scale bar corresponds to 50 µm

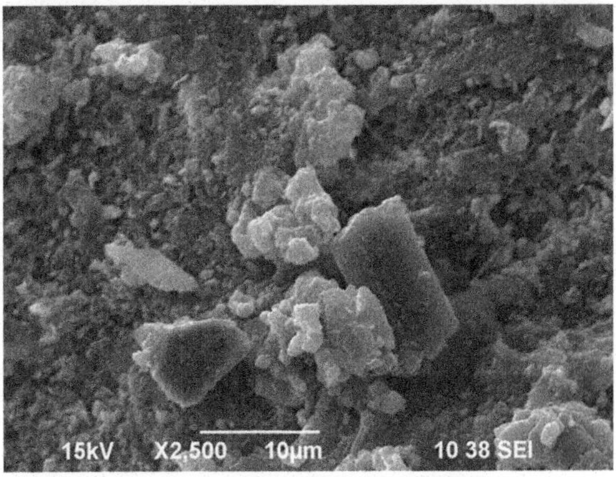

Fig. 3.60 Secondary electron image of dark coating of missile body surface. Accelerating voltage is 15 kV, magnification ×2500, scale bar corresponds to 10 µm

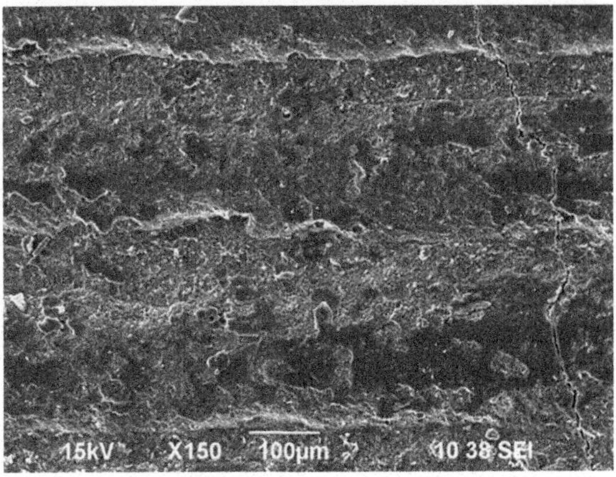

Fig. 3.61 Secondary electron image of green coating of missile body surface. Accelerating voltage is 15 kV, magnification ×150, scale bar corresponds to 100 µm

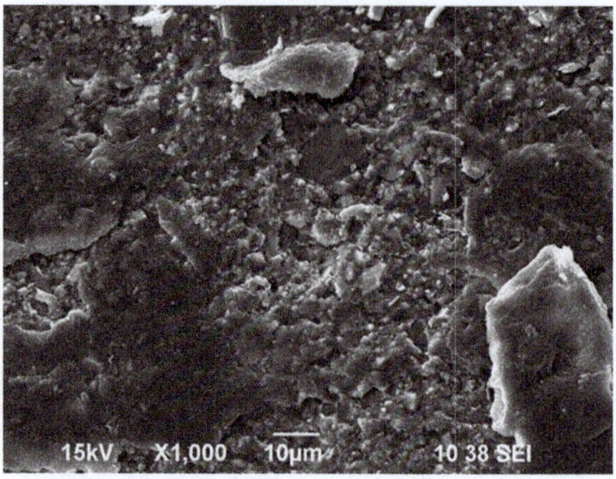

Fig. 3.62 Secondary electron image of the green surface of missile body coating. Accelerating voltage is 15 kV, magnification ×1000, scale bar corresponds to 10μm

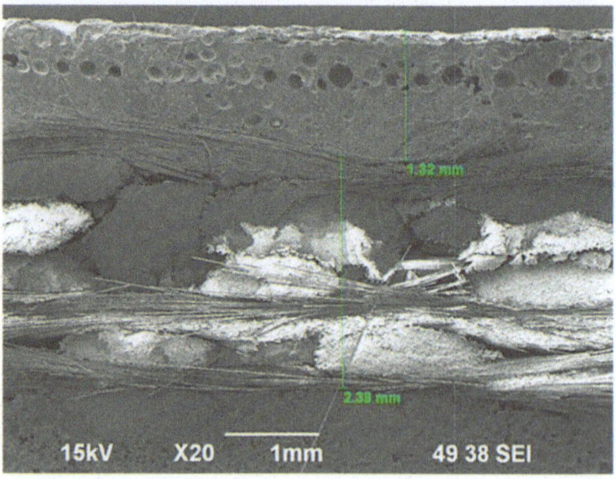

Fig. 3.63 Secondary electron image of a cut of missile body coating. Accelerating voltage is 15 kV, magnification ×20, scale bar corresponds to 1 mm. The sizes of different coating layers are shown

To study the structure and measure chemical element concentrations in the coating bulk, as well as measure coating layer's thicknesses, a section of missile body coating was made. Secondary electron images obtained with a JSM-6490LV scanning electron microscope are presented in Figs. 3.63, 3.64, 3.65, and 3.66.

It can be seen in Figs. 3.63, 3.64, 3.65, and 3.66 that the missile body coating consists of several layers. There is a dark top layer 124 μm thick; under it a 1.32-mm-thick porous layer is located; still lower, there is a layer reinforced with artificial longitudinal and transverse fibers of 2.39 mm thickness. Under the reinforced layer, a foamed material of 7.3 mm thickness is located. The green bottommost layer is 33 μm thick.

Figure 3.67 presents the images of longitudinal and transverse artificial fibers, and Fig. 3.68 shows their thickness.

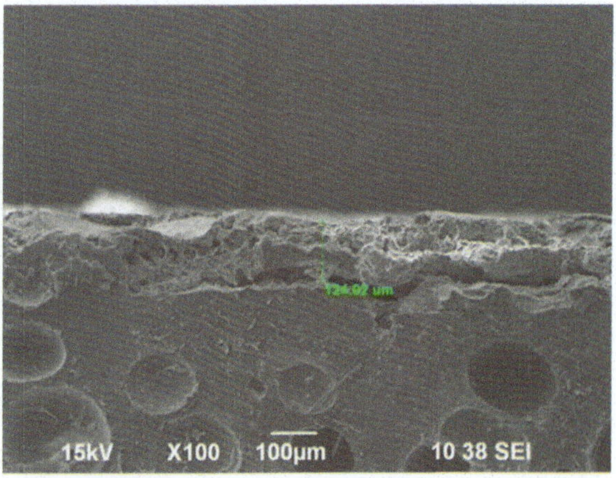

Fig. 3.64 Secondary electron image of a cut of missile body coating. Accelerating voltage is 15 kV, magnification $\times 100$, scale bar corresponds to 100 μm. The size of the dark top coating layer is indicated

Fig. 3.65 Secondary electron image of a cut of missile body coating. Accelerating voltage is 15 kV, magnification $\times 10$, scale bar corresponds to 2 mm (the thickest foamed material layer is located under the reinforced one)

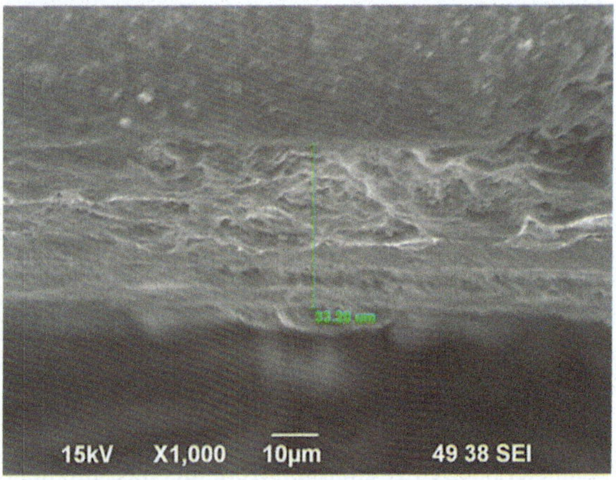

Fig. 3.66 Secondary electron image of a cut of missile body coating. Accelerating voltage is 15 kV, magnification $\times 1000$, scale bar corresponds to 10 μm. The thickness of the green bottommost coating layer is indicated

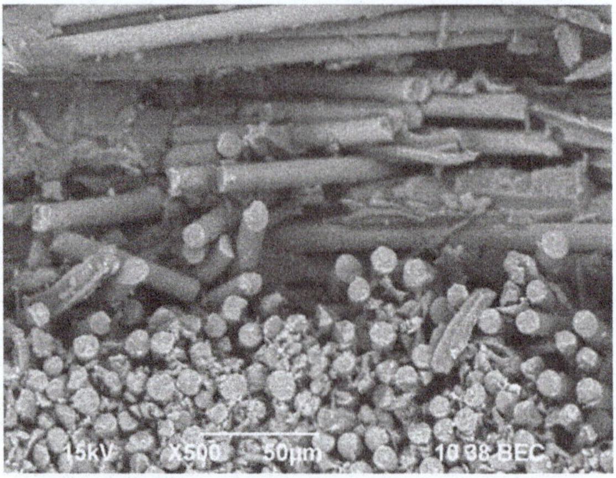

Fig. 3.67 Backscattered electron image of longitudinal and transverse artificial fibers. Accelerating voltage is 15 kV, magnification ×500, scale bar corresponds to 50 μm

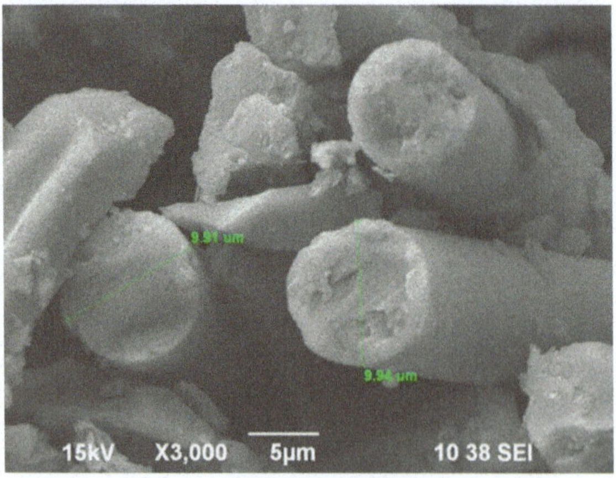

Fig. 3.68 Secondary electron image of artificial fibers. Accelerating voltage is 15 kV, magnification ×3000, scale bar corresponds to 5 μm. The fiber thicknesses (diameters) are indicated (about 10 μm)

It is seen in Fig. 3.63 that the reinforced layer consists of three longitudinal and four transverse interlayers of artificial fiber. The thickness of fiber threads is 10 μm (Fig. 3.68).

Figures 3.69, 3.70, and 3.71 present secondary electron images of a section of missile body in the zone of the thickest foamed layer at different magnifications.

Figure 3.72 gives an example of X-ray microanalysis results in the bulk (in the section) of the dark topmost layer of the missile body. Microanalysis was performed in the indicated rectangular area in the area where particles are present (the measured chemical element concentrations were averaged within this area, which is shown in the secondary electron image and chemical element concentrations are given in mass percentages). The concentrations of all chemical elements were measured on K lines of characteristic X-ray emission. Figure 3.72 also presents the spectrum of characteristic X-ray emission from this measurement area.

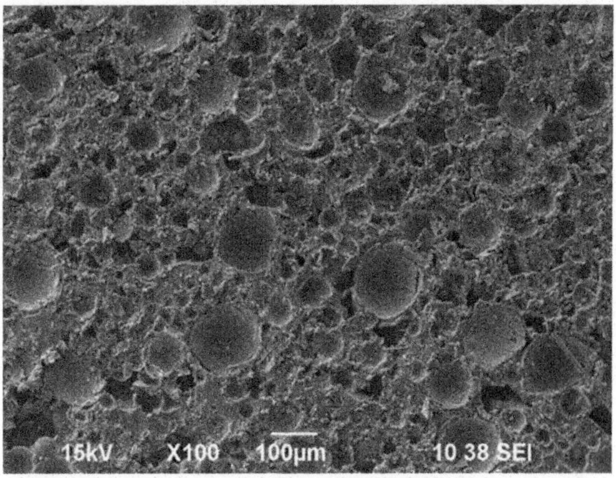

Fig. 3.69 Secondary electron image of section of missile body coating in the zone of the thickest foamed layer. Accelerating voltage is 15 kV, magnification ×100, scale bar corresponds to 100 μm

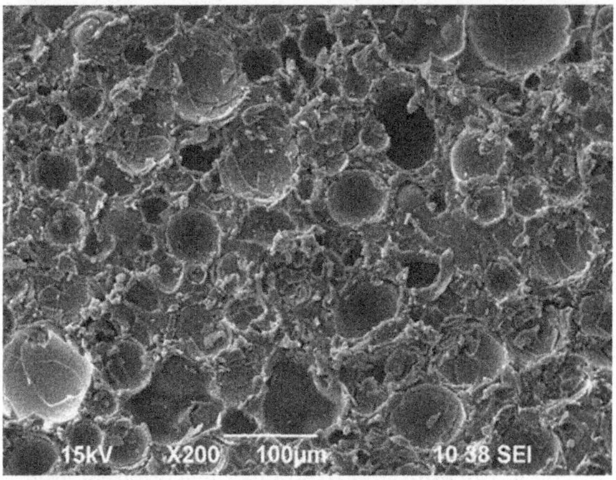

Fig. 3.70 Secondary electron image of a section of missile body coating in zone of the thickest foamed layer. Accelerating voltage is 15 kV, magnification ×200, scale bar corresponds to—100 μm

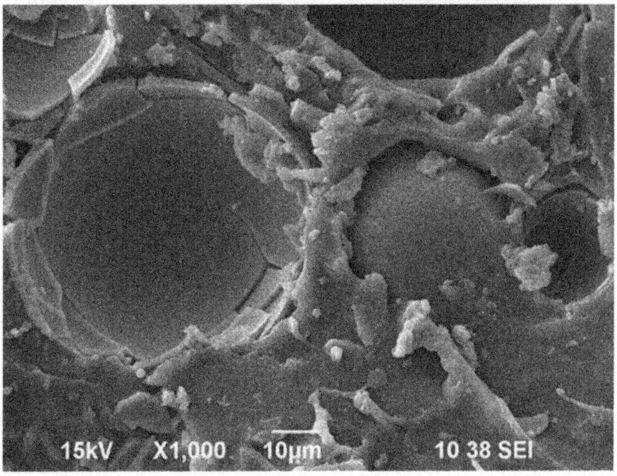

Fig. 3.71 Secondary electron image of a section of missile body coating in zone of the thickest foamed layer. Accelerating voltage is 15 kV, magnification ×1000, scale bar corresponds to—10 μm

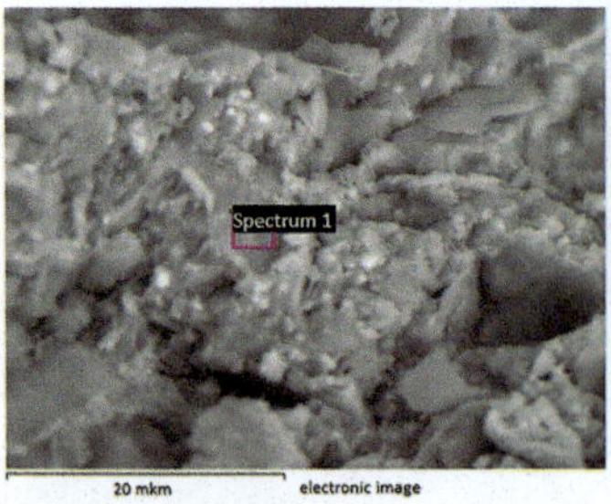

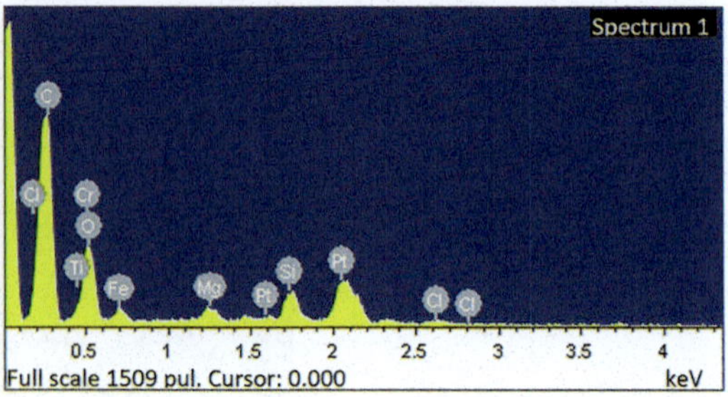

Element	Weight %
C K	41.74
O K	5.19
Si K	1.43
Cl K	2.07
Cr K	14.81
Fe K	34.76
Total	100.00

Fig. 3.72 Results of microanalysis in the bulk (in the cut) of the topmost (dark) layer of missile body coating

Figure 3.73 shows the results of X-ray microanalysis in the bulk (in the section) of the coating topmost (dark) layer of the missile body. Microanalysis was performed in the indicated rectangular area in the matrix (binder) region. The measured concentrations of chemical elements were measured within this area which is shown in the secondary electron image. Chemical element concentrations are given in mass percentages. The concentrations of all chemical elements were measured on K lines of characteristic X-ray emission. Figure 3.73 also presents the spectrum of characteristic X-ray emission from this measurement area.

Figure 3.74 gives an example of X-ray microanalysis results in the bulk (in the section) of the porous layer of missile body coating. Microanalysis was performed in the indicated rectangular area in the zone with particles (the measured concentrations of chemical elements were averaged within this area), which is shown in the secondary electron image. Chemical element concentrations are shown in mass percentages. The concentrations of all chemical elements were measured on K lines of characteristic X-ray emission. Figure 3.74 also presents the spectrum of characteristic X-ray emission from this measurement area.

Figure 3.75 shows an example of X-ray microanalysis in the bulk (section) of the missile body coating layer reinforced with artificial longitudinal fibers. Microanalysis was performed in the indicated rectangular area in the zone with particles (the measured chemical element concentrations were measured within this area), which is shown in the secondary electron image. Chemical element concentrations are shown in mass percentages. The concentrations of all chemical elements were measured on K lines of characteristic X-ray emission. Figure 3.75 also presents the spectrum of characteristic X-ray emission from this measurement area.

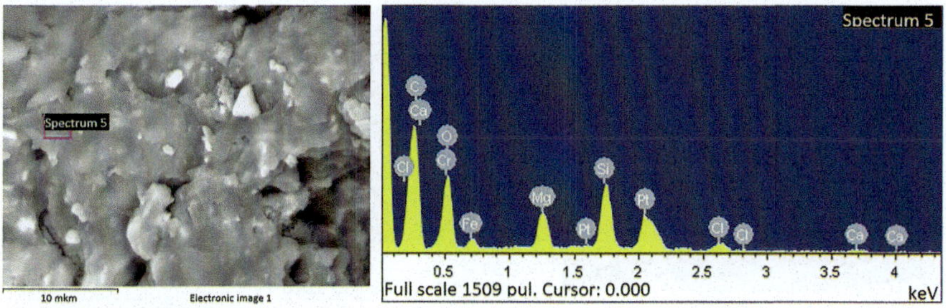

Element	Weight %
C K	59.75
O K	29.22
Mg K	2.37
Si K	3.22
Cl K	0.70
Ca K	0.28
Cr K	0.96
Fe K	3.49
Total	100.00

Fig. 3.73 Results of microanalysis in the bulk (in the cut) of topmost dark layer

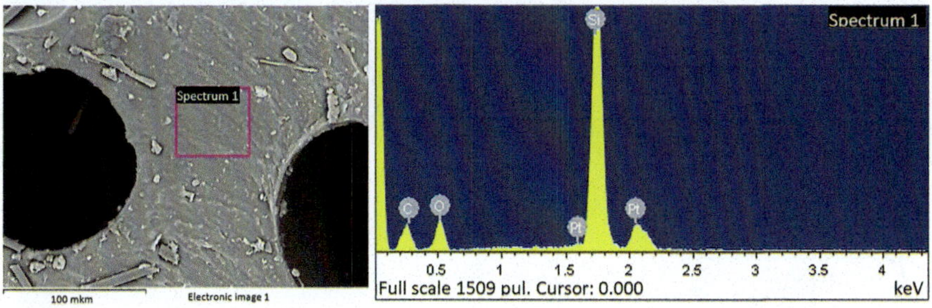

Element	Weight %
C K	49.40
O K	24.45
Si K	26.15
Total	100.00

Fig. 3.74 X-ray microanalysis results in the bulk (section) of the porous layer of missile body coating

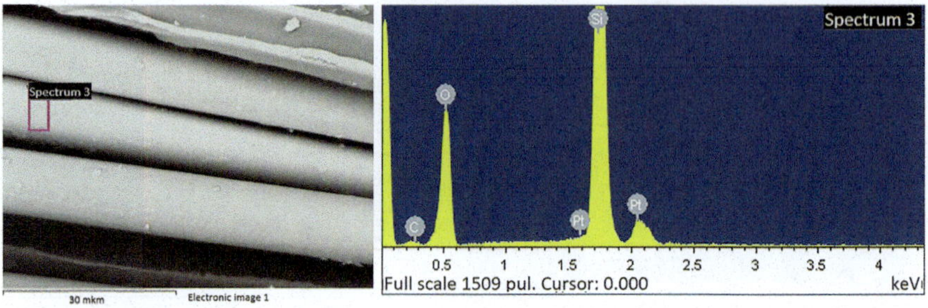

Element	Weight %
C K	11.75
O K	51.47
Si K	36.78
Total	100.00

Fig. 3.75 Results of X-ray microanalysis in the bulk (section) of the missile body coating layer reinforced with longitudinal artificial fibers

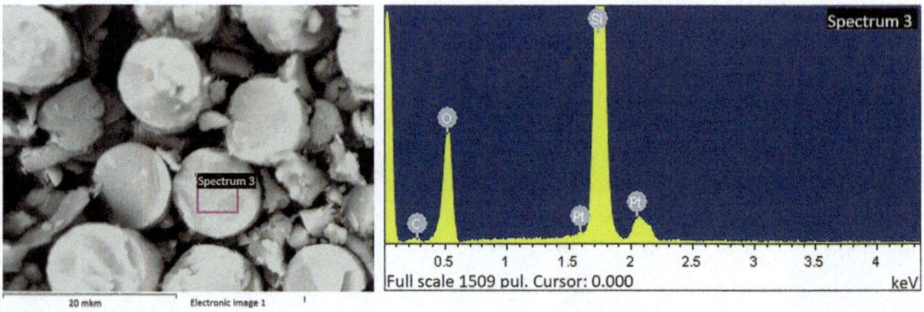

Element	Weight %
C K	9.48
O K	55.34
Si K	35.18
Total	100.00

Fig. 3.76 Results of X-ray microanalysis in the bulk (section) of missile body coating layer reinforced with artificial transverse fibers

Figure 3.76 gives an example of X-ray microanalysis performed in order to determine the chemical composition of fibers in the bulk (section) of the missile body coating layer reinforced with artificial transverse fibers. Microanalysis was performed in the indicated rectangular area in the zone with particles (the measured concentrations of chemical elements were averaged within this area), which is shown in the secondary electron image. Chemical element concentrations are presented in mass percentages. The concentrations of all chemical elements were measured on

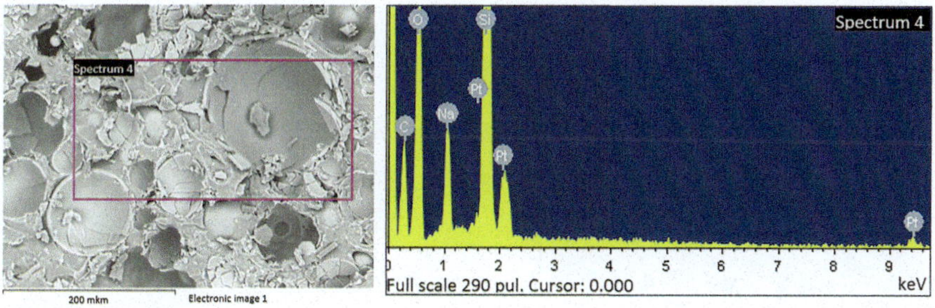

Element	Weight
C K	17.39
O K	44.66
Na K	5.86
Si K	32.08
Total	100.00

Fig. 3.77 Results of X-ray microanalysis in the bulk (section) of the thickest layer of missile body coating

K lines of characteristic X-ray emission. Figure 3.76 also shows the spectrum of characteristic X-ray emission from this measurement area.

Figure 3.77 provides an example of X-ray microanalysis results for the bulk (section) of the thickest layer (the foamed material) of the missile body coating. Microanalysis was performed in the indicated rectangular area in the zone with particles (the measured concentrations of chemical elements were averaged within this area), which is shown in the secondary electron image. Chemical element concentrations are given in mass percentages. The concentrations of all chemical elements were measured on K lines of characteristic X-ray emission from this measurement area.

Figures 3.78, 3.79, and 3.80 show examples of the results of X-ray microanalysis in the bulk (section) of the bottommost (green) layer of missile body coating. Microanalysis was performed in the indicated rectangular area in the zone with particles (the measured chemical element concentrations were averaged within this area), which is shown in the secondary electron image. Chemical element concentrations are given in mass percentages. The concentrations of all chemical elements were measured on K lines of characteristic X-ray emission (except Ba (barium) whose concentration was measured on L line). The figures also present the spectrum of characteristic X-ray emission from this measurement area.

The studies of the chemical composition and the structure of the thermal insulation coating of 9M723 missile body showed that this coating is a composite material and has several layers of different structures and chemical compositions. The topmost layer is dark colored and 124 μm thick. It is followed by a porous layer of 1.32 mm thickness; under this layer, a 2.39 mm layer reinforced with artificial longitudinal and transverse glass fibers is located. Below the reinforced layer, the thickest (7.3

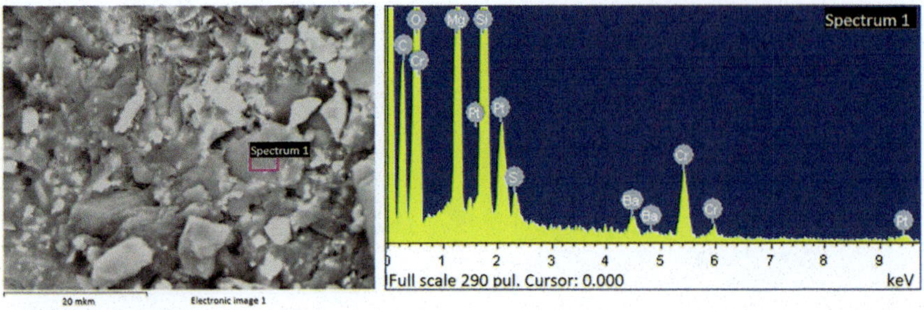

Element	Weight
C K	32.81
O K	45.11
Mg K	7.37
Si K	7.97
S K	0.46
Cr K	4.58
Ba L	1.69
Total	100.00

Fig. 3.78 Results of X-ray microanalysis in the bulk (section) of the bottommost green layer of missile body coating. Matrix (binder) region

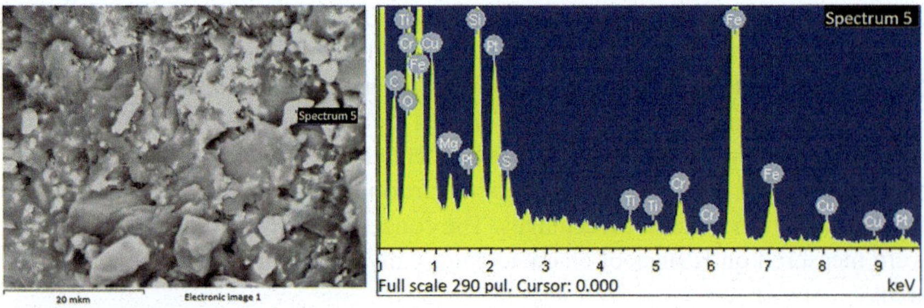

Element	Weight %
C K	28.21
O K	15.81
Mg K	0.79
Si K	3.50
S K	0.49
Ti K	0.41
Cr K	2.01
Fe K	34.77
Cu L	14.00
Total	100.00

Fig. 3.79 Results of X-ray microanalysis in the bulk (section) of green bottommost coating layer of missile body. The study was carried out in the zone that contains particles observed in the bottommost (green) layer

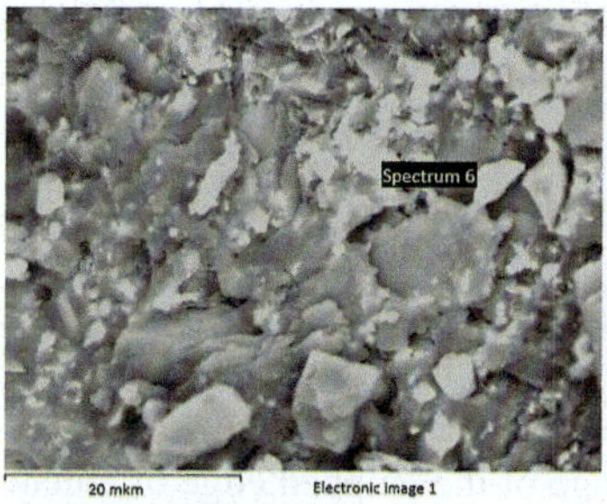

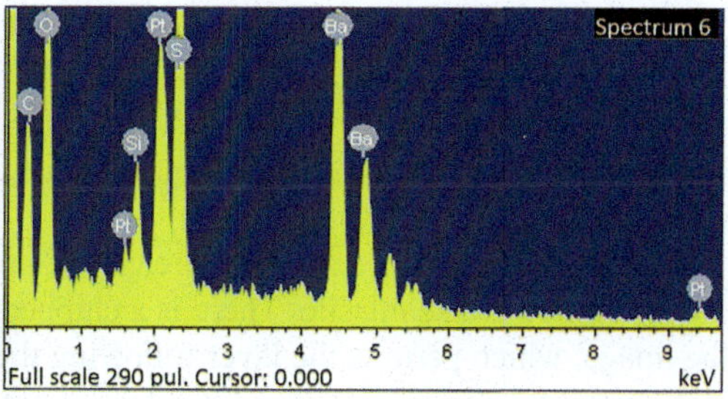

Element	Weight %
C K	49.40
O K	24.45
Si K	26.15
Total	100.00

Fig. 3.80 Results of X-ray microanalysis in the bulk (section) of the bottommost green layer of missile body coating. The study was carried out in the zone that contains particles observed in the bottommost (green) layer

mm) foamed glass layer is located. And, finally, there is the bottommost 33 μm green layer.

The topmost coating layer is a composite with polymer matrix whose chemical composition is 60% C and 29% O with Mg, Si, Cl additives, as well as the filler in the form of particles that contain 40% Fe and 15% Cr.

The porous layer is a polymer that consists of 50% C, 25% O, and 26% Si.

The reinforced layer is a composite with polymer matrix and filler in the form of three longitudinal and four transverse glass fiber interlayers. The glass fiber thickness is about 10 μm. The glass fibers contain 10% C, 51% O, and 37% Si.

The foamed glass layer has homogeneous chemical composition (17% C, 45% O, 32% Si, 6% Na).

The bottommost coating layer is a composite with a polymer matrix consisting of 33% C and 45% O with additives of 7% Mg, 8% Si, and the filler in the form of particles containing 35% Fe and 15% Ba.

Thus, the coating of 9M723 *Iskander-M* missile body is a thermal insulating one due to the presence of a porous layer and a layer of foamed glass. The layer reinforced with glass fiber interlayers increases coating strength characteristics, while the inclusion of particles of Fe (iron), which is a ferromagnetic, makes it possible to consider this coating studied as a heat-insulating and RF absorbent material.

3.3.1.6 Fragment of GPS Antenna Cover of 3M-14 *Kalibr* Cruise Missile

The specimen of antenna cover is made of material consisting of a thin grey layer and a thicker light yellow layer.

Figure 3.81 presents a secondary electron image (SEI) of grey antenna cover of 3M-14 *Kalibr* cruise missile. There is a chipped edge of the grey layer at the top of the image, which permits the layer following the grey one to be seen.

Figure 3.82 presents a secondary electron image of the chipped edge of antenna cover grey surface.

Figure 3.83 shows a secondary electron image of antenna cover light yellow surface, where artificial fabric structure (artificial fabric fiber) can be observed.

Figure 3.84 presents a secondary electron image of antenna cover light yellow surface, which shows artificial fiber size.

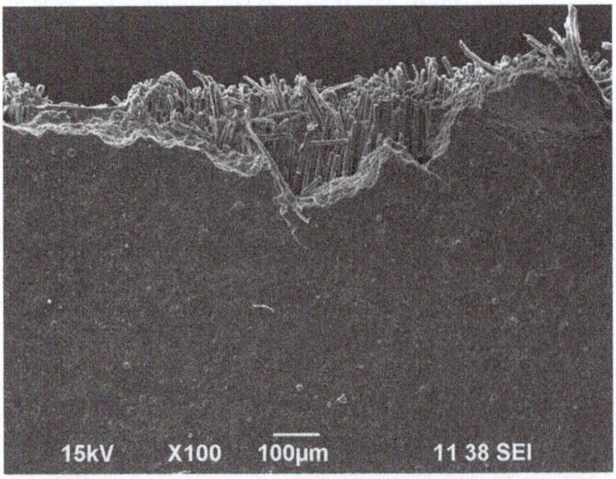

Fig. 3.81 Secondary electron image of antenna cover grey surface. Accelerating voltage is 15 kV, magnification ×100, scale bar corresponds to 100 μm

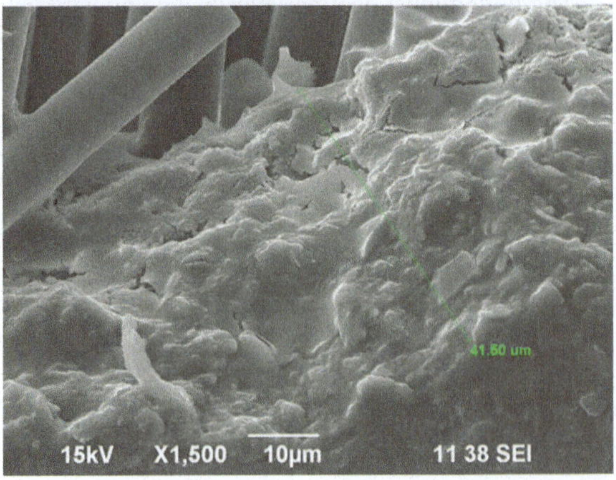

Fig. 3.82 Secondary electron image of the chipped edge of antenna cover grey surface. Accelerating voltage is 15 kV, magnification ×1500, scale bar corresponds to 10 μm

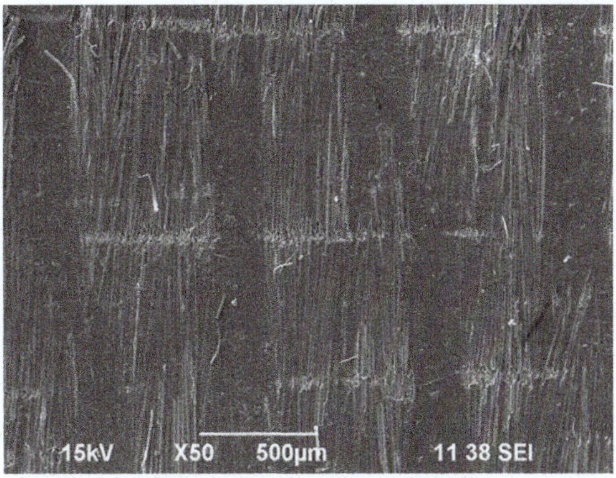

Fig. 3.83 Secondary electron image of antenna cover light yellow surface, where artificial fabric structure (artificial fabric fibers) can be noticed. Accelerating voltage is 15 kV, magnification ×50, scale bar corresponds to 500 μm

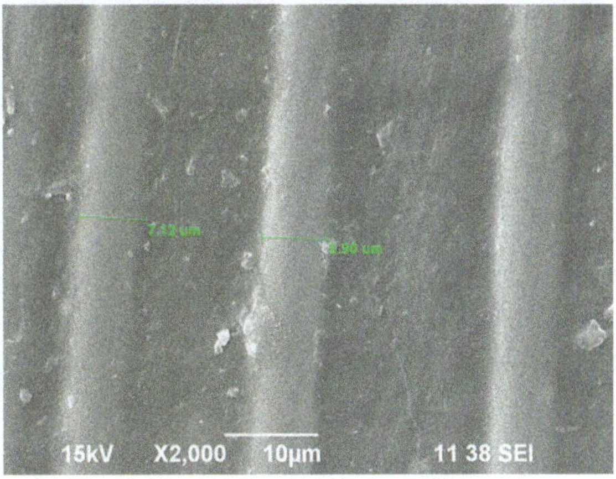

Fig. 3.84 Secondary electron image of antenna cover light yellow surface, where artificial fiber size is shown. Accelerating voltage is 15 kV, magnification ×2000, scale bar corresponds to 10 μm

To study the structure and measure chemical element concentrations in the coating bulk, as well as determine coating thickness, a section of antenna cover specimen was made. Secondary electron images produced with a JSM-6490LV scanning electron microscope are presented in Figs. 3.85 and 3.86.

Figure 3.87 shows an example of secondary electron image of the specimen of antenna cover grey layer 70.02 μm thick. Longitudinal (Fig. 3.88) and transverse (Fig. 3.89) artificial fibers are observed in the specimen bulk (section). The boundary of their intersection can be seen in Fig. 3.90.

Figure 3.91 shows a secondary electron image of the matrix (binder).

Figure 3.92 presents an example of the results of X-ray microanalysis of longitudinal fibers in the bulk (section) of antenna cover specimen. Microanalysis was performed at the point shown in the secondary electron image. Chemical element concentrations are given in mass percentages. The concentrations of all chemical

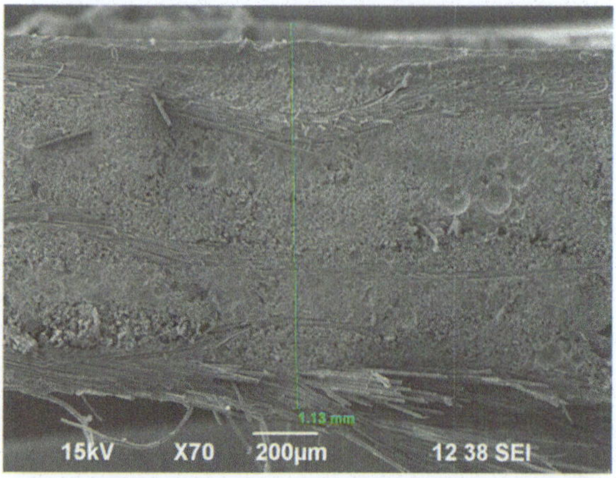

Fig. 3.85 Secondary electron image of a section of antenna cover specimen (thickness of the whole specimen is shown). Accelerating voltage is 15 kV, magnification ×70, scale bar corresponds to 200 µm

Fig. 3.86 Secondary electron image of antenna cover specimen (thickness of the whole specimen is shown). Accelerating voltage is 15 kV, magnification ×80, scale bar corresponds to 200 the specimen µm

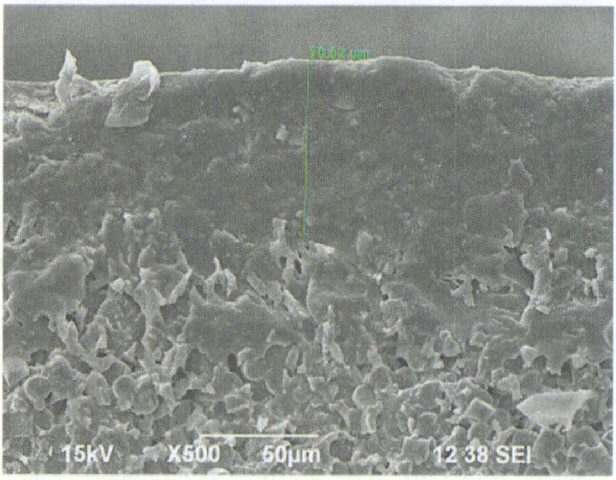

Fig. 3.87 Secondary electron image of grey lever of antenna cover specimen (grey layer thickness is indicated). Accelerating voltage is 15 kV, magnification ×500, scale bar corresponds to 50 µm

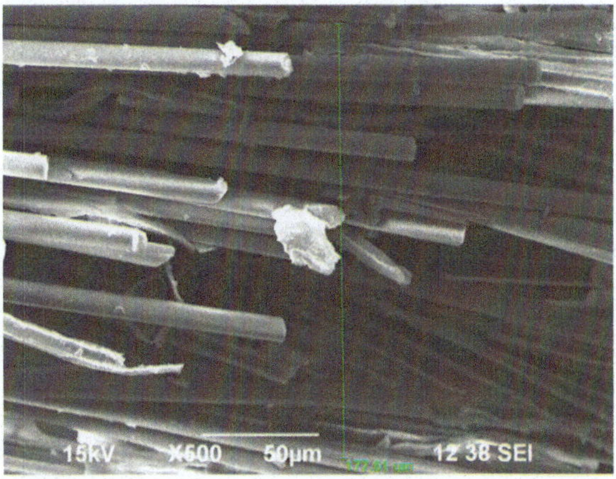

Fig. 3.88 Secondary electron image of artificial longitudinal fibers in the bulk (section) of antenna cover specimen. Accelerating voltage is 15 kV, magnification ×500, scale bar corresponds to 50 μm

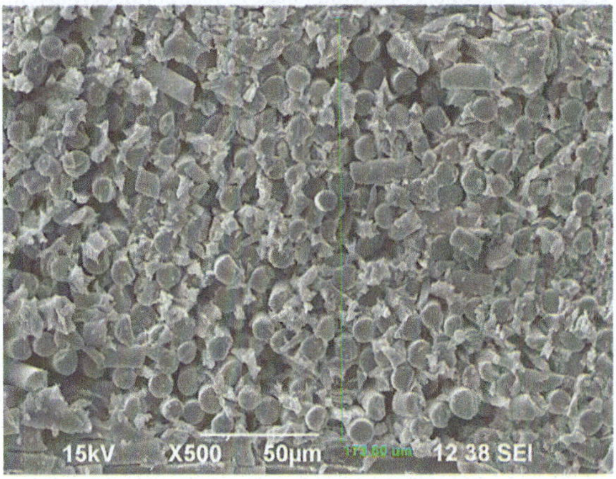

Fig. 3.89 Secondary electron image of a layer of artificial transverse fibers in the bulk (section) of antenna cover specimen. Accelerating voltage is 15 kV, magnification ×500, scale bar corresponds to 50 μm

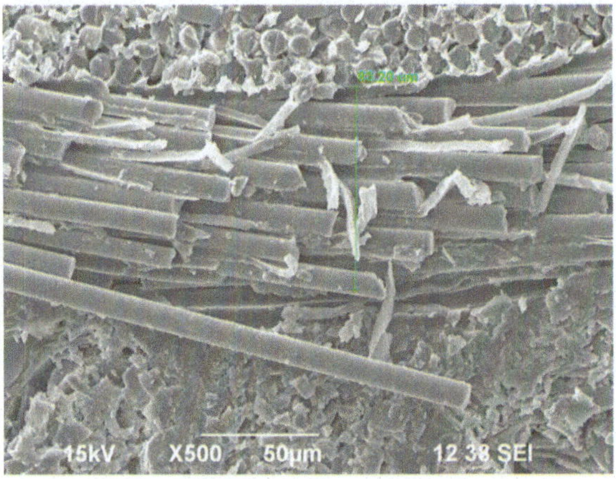

Fig. 3.90 Secondary electron image of the boundary between the layers with longitudinal and transverse fibers. Accelerating voltage is 15 kV, magnification ×500, scale bar corresponds to 50 μm

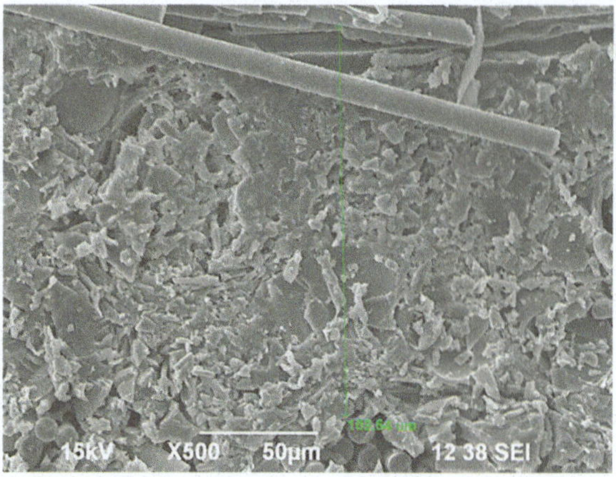

Fig. 3.91 Secondary electron image of matrix (binder) layer. Accelerating voltage is 15 kV, magnification ×500, scale bar corresponds to 50 μm

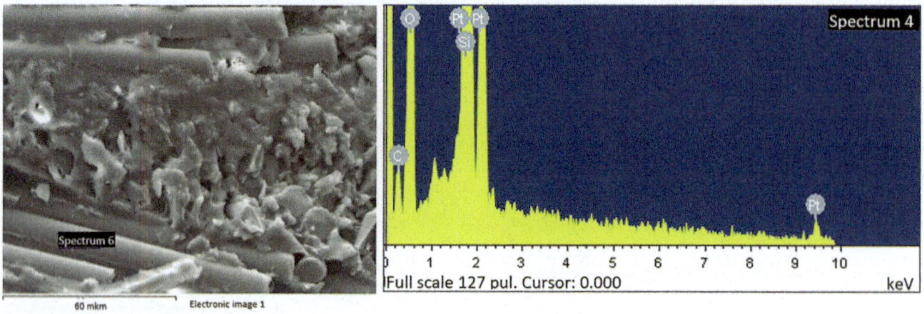

Element	Weight %
C K	16.18
O K	44.63
Si K	39.19
Total	100.00

Fig. 3.92 Results of X-ray microanalysis of longitudinal fiber in the bulk (section) of antenna cover specimen

elements were measured on K lines of characteristic X-ray emission. Figure 3.92 also shows the spectrum of characteristic X-ray emission of this measurement area.

Figure 3.93 gives an example of X-ray microanalysis of transverse fibers in the bulk (section) of antenna cover. Microanalysis was performed in the indicated rectangular area (the measured chemical element concentrations were averaged within this area) which is shown in the secondary electron image. Chemical element concentrations are given in mass percentages. The concentrations of all chemical elements were measured on K lines of characteristic X-ray emission. Figure 3.93 also presents the spectrum of characteristic X-ray emission from this measurement area.

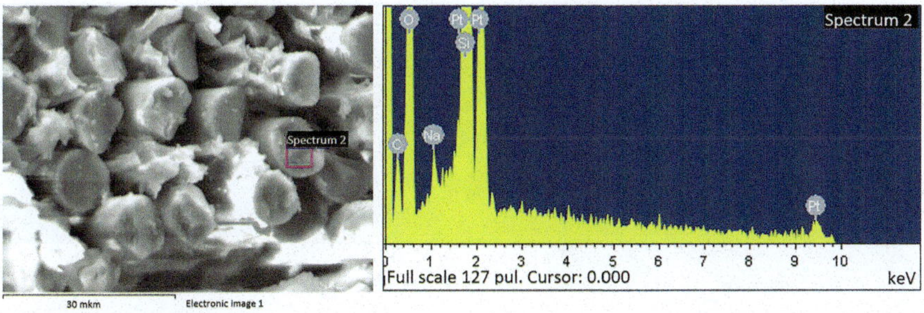

Element	Weight %
C K	18.42
O K	40.92
Na K	0.23
Si K	40.43
Total	100.00

Fig. 3.93 Results of X-ray microanalysis of transverse fiber in the bulk (section) of antenna cover

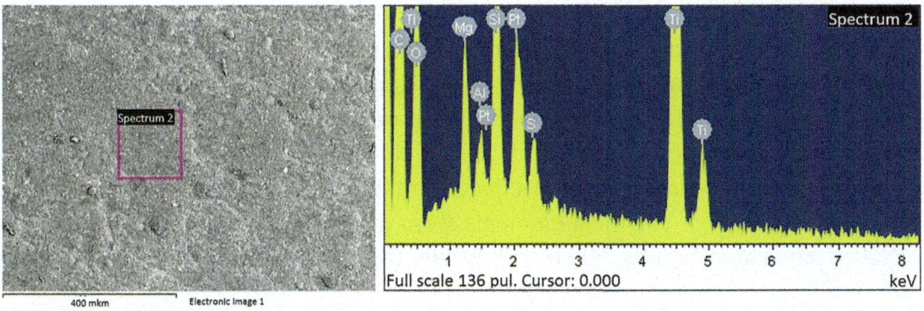

Element	Weight %
C K	52.97
O K	27.30
Mg K	1.57
Al K	0.47
Si K	2.27
S K	0.71
Ti K	14.72
Total	100.00

Fig. 3.94 Results of X-ray microanalysis of the grey layer of antenna cover specimen

Figure 3.94 shows an example of the results of X-ray microanalysis of grey layer of antenna cover specimen. Microanalysis was performed in the indicated rectangular area (the measured chemical element concentrations were averaged within this area) shown in the secondary electron image. Chemical element concentrations are given in mass percentages. The concentrations of all chemical elements were measured on K lines of characteristic X-ray emission. Figure 3.94 also presents the spectrum of characteristic X-ray emission from this measurement area.

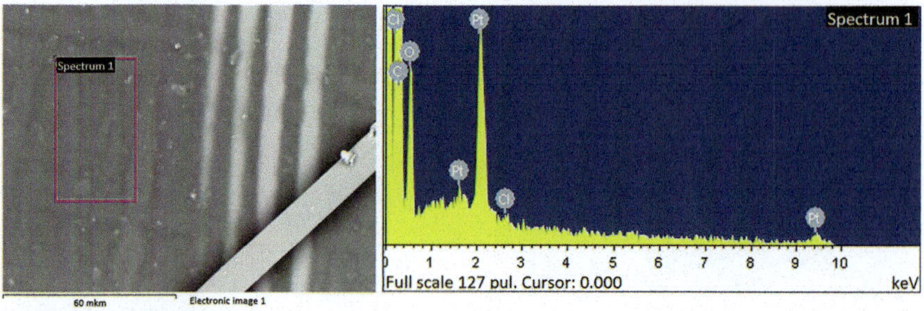

Element	Weight %
C K	86.02
O K	13.84
Cl K	0.14
Total	100.00

Fig. 3.95 Results of X-ray microanalysis of the matrix (composite binder) of the main (the thickest) layer of antenna cover specimen

Figure 3.95 gives an example of X-ray microanalysis results for the matrix (composite binder) of the main (the thickest) layer of antenna cover specimen. Microanalysis was performed at the indicated rectangular area (the measured chemical element concentrations were averaged within this area) which is shown in the secondary electron image. Chemical element concentrations are given in mass percentages. The concentrations of chemical elements were measured on K lines of characteristic X-ray emission. Figure 3.95 also presents the spectrum of characteristic X-ray emission from this measurement area.

The analysis of the results of studying the structure and chemical composition of the antenna cover specimen showed that the specimen is made of the material consisting of two layers of different structures and different chemical compositions, viz., of a thin (grey) layer of 70 μm thickness and the main (the thickest) 1060-μm-thick layer of antenna cover specimen. The specimen main layer is a composite material reinforced (for improving its strength characteristics) with three (Fig. 3.86) interlayers of longitudinal and transverse glass fibers about 7 μm thick. The Si (silicon) concentration in the fibers is as high as 40 mass percentages. The interlayers of longitudinal and transverse fibers are 250 μm thick. The distance between the interlayers that is filled with matrix (binder) material is 150 μm. The matrix (binder) of this composite is, in terms of its chemical composition, a polymer that consists of the following chemical elements: 86.02% C (carbon), 13.84 O (oxygen), and 0.14% Cl (chlorine). It is also highly probable that H (hydrogen) is present in the matrix; however, the study did not reveal its presence, as the equipment used permitted measurements of chemical element concentrations starting from $Z = 6$.

The study of the structures and chemical compositions of the surface and the bulk (section) of antenna cover specimen did not reveal inclusions that could absorb or scatter electromagnetic radiation. Therefore, the specimen of GPS antenna cover of 3M-14 cruise missile of *Kalibr* system is an RF transparent material.

3.3.1.7 Fragment of Radar Homing Head Fairing of 3M-14 *Kalibr* Cruise Missile

The examined fragment of the radar homing head fairing of 3M-14 *Kalibr* cruise missile is brown on the one side and black on the other.

Figure 3.96 shows a secondary electron image (SEI) of the surface of fairing fragment of radar homing head of 3M-14 *Kalibr* cruise missile, which is brown.

Figure 3.97 presents a secondary electron image of fairing fragment black surface.

To study the structure and measure chemical element concentrations in the specimen bulk, as well as measure layer and interlayer thicknesses that are observed in the section, a specimen of fairing section was prepared. Secondary electron images of the section that were obtained with a JSM-6490LV scanning electron microscope are shown in Figs. 3.98, 3.99, 3.100, and 3.101.

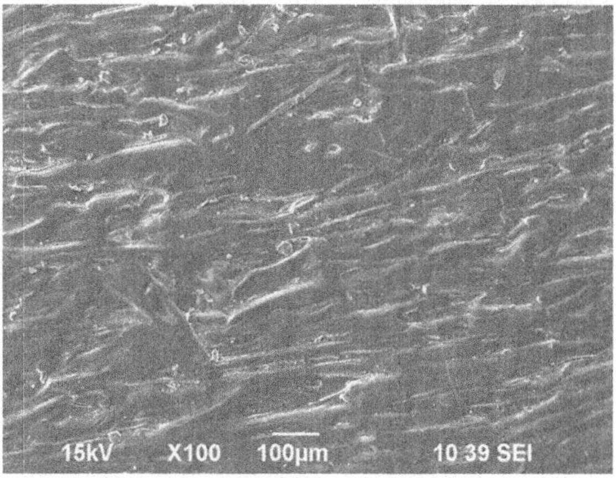

Fig. 3.96 Secondary electron image of fairing fragment brown surface. Accelerating voltage is 15 kV, magnification ×100, scale bar corresponds to 100 μm

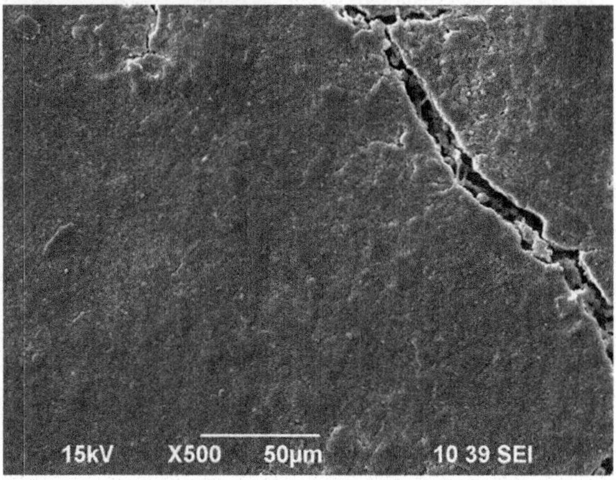

Fig. 3.97 Secondary electron image of fairing fragment black surface. Accelerating voltage is 15 kV, magnification ×500, scale bar corresponds to 50 μm

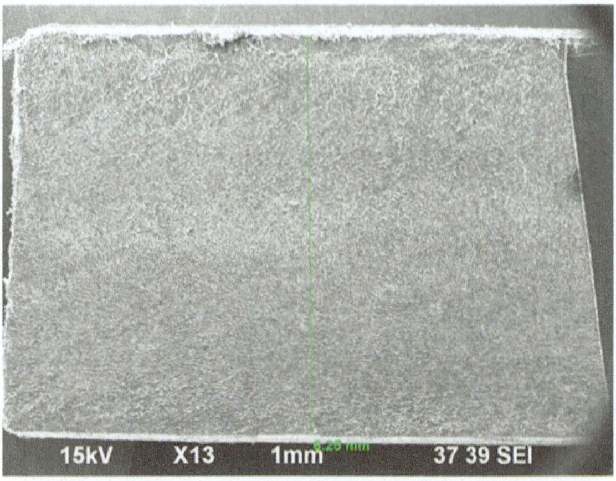

Fig. 3.98 Secondary electron images of fairing section (overall thickness of the specimen is indicated). Accelerating voltage is 15 kV, magnification ×13, scale bar corresponds to 1 mm

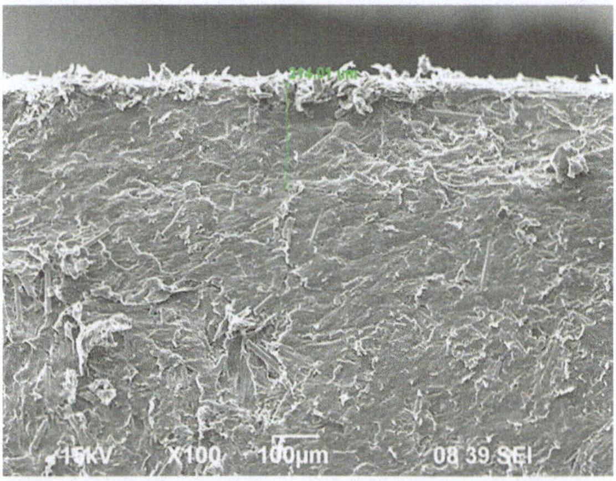

Fig. 3.99 Secondary electron image of fairing fragment. Accelerating voltage is 15 kV, magnification ×100, scale bar corresponds to 100 μm. Brown layer thickness is indicated

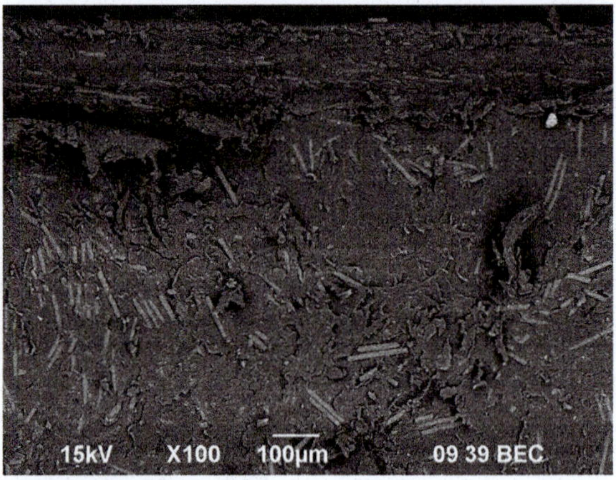

Fig. 3.100 Backscattered electron image of a section of fairing fragment in the thickest layer region. Accelerating voltage is 15 kV, magnification ×100, scale bar corresponds to 100 μm. Individual randomly oriented tubes are observed

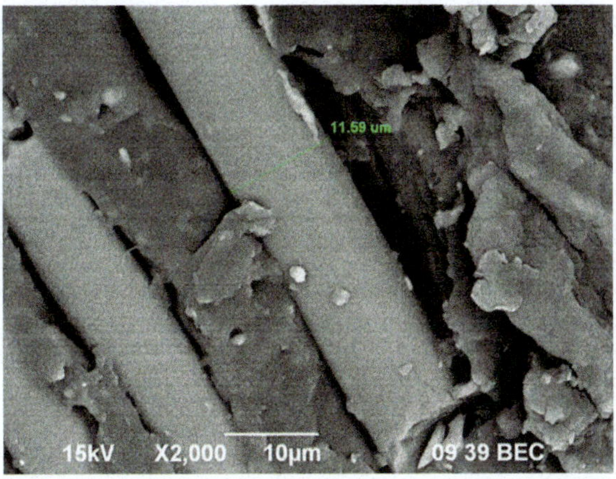

Fig. 3.101 Backscattered electron image of a section of fairing fragment in the thickest layer region. Accelerating voltage is 15 kV, magnification ×2000, scale bar corresponds to 10 µm. The thicknesses of tubes and fiber pieces are indicated

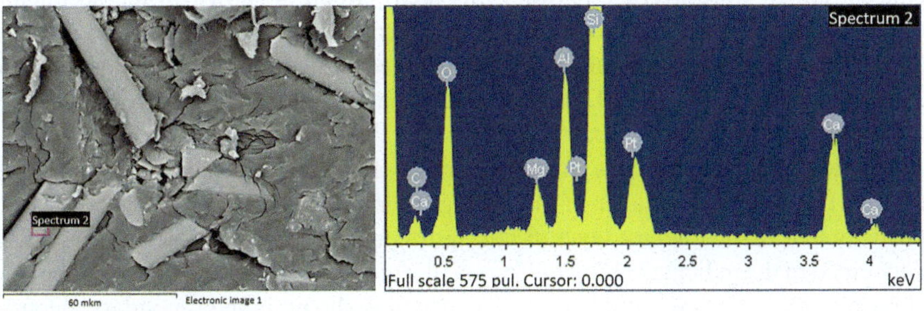

Element	Weight %
C K	13.46
O K	48.69
Mg K	2.16
Al K	6.40
Si K	20.02
Ca K	9.28
Total	100.00

Fig. 3.102 Results of X-ray microanalysis in the bulk (section) of the thickest layer of fairing fragment on tube surface

It can be seen in Figs. 3.100 and 3.101 that individual randomly oriented tubes and fiber pieces of 30–100 µm length and 11.5 µm thickness are observed in the bulk of the thickest layer of the radar homing head fairing fragment.

Figure 3.102 presents an example of X-ray microanalysis results obtained in the bulk (section) of the thickest layer of fairing fragment on tube surface. Microanalysis was performed in the indicated rectangular area (the measured chemical element

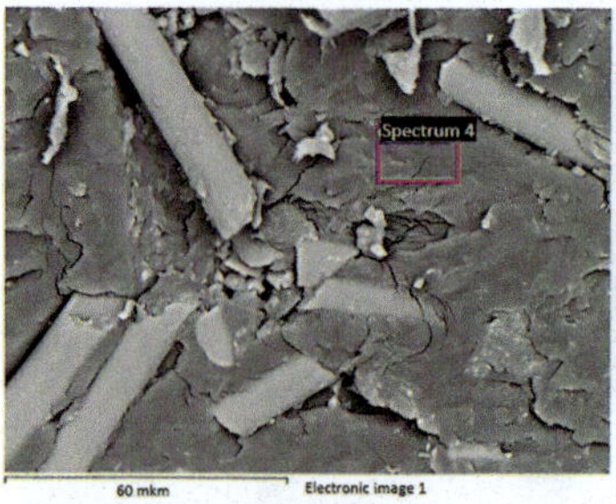

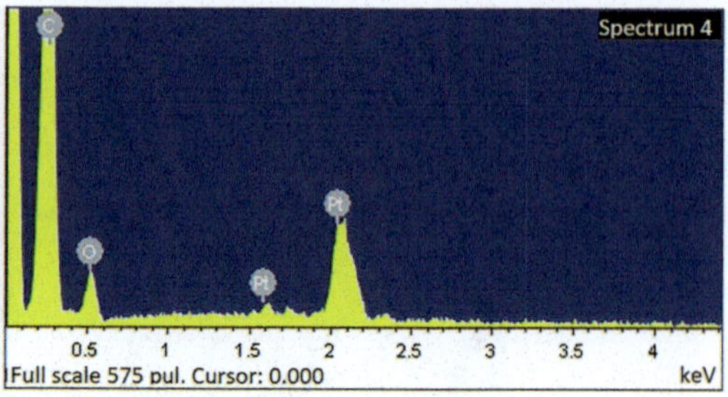

Element	Weight %
C K	82.79
O K	17.21
Total	100.00

Fig. 3.103 Results of X-ray microanalysis in the bulk (section) of the thickest fairing fragment layer in the region of matrix (composite binder in the thickest layer of fairing fragment specimen)

concentrations were averaged within this area) which is shown in the secondary electron image. Chemical element concentrations are given in mass percentages. The concentrations of all chemical elements were measured on *K* lines of characteristic X-ray emission. Figure 3.102 also presents the spectrum of characteristic X-ray emission from this measurement area.

Figure 3.105 demonstrates an example of X-ray microanalysis results in the bulk (section) of the thickest layer of the fairing fragment in the region of matrix (composite binder in the thickest layer of fairing fragment specimen). Microanalysis was performed in the indicated rectangular area (the measured concentrations of chemical elements were averaged within this area) which is shown in the secondary electron image. Chemical element concentrations are given in mass percentages. The concentrations of all chemical elements were measured on *K* lines of characteristic X-ray emission (except those of Zn (zinc) and Pb (lead); Zn concentration was measured on L line and that of Pb—on M line). Figure 3.103 also presents the spectrum of characteristic X-ray emission from this measurement area (Fig. 3.104).

The analysis of the results of studying the structure and chemical composition of a fragment of radar homing head fairing of 3M-14 *Kalibr* missile has shown that this fragment consists of three layers of different structures and chemical compositions. The topmost layer is 214 μm. The main (thickest) layer of the fragment of the radar homing head fairing is 6250 μm thick, while the bottom layer is 15 μm thick. The main layer of the fragment is a composite material reinforced (to improve its strength characteristics) with randomly oriented tubes and pieces of, probably, glass fibers of 30–100 μm length and 11.5 μm thickness. Si (silicon) content in the fiber pieces is as high as 30 mass percent. Chemically, the matrix—this composite binder—is a polymer consisting of the following chemical elements: 83.09% C (carbon) and 16.91% O (oxygen); very likely, H (hydrogen) is also present but its concentration was not measured due to experimental limitations.

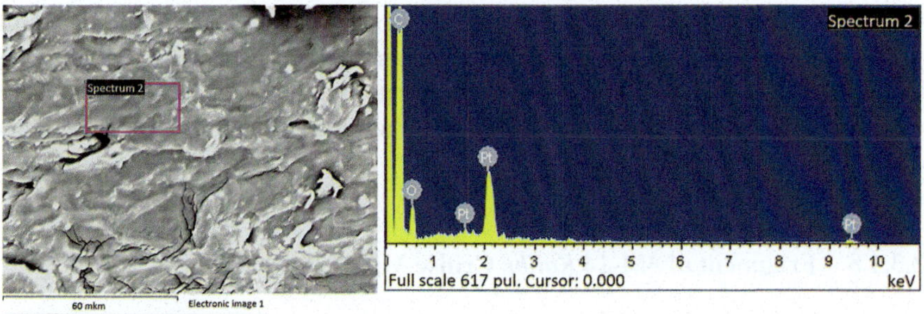

Element	Weight %
C K	83.09
O K	16.91
Total	100.00

Fig. 3.104 Results of X-ray microanalysis in the bulk (section) of the brown topmost layer of the fairing fragment

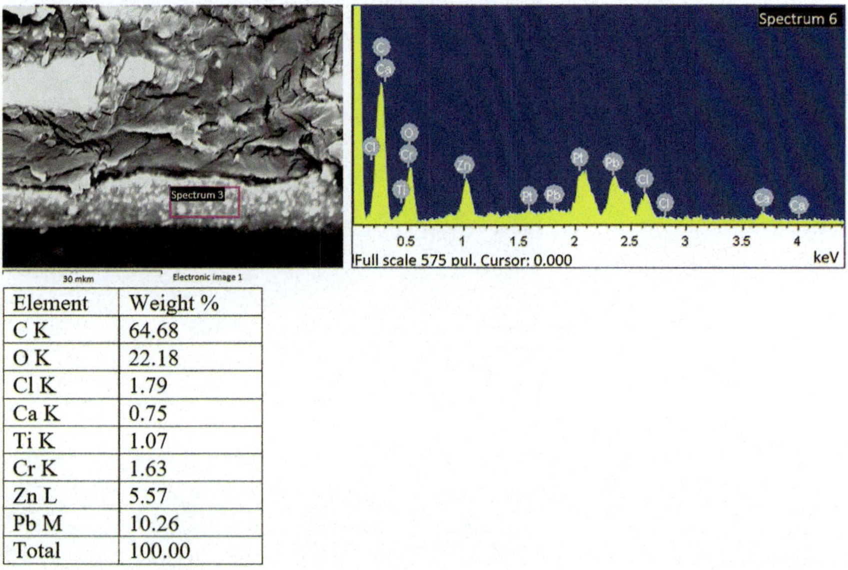

Element	Weight %
C K	64.68
O K	22.18
Cl K	1.79
Ca K	0.75
Ti K	1.07
Cr K	1.63
Zn L	5.57
Pb M	10.26
Total	100.00

Fig. 3.105 Results of X-ray microanalysis in the bulk (section) of the bottommost (dark colored) layer of the fairing fragment

While examining the structure and chemical composition of the surface and bulk (section) of the fragment of radar homing head fairing of 3M-14 *Kalibr* cruise missile, the inclusions that could cause absorption or scattering of electromagnetic radiation were not detected. Hence, the fragment of radar homing head fairing of 3M-14 *Kalibr* cruise is an RF transparent material.

3.3.1.8 Fragment of 3M-14 *Kalibr* Cruise Missile Rudder

Figure 3.106 presents a secondary electron image (SEI) of the black surface of 3M-14 *Kalibr* cruise missile rudder.

Figures 3.107 and 3.108 show secondary electron (Fig. 3.107) and backscattered electron (Fig. 3.108) images of the rudder fragment bottom surface.

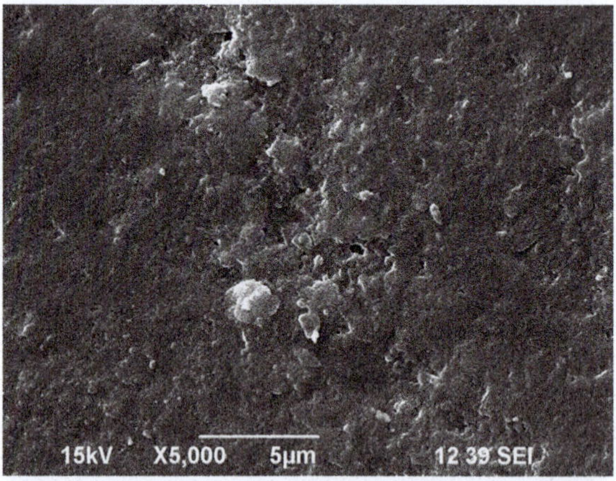

Fig. 3.106 Secondary electron image of a fragment of rudder black surface. Accelerating voltage is 15 kV, magnification ×5000, scale bar corresponds to 5 μm

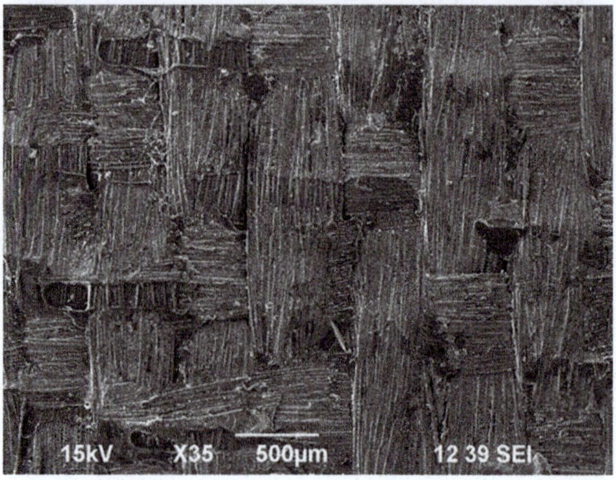

Fig. 3.107 Secondary electron image of the bottom surface of rudder fragment. Accelerating voltage is 15 kV, magnification ×35, scale bar corresponds to 500 μm

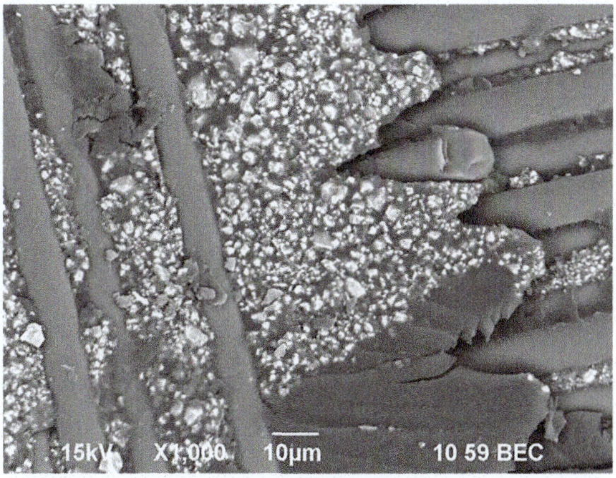

Fig. 3.108 Backscattered electron image (BEC) of the rudder fragment bottom surface. Accelerating voltage is 15 kV, magnification ×1000, scale bar corresponds to 10 μm

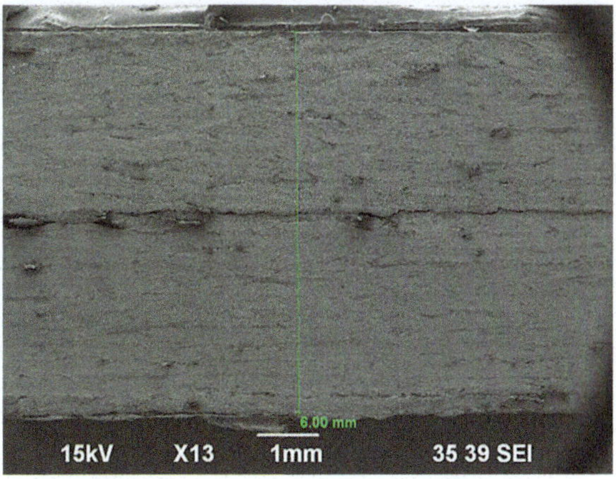

Fig. 3.109 Secondary electron image of the whole rudder fragment section. Accelerating voltage is 15 kV, magnification ×13, scale bar corresponds to 1 mm

It can be seen in Fig. 3.107 that the bottom surface of rudder fragment is an artificial fabric layer; in Fig. 3.108, inclusions (particles) are observed. As further studies have shown, these particles contain Fe (iron) in high concentration.

To examine the structure of and measure chemical element concentrations in the bulk of the fragment of 3M-14 *Kalibr* cruise missile rudder, as well as measure specimen layer thicknesses, a fragment section was made. Secondary electron and backscattered electron images produced with a JSM-6490LV electron microscope are shown in Figs. 3.109, 3.110, 3.111, 3.112, and 3.113.

It can be seen in Figs. 3.111 and 3.112 that artificial fabric interlayers with fiber about 11 μm thick are observed in the region of the main layer of rudder fragment section. The studies of the whole main layer of rudder fragment section have shown the presence of 25 such interlayers in the bulk of the main layer of rudder fragment section. Inclusions in the form of particles among fabric fibers are observed in

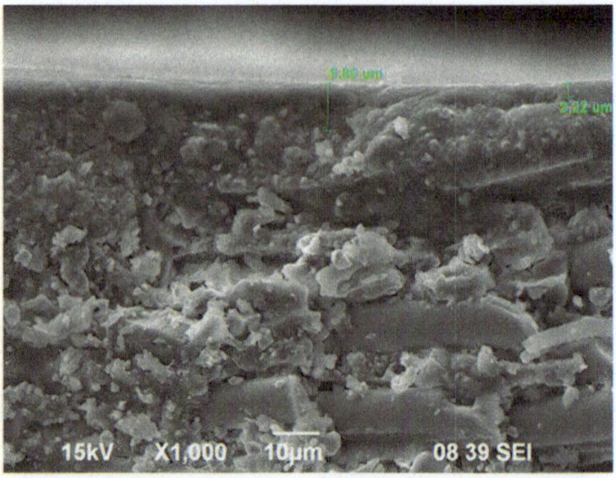

Fig. 3.110 Secondary electron image of the top layer of rudder fragment section. Accelerating voltage is 15 kV, magnification ×1000, scale bar corresponds to 10 μm

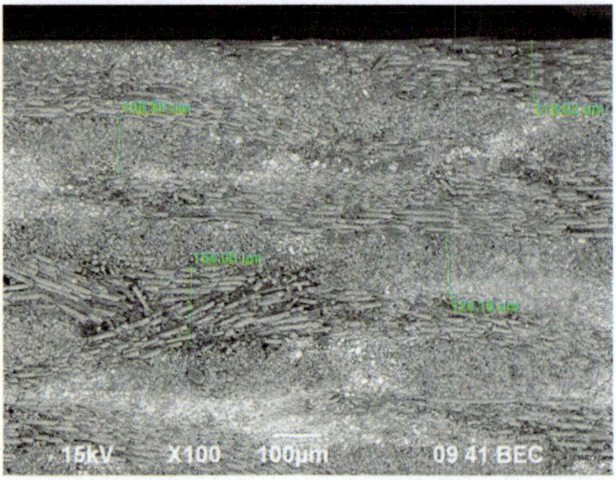

Fig. 3.111 Backscattered electron image of the main layer (about 6 mm thick) of rudder fragment section. Accelerating voltage is 15 kV, magnification ×100, scale bar corresponds to 100 μm

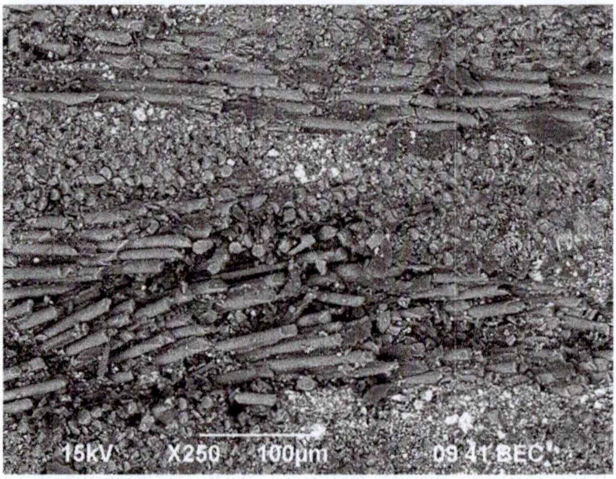

Fig. 3.112 Backscattered electron image of the main layer (about 6 mm thick) of rudder fragment section. Accelerating voltage is 15 kV, magnification ×250, scale bar corresponds to 100 μm

Fig. 3.113 Backscattered electron image of artificial fabric fibers in the region of the main layer of rudder fragment section. Accelerating voltage is 15 kV, magnification ×2000, scale bar corresponds to 10 μm. Fiber thickness (diameter) is shown

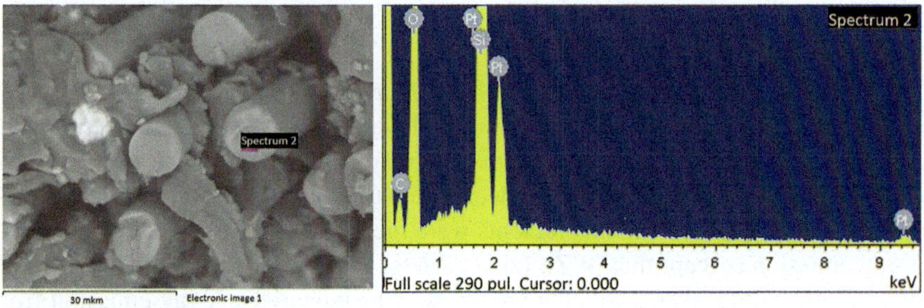

Element	Weight %
C K	15.24
O K	50.18
Si K	34.58
Total	100.00

Fig. 3.114 Results of X-ray microanalysis of artificial fabric fibers in the region of the main layer of rudder fragment section

Fig. 3.112 (the same as in Fig. 3.108). As further studies have shown, these particles contain Fe (iron) in high concentrations.

Figure 3.114 presents an example of the results of X-ray microanalysis of artificial fabric fibers in the region of the main layer of rudder fragment section. The microanalysis was performed at the point shown in the secondary electron image. Chemical element concentrations are given in mass percentages. The concentrations of all chemical elements were measured on K lines of characteristic X-ray emission. Figure 3.114 also presents the spectrum of characteristic X-ray emission from this measurement area.

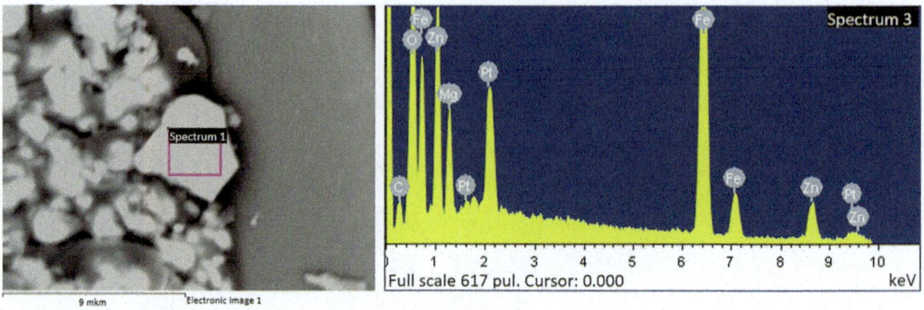

Element	Weight %
C K	9.60
O K	25.01
Mg K	3.62
Fe K	44.78
Zn L	16.99
Total	100.00

Fig. 3.115 Results of X-ray microanalysis of the particles observed among fabric fibers in the region of the main layer of rudder fragment section. Particles of 0.5–10 μm size are observed

Figure 3.115 shows an example of X-ray microanalysis of the particles observed among fabric fibers in the area of the main layer of rudder fragment section. The microanalysis was performed in the rectangular area shown in the backscattered electron image. Chemical element concentrations are given in mass percentages. The concentrations of all chemical elements were measured on K lines of characteristic X-ray emission (except that of Zn (zinc), whose concentration was measured on L line). Figure 3.115 also shows the spectrum of characteristic X-ray emission from this measurement area.

It can be seen in Figs. 3.116 and 3.117 that Fe (iron) particles accumulate on artificial fabric interlayers. The number of iron interlayers is the same as that of artificial fabric interlayers, viz., about 25. Besides, Fig. 3.117 confirms that artificial fabric fibers consist mainly of Si and O.

Figure 3.118 presents an example of the results of X-ray microanalysis in the area of matrix (binder of the main layer composite in the rudder fragment section). The microanalysis was performed in the indicated rectangular area (the measured chemical element concentrations were averaged within this area) which is shown in the secondary electron image. Chemical element concentrations are given in mass percentages. The concentrations of all chemical elements were measured on K lines of characteristic X-ray emission. Figure 3.118 also shows the spectrum of characteristic X-ray emission from this measurement area.

The analysis of the results of studying the structure and chemical composition of the rudder fragment of 3M-14 *Kalibr* cruise missile has shown that this specimen is a composite consisting of two layers different in structure and chemical composition, namely, a thin top layer (10 μm thick) and the main layer (6 mm thick). The main composite layer is a polymer (matrix) reinforced for improving its strength and

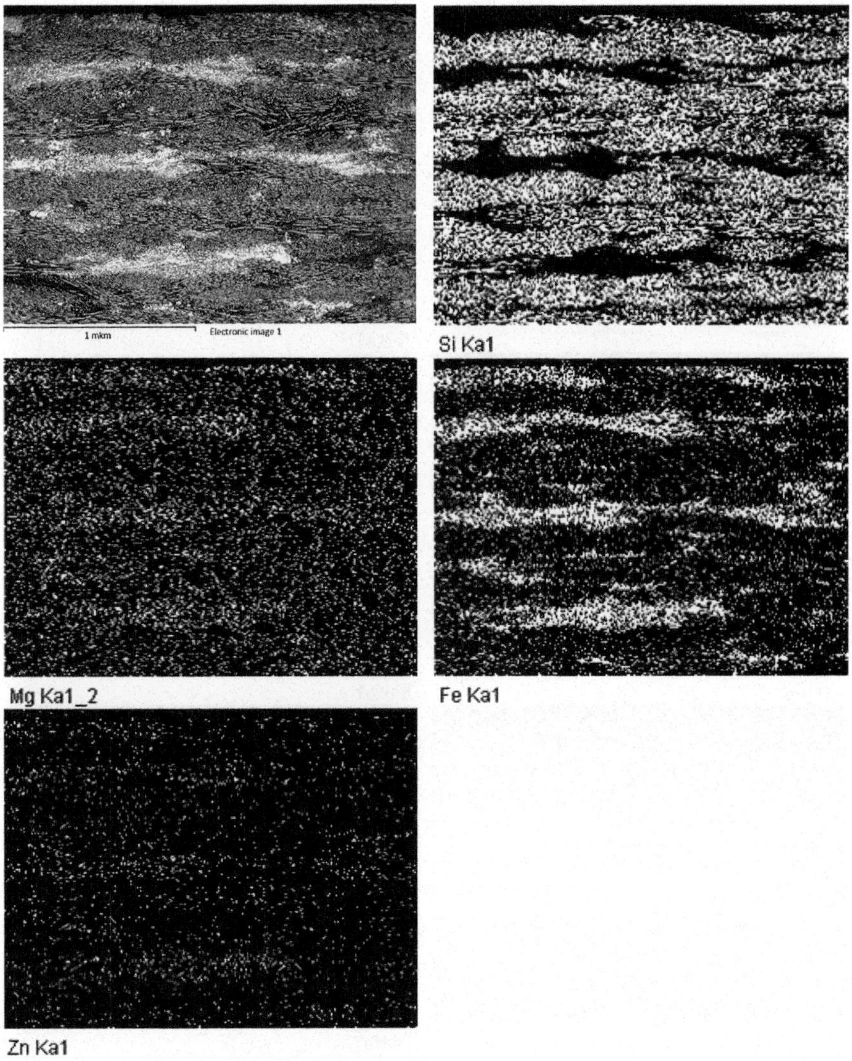

Fig. 3.116 Backscattered electron image in the region of the main layer of rudder fragment section with 1 mm scale bar, along with the images of the same area in characteristic X-ray emission (spatial distribution of chemical elements over the section of the examined specimen). White dots correspond to the presence of the chemical element indicated under each image

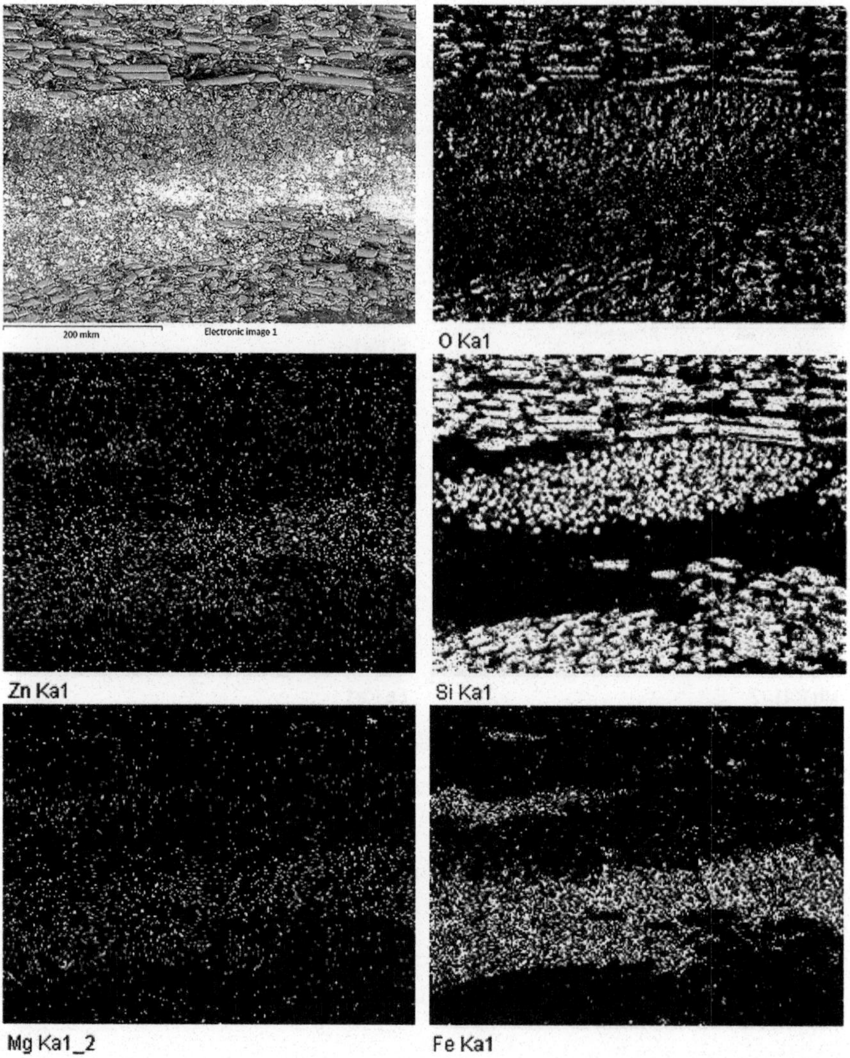

Fig. 3.117 Backscattered electron image in the area of the main layer of rudder fragment section with 200 μm scale bar, along with the images of the same area shown in characteristic X-ray emission (spatial distribution of chemical elements over the examined specimen section). White dots correspond to the presence of the chemical element indicated under each image

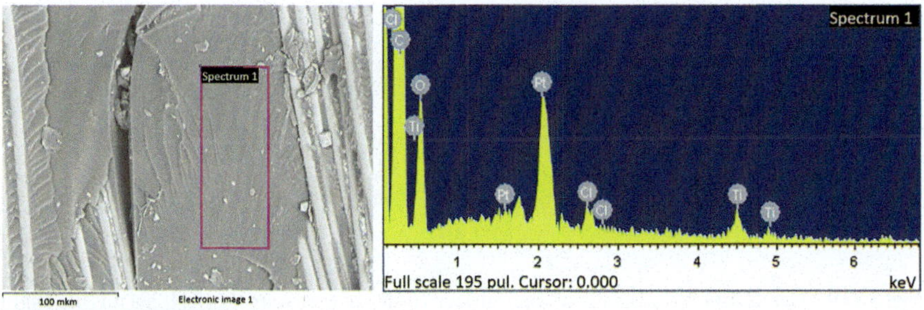

Element	Weight %
C K	76.28
O K	22.13
Cl K	0.55
Ti K	1.04
Total	100.00

Fig. 3.118 Results of X-ray microanalysis in the area of matrix (the main layer composite binder in the rudder fragment section)

heat resistance characteristics by glass fabric interlayers (about 25 interlayers in total), with glass fibers' thickness about 11 μm. The matrix—binder—consists of the following chemical elements: 76.28% C (carbon), 22.13% O (oxygen), 0.55% Cl (chlorine), 1.04% Ti (titanium); in high probability, it contains H (hydrogen) as well. It should be noted that hydrogen concentration was not measured in this study (the equipment used permits measurements of chemical elements starting with $Z = 6$). Chemical composition of the thin top layer is close to that of the main layer matrix. In artificial fabric interlayers, there are particles of 0.5–100 μm size with high (up to 45%) Fe (iron) concentration. The number of iron-containing interlayers is the same as that of artificial fabric interlayers, viz., about 25.

Thus, due to reinforcing with glass fabric, the rudder fragment of 3M-14 *Kalibr* cruise missile has high strength and heat resistance characteristics, while the presence of particle interlayers (related to those of glass fabric) with high iron concentration (iron is a ferromagnetic) permits us to classify the rudder fragment composite as a magnetic RF absorbent material.

3.3.2 Results of Studying Ballistic Missile Fragments

3.3.2.1 Fragment of Outer Part of *Hwasong-11* (KN-23) (South Korea) Missile Engine Nozzle

The fragment of the outer part of *Hwasong-11* (KN-23) missile engine nozzle (further—X3 specimen) is a structure where a 4-mm-thick polymer layer is applied

Fig. 3.119 Metal surface (left) and polymer surface (right) of *Hwasong-11* (KN-23) missile engine nozzle

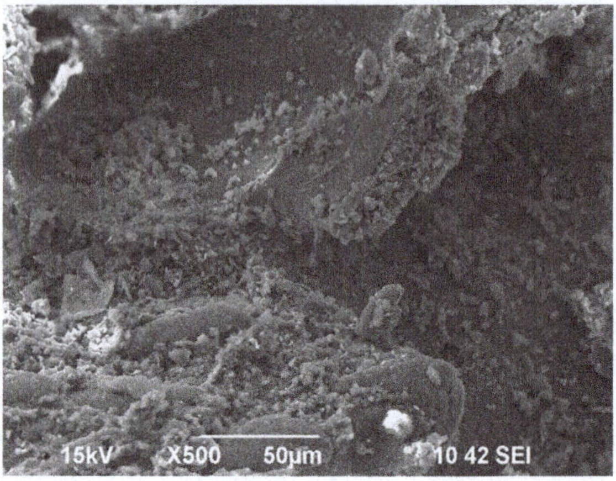

Fig. 3.120 Secondary electron image of specimen polymer layer. Accelerating voltage is 15 kV. Magnification is $\times 500$, scale bar corresponds to 50 μm

on a metal part of 2.5 mm thickness. Figure 3.119 shows the metal surface of X3 specimen (left) and a polymer surface (right).

The results of examining the metal part are presented in Sect. 2.3.2.1. As to the specimen side with a polymer applied, Figs. 3.120 and 3.121 show the secondary electron images of the layer surface and section, respectively, at different magnifications.

Figures 3.122 and 3.123 show secondary electron images of the surface of X3 specimen polymer layer at different magnifications.

It can be seen in Figs. 3.122 and 3.123 that there are damaged surface areas on X3 polymer layer. Fiber fragments can be observed in them.

Figure 3.124 shows the image of the abovementioned fiber fragments. The image has been produced in the region of damaged surface of specimen composite layer.

Figure 3.125 gives an example of the results of X-ray microanalysis of the fiber fragments observed in the regions of damaged specimen composite surface. The microanalysis was performed at the indicated point shown in the backscattered electron image. Chemical element concentrations are presented in mass percentages. The concentrations of all chemical elements were measured on K lines of characteris-

Fig. 3.121 Secondary electron image of specimen polymer layer section. Accelerating voltage is 15 kV. Magnification ×1500, scale bar corresponds to 10 µm

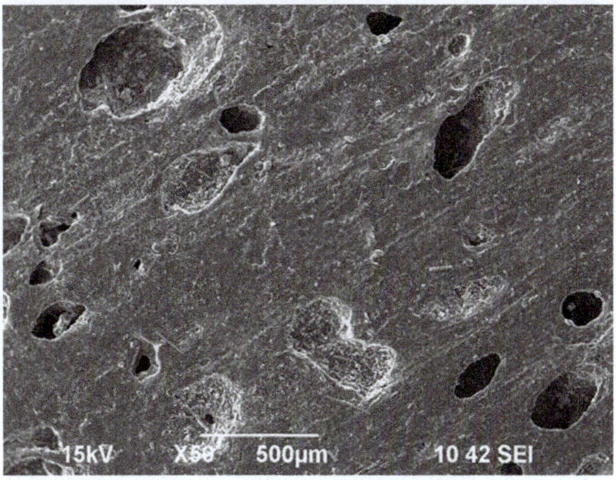

Fig. 3.122 Secondary electron image of polymer layer surface of the specimen. Accelerating voltage is 15kV. Magnification × 50, scale bar corresponds to 500 µm

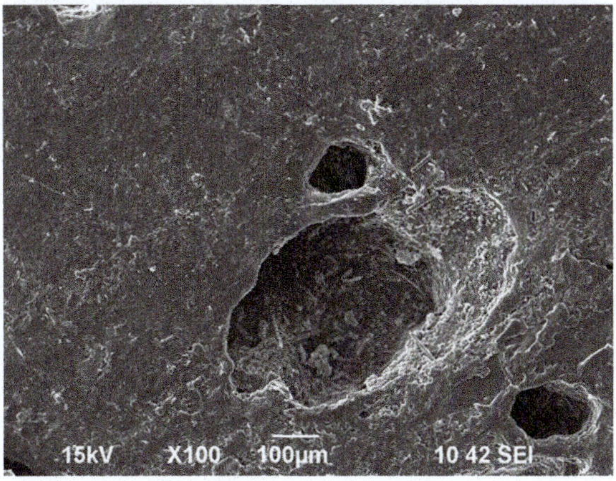

Fig. 3.123 Secondary electron image of polymer layer surface of the specimen. Accelerating voltage is 15kV. Magnification ×100, scale bar corresponds to 100 µm

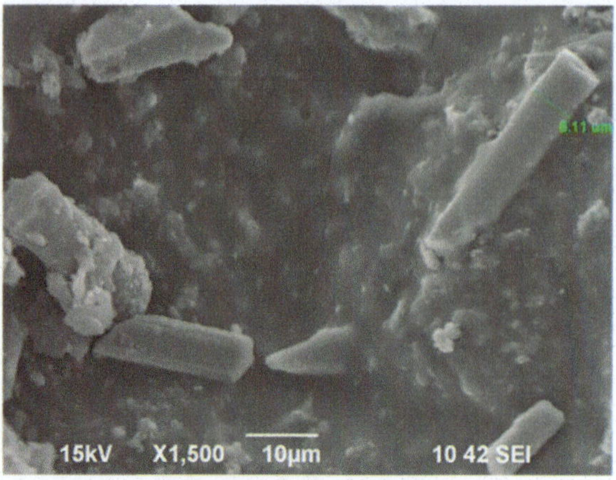

Fig. 3.124 Secondary electron image of fiber fragments. The image is produced in the region of damaged surface of the specimen composite layer. Accelerating voltage is 15 kV, magnification ×1500, scale bar corresponds to 10 μm. The image shows the size of fiber fragments

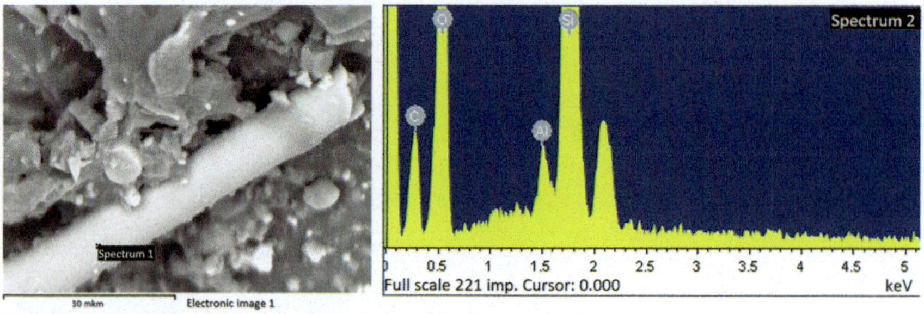

Element	Weight %
C K	26.81
O K	38.10
Al K	0.72
Si K	34.38
Total	100.00

Fig. 3.125 Results of X-ray microanalysis of fiber fragments observed in the damaged surface areas of specimen composite

tic X-ray emission. Figure 3.125 also presents the spectrum of characteristic X-ray emission from this measurement point.

Figure 3.126 illustrates the results of X-ray microanalysis of the specimen polymer layer material. The microanalysis was performed in the indicated rectangular area. The measured concentrations of chemical elements were averaged within this area which is shown in the backscattered electron image. Chemical element concentrations are shown in mass percentages. The concentrations of all chemical elements were

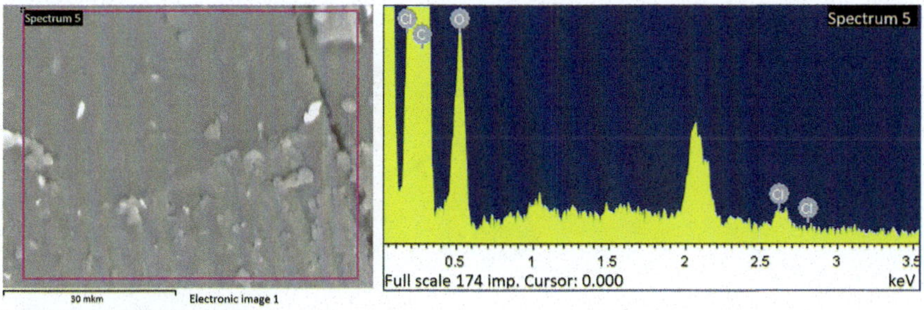

Element	Weight %
C K	80.86
O K	18.67
Cl K	0.48
Total	100.00

Fig. 3.126 Results of X-ray microanalysis of matrix (binder) layer of the specimen polymer composite

measured on K lines of characteristic X-ray emission. Figure 3.126 also presents the spectrum of characteristic X-ray emission from this measurement area.

Thus, the polymer layer of X3 specimen has the following chemical composition of the matrix (binder): 80.86% C (carbon), 18.67% O (oxygen), 0.48% Cl (chlorine). Silicon (Si) (up to 34%) in the concentration of up to 34% is present in the chemical composition of fiber fragments of 6 μm diameter.

It should be emphasized that fiber fragments are only observed in the damaged regions of polymer surface; they are not observed in the specimen bulk. So, most likely, the presence of fiber fragments in the damaged areas of polymer surface is due to nozzle functioning (e.g., due to high temperatures) and separation of fiber pieces from specimen layers (possibly, the widest brown layer) followed by their deposition on the specimen surface.

3.3.2.2 Fragment of Inner Part of *Hwasong-11* (KN-23) Engine Nozzle, North Korea

The fragment of the inner part of *Hwasong-11* (*KN-23*) missile engine nozzle (further—X3 specimen) is a multi-layer structure (four layers in total), each of those being a composite. The specimen thickness varies from 20 to 24 mm, mainly due to a change in thickness of the widest (brown) layer which is located under the topmost black layer. Figure 3.127 shows the image of X6 specimen.

The topmost black layer (see Fig. 3.127) of the specimen is partially damaged. Fragments of artificial fabric with mutually perpendicular fibers are observed on its surface. Figure 3.128 presents a secondary electron image of the topmost surface layer.

Fig. 3.127 Image of the fragment of the inner part of *Hwasong-11* (*KN-23*) missile engine nozzle

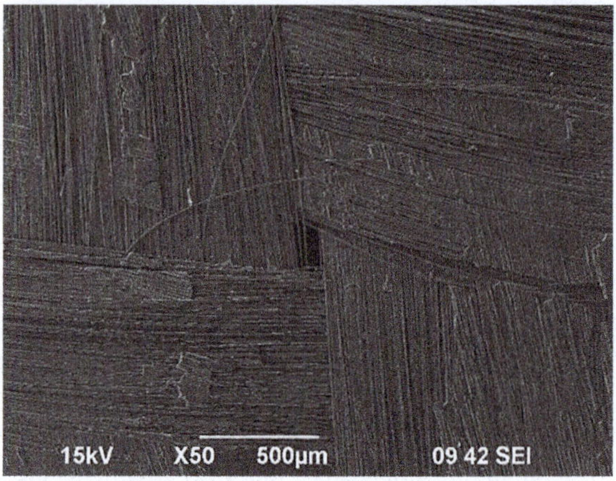

Fig. 3.128 Secondary electron image of the specimen's topmost layer. Accelerating voltage is 15 kV. Magnification is ×50, scale bar corresponds to 500 μm

Figures 3.129 and 3.130 show secondary electron images of the topmost black X6 specimen section at different multiplications. It can be seen in Fig. 3.129 that mutually perpendicular artificial fabric fibers are present in the layer bulk. Figure 3.130 shows that fiber size (diameter) is nearly equal to 7 μm.

Figures 3.131 and 3.132 present the images of the specimen's widest brown section at different magnifications, which is located outside of the black layer (see Fig. 3.127). The image in Fig. 3.131 is the backscattered electron one, and that in Fig. 3.132—a secondary electron image. It can be seen in Fig. 3.131 that artificial fabric fibers are present in this layer bulk. Figure 3.132 shows that fiber size (diameter) is approximately 6 μm.

Figure 3.133 shows the backscattered electron image of the boundary area of the specimen section at the top of which a brown layer can be seen, as well as the yellow one next to it (see Fig. 3.127). Artificial fabric fibers can be seen in the bulk of both layers. In the yellow layer, fibers are brighter, which points to the presence of chemical elements with higher average atomic numbers as compared to the average atomic numbers of the chemical elements present in brown layer fibers (the contrast of backscattered electron images is due to the difference of average atomic numbers in different areas of the studied specimen).

Fig. 3.129 Secondary electron image of the specimen's topmost black layer, in whose bulk mutually perpendicular fibers of artificial fabric are observed. Accelerating voltage is 15 kV. Magnification ×500, scale bar corresponds to 50 μm

Fig. 3.130 Secondary electron image of the specimen's topmost black layer section in whose bulk artificial fabric fibers of 7 μm diameter are observed. Accelerating voltage is 15 kV. Magnification is ×2000, scale bar corresponds to 10 μm

Fig. 3.131 Backscattered electron image of section of the widest brown layer located under the black one. Accelerating voltage is 15kV. Magnification is ×500, scale bar corresponds to 50 μm

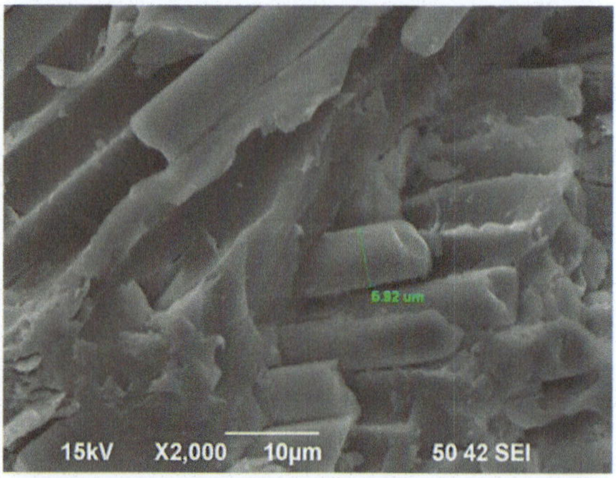

Fig. 3.132 Secondary electron image of section of the specimen's widest brown layer located under the black one. Accelerating voltage is 15kV, multiplication × 2000, scale bar corresponds to 10 μm. The image shows the fiber size—approximately 6 μm

Fig. 3.133 Backscattered electron image of the specimen section boundary area at the top of which a brown layer can be seen (the layer under the black one), as well as the following yellow layer (see Fig. 3.127). Accelerating voltage is 15 kV. Magnification is ×100, scale bar corresponds to 100 μm

Figure 3.134 shows a backscattered electron image of X6 specimen penultimate layer (the yellow one) (see Fig. 3.127), which follows the brown layer. Several artificial fabric layers can be seen in the bulk of this layer.

Figure 3.135 presents backscattered electron image of section of the specimen penultimate (yellow) layer (see Fig. 3.127), which follows the brown one and in which artificial fabric fibers of approximately 8 μm size (diameter) are observed.

Figure 3.136 demonstrates backscattered electron image of section of the fourth (the last, bottommost) specimen layer, which is white (see Fig. 3.127). The image shows numerous light-colored inclusions (particles).

Figure 3.137 presents the backscattered electron image of the fracture surface of fourth (last, bottom) white layer of the specimen (see Fig. 3.127). The fracture was produced for studying the structure and chemical composition of the fourth (last, bottommost) white layer of the specimen directly in the bulk of this layer, without mechanical and other effects of the instruments used to produce sections. Numerous

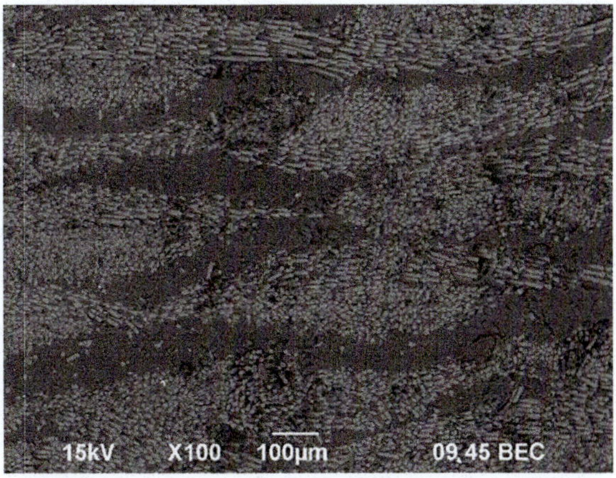

Fig. 3.134 Backscattered electron image of the penultimate (yellow) specimen layer section (see Fig. 3.127), which follows the brown one. Several artificial fabric layers can be seen in the bulk of this one. Accelerating voltage is 15 kV. Magnification is ×100, scale bar corresponds to 100 μm

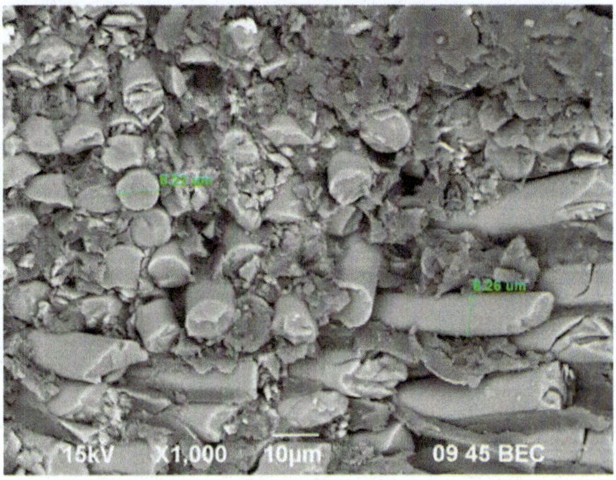

Fig. 3.135 Backscattered electron image of the penultimate (yellow) specimen layer slice (see Fig. 3.127), which follows the brown one. Artificial fabric fibers of approximately 8 μm size (diameter) are observed in this layer. Accelerating voltage is 15kV. Magnification ×1000, scale bar corresponds to 10 μm

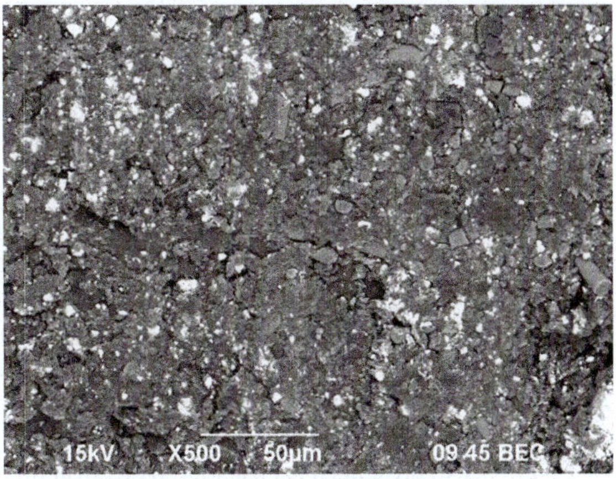

Fig. 3.136 Backscattered electron image of section of the specimen fourth (last, bottommost) white layer (see Fig. 3.127). Accelerating voltage is 15 kV, magnification ×500, scale bar corresponds to 50 μm

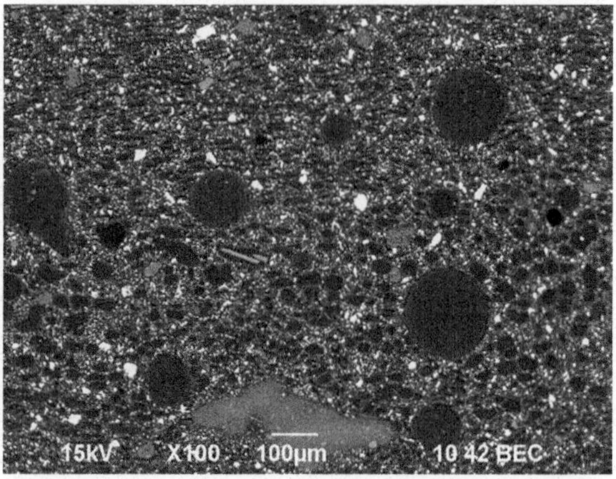

Fig. 3.137 Backscattered electron image of fracture surface of the specimen fourth (last, bottommost) white layer (see Fig. 3.127). Accelerating voltage is 15 kV. Magnification ×100, scale bar corresponds to 100 μm

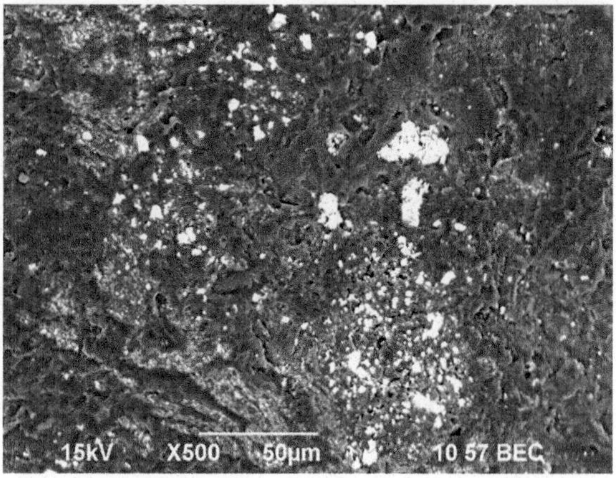

Fig. 3.138 Backscattered electron image of the surface of the fourth (the last, bottom) white layer of the specimen (see Fig. 3.127). Accelerating voltage is 15 kV. Magnification is 500, scale bar corresponds to 50 μm

light-colored inclusions (particles) can also be seen in the bulk of the abovementioned layer.

Figure 3.138 shows the image of the surface of the fourth (last, bottommost) white layer of the specimen (see Fig. 3.127). Light-colored inclusions (particles) are also observed on the layer surface.

Figure 3.139 illustrates the results of X-ray microanalysis of the fragments of artificial fabric fibers, which are observed in the bulk of specimen topmost layer, which is black. The microanalysis was performed at the indicated point shown in the backscattered electron image. Chemical element concentrations are given in mass percentages. The concentrations of all chemical elements were measured on K lines of characteristic X-ray emission. The spectrum of characteristic X-ray emission from this measurement point is also shown in this figure.

Figure 3.140 presents a backscattered electron image with 30 μm scale bar of the black area of the topmost specimen layer, where fibers forming the artificial fabric

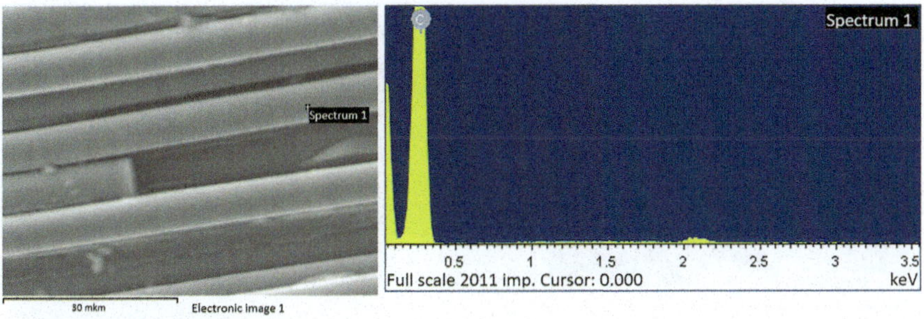

Element	Weight %
C K	100.00
Total	100.00

Fig. 3.139 Results of X-ray microanalysis of the fragments of artificial fabric fibers observed in the bulk of the specimen topmost (black) layer

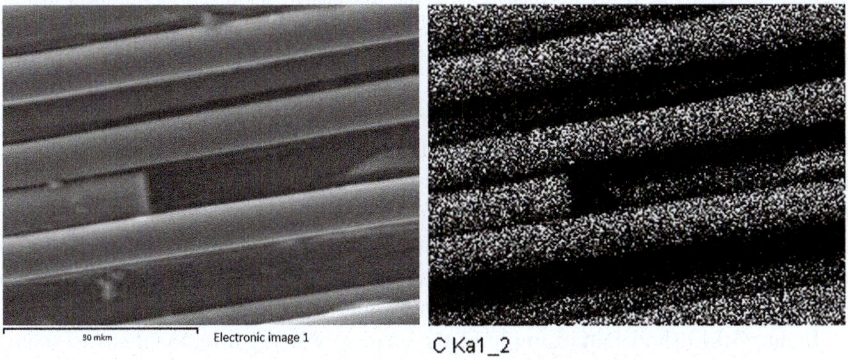

Fig. 3.140 Backscattered electron image with 30 μm scale bar of the area of the topmost (black) layer, where fibers forming the artificial fabric can be seen, as well as the X-ray characteristic emission images of the same area (the spatial distribution of chemical elements over the area examined). White points correspond to the presence of the chemical element indicated under the image

can be seen, as well as the X-ray characteristic emission images of the same area (the spatial distribution of chemical elements over the examined area). White points correspond to the presence of the chemical element indicated under the image.

Figure 3.141 shows an example of the results of X-ray microanalysis of the fragments of artificial fabric fibers observed in the bulk of the specimen widest layer (brown), which is located below the black one. The microanalysis was performed at the indicated point shown in the backscattered electron image. Concentrations of chemical elements are given in mass percentages. The concentrations of all chemical elements were measured on K lines of characteristic X-ray emission. Figure 3.141 also shows the spectrum of characteristic X-ray emission from this measurement point.

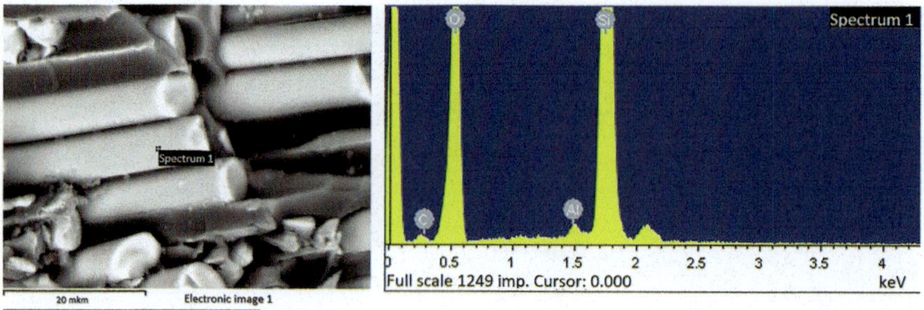

Element	Weight %
C K	8.97
O K	57.61
Al K	0.66
Si K	32.76
Total	100.00

Fig. 3.141 Results of X-ray microanalysis of the fragments of artificial fabric fibers that are observed in the bulk of the widest brown layer of X6 specimen

Figure 3.142 gives an example of the results of X-ray microanalysis of artificial fabric fiber fragments observed in the bulk of the penultimate specimen layer (yellow, following the brown one). The microanalysis was performed at the indicated point shown in the backscattered electron image. Chemical element concentrations are given in mass percentages. The concentrations of all chemical elements were measured on K lines of characteristic X-ray emission. Figure 3.142 also presents the spectrum of characteristic X-ray emission from this measurement point.

Figure 3.143 shows an example of the results of X-ray microanalysis of matrix (binder) substance of the specimen black polymer top layer. The microanalysis was performed in the indicated rectangular area. The measured concentrations of chemical elements were averaged within this area shown in the backscattered electron image. Chemical element concentrations are given in mass percentages. The concentrations of all chemical elements were measured on K lines of characteristic X-ray emission. Figure 3.143 also presents the spectrum of characteristic X-ray emission from this measurement area.

Figure 3.144 demonstrates an example of the results of X-ray microanalysis of matrix (binder) material of the specimen polymer (widest) layer, which is brown and is located under the black one. The microanalysis was performed in the indicated rectangular area. The measured concentrations of chemical elements were averaged within this area shown in the backscattered electron image. Chemical element concentrations are given in mass percentages. The concentrations of all chemical elements were measured on K lines of characteristic X-ray emission. Figure 3.144 also shows the spectrum of characteristic X-ray emission from this measurement area.

Figure 3.145 shows an example of the results of X-ray microanalysis of matrix (binder) material of the polymer (penultimate, yellow) layer, which follows the widest (brown) specimen layer. The microanalysis was performed in the indicated rectan-

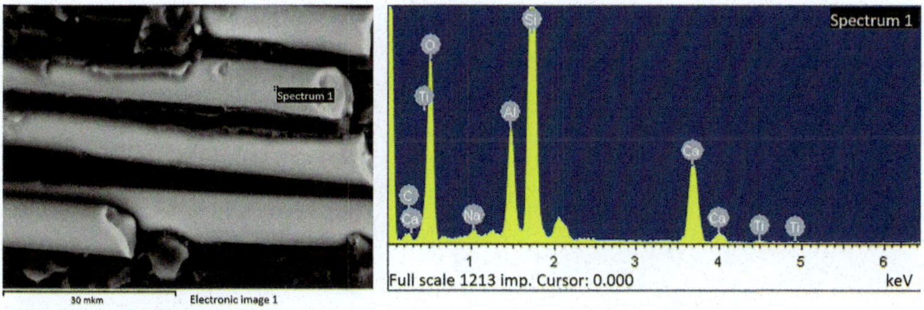

Element	Weight %
C K	6.76
O K	52.07
Al K	6.55
Si K	22.12
Ca K	12.50
Total	100.00

Fig. 3.142 The results of X-ray microanalysis of the fragments of artificial fabric fibers that are observed in the bulk of the penultimate specimen layer (yellow), which follows the brown one

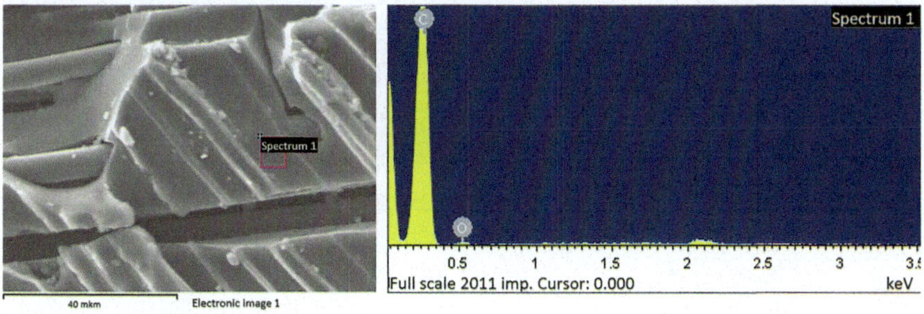

Element	Weight %
C K	96.68
O K	3.32
Total	100.00

Fig. 3.143 Results of X-ray microanalysis of matrix (binder) material of the polymer topmost layer (black) of X6 specimen

gular area. The measured concentrations of chemical elements were averaged within this area shown in the backscattered electron image. Chemical element concentrations are given in mass percentages. The concentrations of all chemical elements were measured on K lines of characteristic X-ray emission. Figure 3.145 also demonstrates the spectrum of characteristic X-ray emission from this measurement area.

Figure 3.146 gives an example of the results of X-ray microanalysis for a section of the fourth (the last, bottommost) specimen layer (see Fig. 3.127), which is white. The microanalysis was performed in the indicated rectangular area. The measured concentrations of chemical elements were averaged within this area shown in the

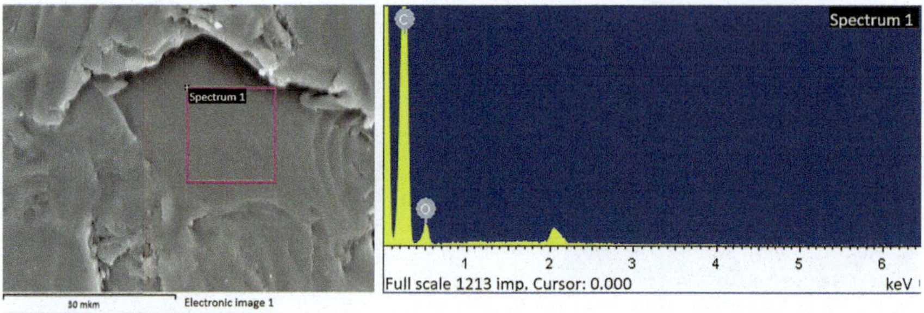

Element	Weight %
C K	84.76
O K	15.24
Total	100.00

Fig. 3.144 Results of X-ray microanalysis of matrix (binder) material of the widest polymer (penultimate) specimen layer, which is brown and is located under the black one

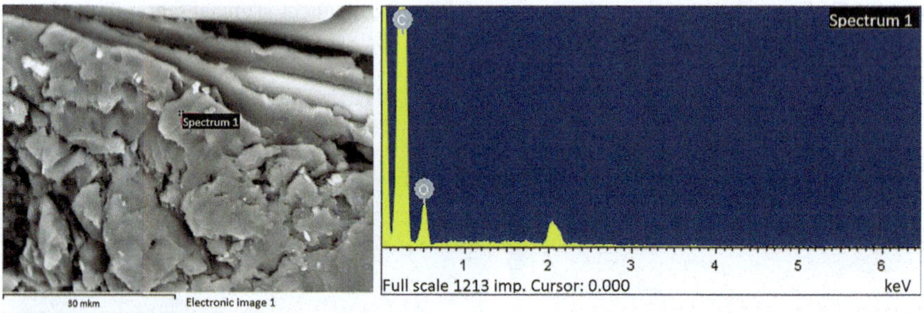

Element	Weight %
C K	82.60
O K	17.40
Total	100.00

Fig. 3.145 Results of X-ray microanalysis of matrix (binder) material of the polymer (penultimate, yellow) layer, which follows the brown one

backscattered electron image. Chemical element concentrations are given in mass percentages. The concentrations of all chemical elements were measured on K lines of characteristic X-ray emission (except that of Zr (zircon), whose concentration was measured on L line). Figure 3.146 also shows the spectrum of characteristic X-ray emission from this measurement area.

Figure 3.147 gives an example of the results of X-ray microanalysis of light-colored inclusions (particles) that are observed in the bulk of the fourth (the last, bottommost) white specimen layer. Microanalysis was performed at the indicated point shown in the backscattered electron image. Chemical element concentrations are given in mass percentages. The concentrations of all chemical elements were measured on K lines of characteristic X-ray emission (except that of Zr (zircon), whose

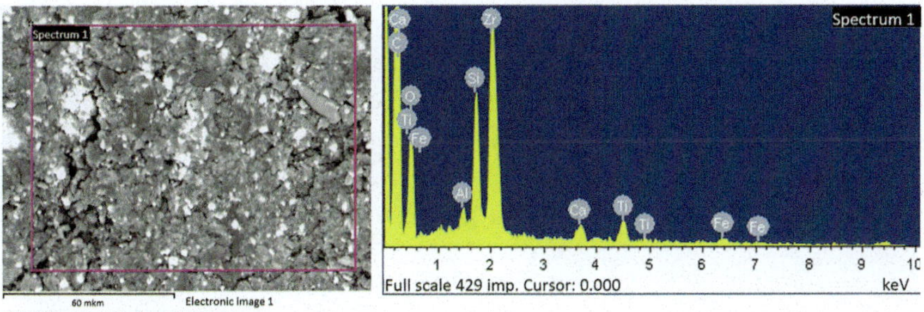

Element	Weight %
C K	64.56
O K	17.82
Al K	0.50
Si K	2.99
Ca K	0.86
Ti K	2.04
Fe K	0.89
Zr L	10.33
Total	100.00

Fig. 3.146 Results of X-ray microanalysis of a section of the fourth (the last, bottommost) white specimen layer

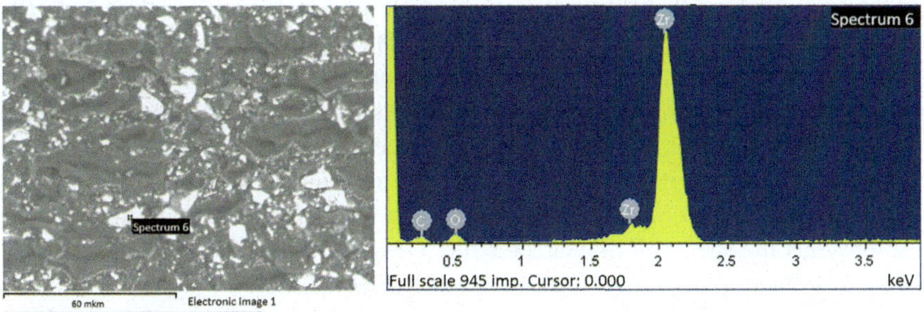

Element	Weight %
C K	17.34
O K	7.75
Zr L	74.91
Total	100.00

Fig. 3.147 Results of X-ray microanalysis of light-colored inclusions (particles) observed in the bulk of the fourth (the last, bottommost) white layer of the specimen

concentration was measured on L line). Figure 3.147 also presents the spectrum of characteristic X-ray emission from this measurement point.

It should be stressed that inclusions (particles) with high Zr (zirconium) concentration (up to 75%) are observed in the studies of the section and fracture surface, as well as surface of the fourth (the last, bottommost) white layer of the specimen.

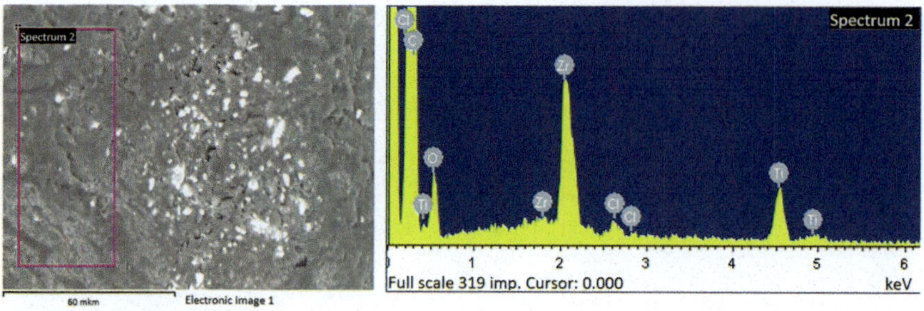

Element	Weight %
C K	70.49
O K	17.69
Cl K	0.46
Ti K	4.13
Zr L	7.24
Total	100.00

Fig. 3.148 Results of X-ray microanalysis of the surface (in the region where there are not many particle-like inclusions) of the fourth (last, bottommost) white specimen layer

Figure 3.148 gives an example of the results of X-ray microanalysis of the surface (in the region where there are not many particle-like inclusions) of the fourth (the last, bottommost) white layer of the specimen. The microanalysis was performed in the indicated rectangular area. The measured concentrations of chemical elements were averaged within this area shown in the backscattered electron image. Chemical element concentrations are given in mass percentages. The concentrations of all chemical elements were measured on K lines of characteristic X-ray emission (except that of Zr (zirconium), whose concentration was measured on L line). Figure 3.148 also presents the spectrum of characteristic X-ray emission from this measurement area.

Thus, specimen X6 (the inner part of the nozzle) is a multi-layered structure that consists of four layers, each of those being a composite. The specimen thickness varies from 20 to 24 mm, mainly, due to a change in thickness (from 9 to 13 mm) of the widest brown layer located under the black one.

Carbon fabric is used as the filler in the first layer. In the next two layers, the filler consists of glass fabric. Glass fabric fibers in these two layers are of different sizes (diameters) and chemical compositions (see the table below). In the last, fourth layer, which has a porous structure, the filler consists of inclusions (particles) with zirconium concentration of up to 75%. Such particles are present both on the surface and in the bulk of the layer.

Below, the table presents the main characteristics of all layers of specimen X6. The layers in the table are denoted according to Fig. 3.127 (Table 3.1).

Table 3.1 Summarized results of specimen X6 studies

Layer	Thickness	Main characteristics
Topmost black layer	5 mm	Composite. Matrix: 96.68% C (carbon); 3.32% O (oxygen). Filler: artificial carbon fabric with fiber diameters of 7 μm and carbon concentration of up to 100%
The widest brown layer	9–13 mm	Composite. Matrix: 84.76% C (carbon); 15.24% O (oxygen). Filler: glass fabric with fiber diameters of 6 μm. The principal chemical elements of fibers: 32.76% Si (silicon), 57.61% O (oxygen), 0.66% Al (aluminum)
Penultimate yellow layer	4 mm	Composite. Matrix: 82.60% C (carbon), 17.40% O (oxygen). Filler: glass fabric with fiber diameters of 8 μm. Principal chemical elements of fibers: 22.12% Si (silicon), 52.07% O (oxygen), 6.55% Al (aluminum), 12.50% Ca (calcium)
The last bottommost white layer	2 mm	Composite. Matrix: 70.49% C (carbon), 17.69% O (oxygen), 0.46% Cl (chlorine), 4.13% Ti (titanium). Filler: inclusions (particles) with Zr (zirconium) concentration of up to 75%

3.3.3 Results of Studying Fragments of Hypersonic Air-to-Surface Ballistic Missiles

3.3.3.1 Fragment of Heat Insulation Layer of *Zircon* Hypersonic Missile (RF) Body

A fragment of heat insulation layer of *Zircon* hypersonic missile body (further—specimen 10.1) is a composite material consisting of several layers. The topmost layer is a black porous material of 3 mm thickness. Under it, there is the main (the widest) layer 8 mm thickness which, in turn, consists of several fiber reinforced interlayers. Then there is a dark-colored bottommost layer (its width could not be determined since specimen 10.1 had been heavily damaged).

Figure 3.149 demonstrates the fragment of the heat insulation layer of *Zircon* hypersonic missile body.

Figure 3.150 shows secondary electron image of a section of specimen 10.1 topmost porous composite layer.

Figure 3.151 presents a secondary electron image of a section of the main (the widest) layer of composite 10.1 fiber reinforced composite. The section plane is perpendicular to the specimen surface. Fiber size is indicated.

Figures 3.152 and 3.15 present secondary electron images at different magnification of the section of the main (the widest) layer of the composite of fiber reinforced specimen 10.1. The section plane is parallel to the specimen surface. Figure 3.153 indicates fiber size.

Fig. 3.149 Image of specimen 10.1

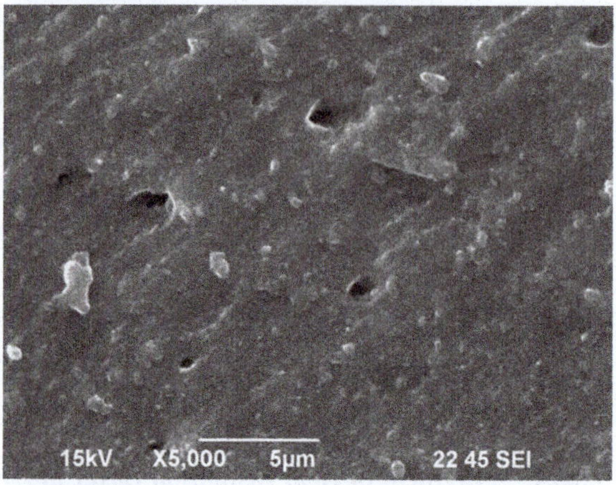

Fig. 3.150 Secondary electron image of topmost porous layer of specimen 10.1 composite. Accelerating voltage is 15 kV. Magnification is ×5000, scale bar corresponds to 5 μm

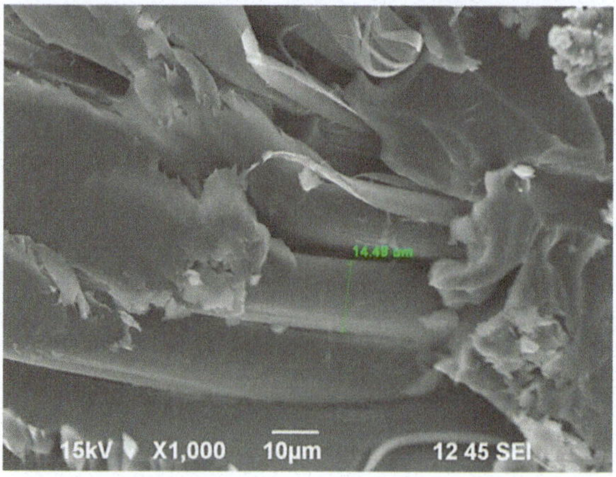

Fig. 3.151 Secondary electron image of section of the main (the widest) layer of fiber reinforced specimen 10.1. Fiber size is indicated. Accelerated voltage is 15 kV. Magnification is ×1000, scale bar corresponds to 10 μm

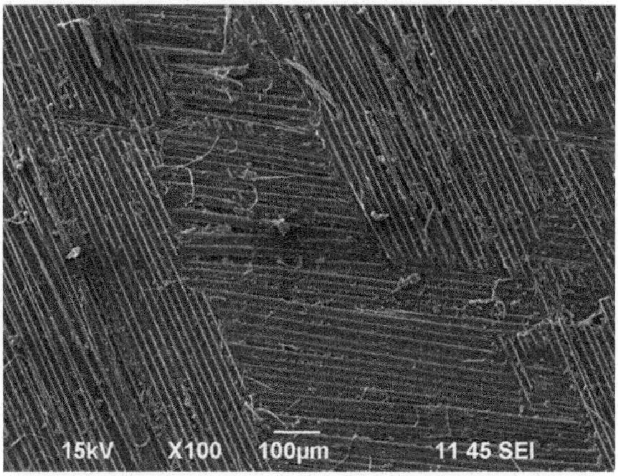

Fig. 3.152 Secondary electron image of section of the main (the widest) layer composite of fiber reinforced specimen 10.1. The section plane is parallel to the specimen surface. Accelerating voltage is 15kV. Magnification × 100, scale bar corresponds to 100 μm

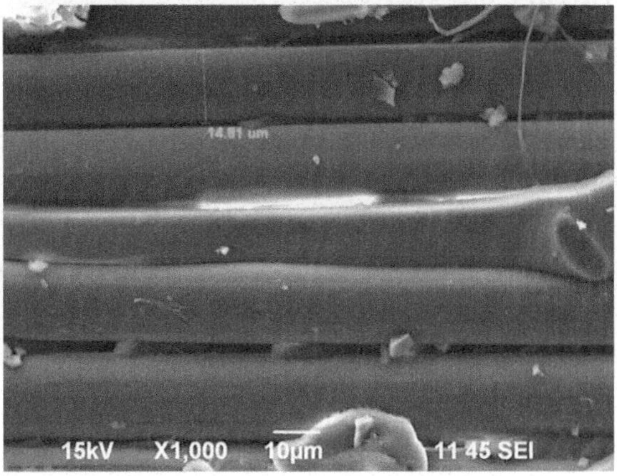

Fig. 3.153 Secondary electron image of a section of the main (the widest) layer of the composite of fiber reinforced specimen 10.1. The section plane is parallel to the specimen surface. Accelerating voltage is 15 kV. Magnification ×1000, scale bar corresponds to 10 μm. Fiber size is shown

Figure 3.154 shows a backscattered electron image of the surface of the dark-colored bottommost layer of specimen 10.1. Fiber segments of smaller diameters than those observed in the sections of the main layer can be seen on the surface. As demonstrated by further studies, these fibers differ from those of the widest layer both in diameter and in chemical composition.

Figure 3.155 gives an example of the results of X-ray microanalysis of the material of the topmost porous layer of specimen 10.1. Microanalysis was performed in the indicated rectangular area. The measured concentrations of chemical elements were averaged within this area shown in the backscattered electron image. Chemical element concentrations are given in mass percentages. The concentrations of all chemical elements were measured on K lines of characteristic X-ray emission (except that of Zn (zinc), whose concentration was measured on L line). Figure 3.155 also demonstrates the spectrum of characteristic X-ray emission from this measurement area.

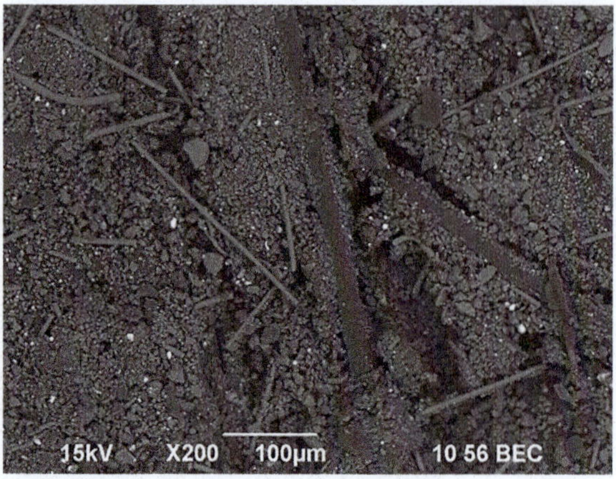

Fig. 3.154 Backscattered electron image of the surface of the dark-colored bottommost layer of specimen 10.1 Accelerating voltage is 15 kV. Magnification is ×200, scale bar corresponds to 100 µm

Element	Weight %
C K	91.71
O K	4.65
P K	0.38
S K	1.23
Zn L	2.03
Total	100.00

Fig. 3.155 Results of X-ray microanalysis of the topmost porous layer material

Figure 3.156 demonstrates an example of the results of X-ray microanalysis of fibers observed in the bulk of the main (the widest) layer of specimen composite. The microanalysis was performed at the indicated point shown in the backscattered electron image. Chemical element concentrations are given in mass percentages. The concentrations of all chemical elements were measured on K lines of characteristic X-ray emission. Figure 3.156 also presents the spectrum of characteristic X-ray emission from this measurement point.

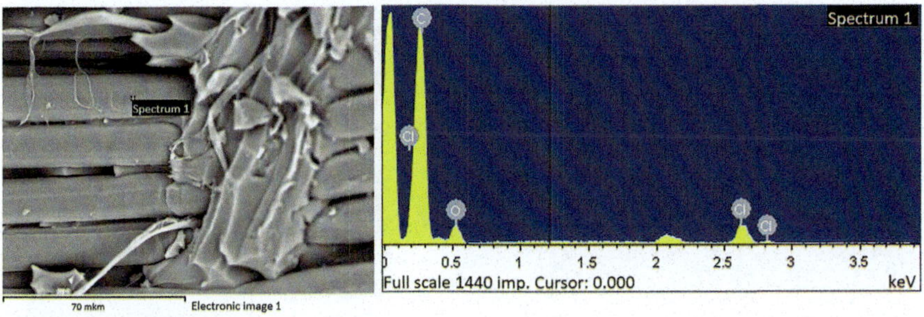

Element	Weight %
C K	83.15
O K	13.91
Cl K	2.95
Total	100.00

Fig. 3.156 Results of X-ray microanalysis of fibers observed in the bulk of the main (the widest) layer of specimen 10.1 composite

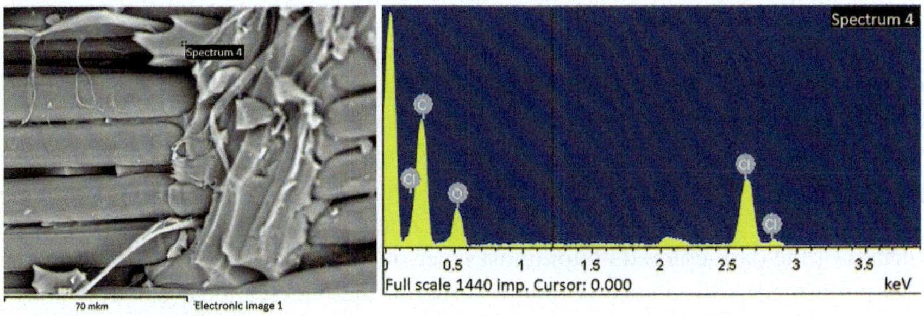

Element	Weight %
C K	71.07
O K	20.49
Cl K	8.44
Total	100.00

Fig. 3.157 Results of X-ray microanalysis of matrix (binder) material of the main (the widest) layer of specimen composite

Figure 3.157 shows an example of the results of X-ray microanalysis of matrix (binder) material of the main (the widest) layer of specimen composite. The microanalysis was performed at the indicated point shown in the backscattered electron image. Chemical element concentrations are given in mass percentages. The concentrations of all chemical elements were measured on K lines of characteristic X-ray emission. Figure 3.157 also presents the spectrum of characteristic X-ray emission from this measurement area.

Figure 3.158 demonstrates an example of the results of X-ray microanalysis of the fibers observed in the bulk of the main (the widest) layer of composite 10.1. Micro-

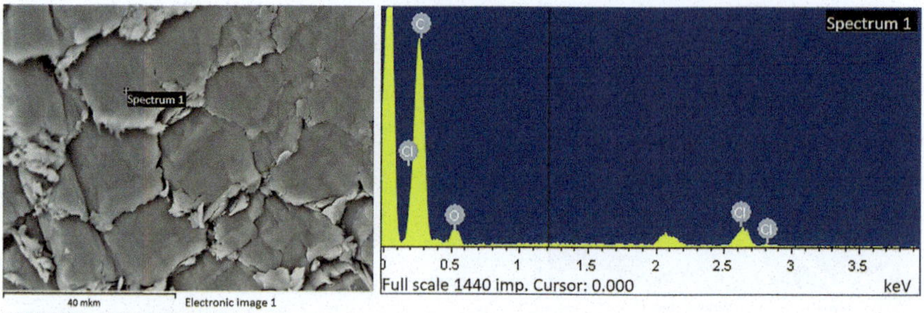

Element	Weight %
C K	83.43
O K	13.60
Cl K	2.97
Total	100.00

Fig. 3.158 Results of X-ray microanalysis of the fibers observed in the bulk of the main (the widest) layer of specimen composite. The measurement point is located at fiber end

analysis was performed at the indicated point shown in the backscattered electron image. The measurement point is located at fiber end. Chemical element concentrations are given in mass percentages. The concentrations of all chemical elements were measured on K lines of characteristic X-ray emission from this measurement point.

Figure 3.159 gives an example of the X-ray microanalysis of fibers observed on the surface of the dark-colored bottommost layer of specimen 10.1. The microanalysis was performed at the indicated point shown in the backscattered electron image. Chemical element concentrations are given in mass percentages. The concentrations of all chemical elements were measured on K lines of characteristic X-ray emission. Figure 3.159 also shows the spectrum of characteristic X-ray emission from this measurement point.

Figure 3.160 shows an example of X-ray microanalysis of the surface of the dark-colored bottommost layer of specimen 10.1. The microanalysis was performed in the indicated rectangular area. The measured concentrations of chemical elements were averaged within this area, which is shown in the backscattered electron image. Chemical element concentrations are given in mass percentages. The concentrations of all chemical elements were measured on K lines of characteristic X-ray emission. Figure 3.160 also shows the spectrum of characteristic X-ray emission from this measurement area.

Thus, specimen 10.1 is a composite material consisting of several layers. The topmost layer is a black porous material of 3 mm thickness. Under it is the main (the widest) layer of 8 mm thickness, which, in turn, consists of several interlayers reinforced with carbon fibers of approximately 15 μm diameter, which are located in the planes parallel to the plane of specimen surface. The bottom layer, whose width could not be determined due to its heavy damage, is dark colored.

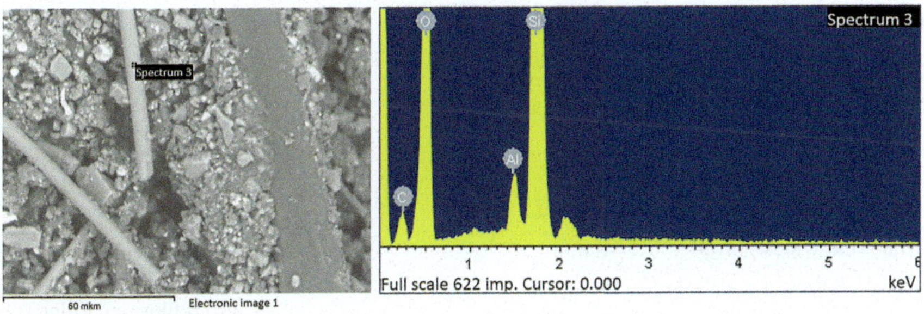

Element	Weight %
C K	16.34
O K	50.11
Al K	1.53
Si K	32.01
Total	100.00

Fig. 3.159 Results of X-ray microanalysis of fibers observed on the surface of the dark-colored bottommost layer of specimen 10.1

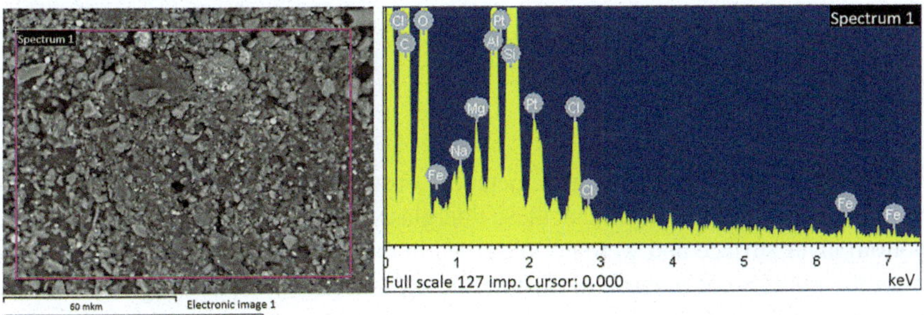

Element	Weight %
C K	57.85
O K	28.10
Na K	0.35
Mg K	0.54
Al K	2.33
Si K	8.81
Cl K	1.34
Fe K	0.69
Total	100.00

Fig. 3.160 Results of X-ray microanalysis of the material of the specimen dark-colored bottommost layer

Table 3.2 Summarized results of specimen 10.1 studies

Layer	Thickness	Principal characteristics
Black topmost layer	3 mm	Polymer of the following chemical composition: 91.71% C (carbon), 4.65% O (oxygen), 0.38% P (phosphorus), 1.23% S (sulfur), 2.03% Zn (zinc)
Main (the widest) layer	8 mm	Composite. Matrix: 71.07% C (carbon), 20.49% O (oxygen), 8.44% Cl (chlorine). Filler: carbon fibers of 15 μm diameter. Fiber chemical composition: 83.15% C (carbon), 13.91% O (oxygen), 2.95% Cl (chlorine)
Dark bottommost layer	Layer thickness could not be determined due to specimen heavy damage.	Layer chemical composition: 57.85% C (carbon), 28.10% O (oxygen), 1.34% Cl (chlorine), 35% Na (sodium), 0.54% Mg (magnesium), 2.33% Al (aluminum), 8.81% Si (silicon), 0.69% Fe (iron)

Fiberglass segments of 6 μm diameter are observed on the surface of the bottommost layer. It should be noted that these fiberglass segments were only noticed on the bottom layer surface and were not observed in other layers, either on their surfaces or in the bulk. So, it is more likely that the presence of fiberglass segments on the bottommost layer surface is due to their adhesion to the surface of this layer when breaking away from other parts of the missile when it was damaged (in the presence of high temperatures and mechanical loads), whose constituent part is the specimen.

Below, the table shows the principal characteristics of all specimen layers. The layers are designated conditionally (the black porous layer is designated as the topmost layer) (Table 3.2).

3.3.3.2 Fragment of Nozzle Unit Insert of *Zircon* Hypersonic Missile (RF)

The examined specimen is a composite material in which fibers are used as the filler.

Figure 3.161 shows the nozzle unit insert of *Zircon* hypersonic missile (further—specimen 10.2).

Figure 3.162 presents a secondary electron image of an area of specimen 10.2 outer surface. Figure 3.163 demonstrates fiber bundles, which consist of large numbers of individual fibers.

Fig. 3.161 Image of specimen 10.2

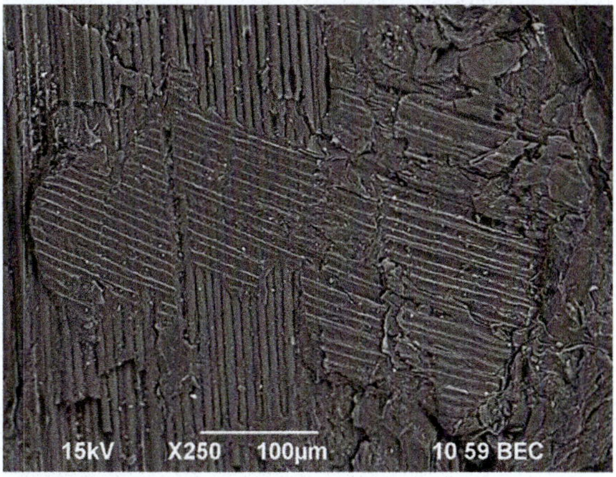

Fig. 3.162 Backscattered electron image of an area of specimen 10.2 outer surface. Accelerating voltage is 15kV, magnification × 250, scale bar corresponds to 100 μm

Figure 3.163 shows a secondary electron image of a fiber bundle. Bundle size is indicated—550 μm.

Figure 3.164 demonstrates a secondary electron image of individual fibers. It indicates the size of an individual fiber—7.5 μm.

Figure 3.165 shows a secondary electron image of fiber bundle ends on specimen 10.2 section.

Figure 3.166 presents a secondary electron image of individual fiber ends on specimen 10.2 section.

Figure 3.167 gives an example of the results of fiber X-ray microanalysis on specimen 10.2 section. The microanalysis was performed at the point that is located at fiber end and is shown in the backscattered electron image. Chemical element

Fig. 3.163 Secondary electron image of a fiber bundle. The bundle size—550 μm—is indicated. Accelerating voltage is 15 kV, magnification ×150, scale bar corresponds to 100 μm

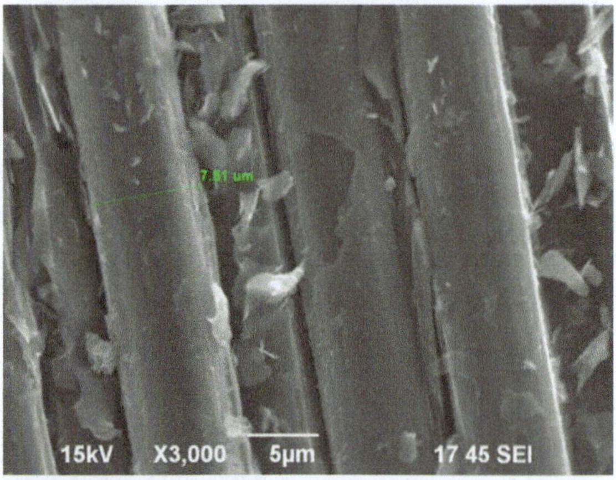

Fig. 3.164 Secondary electron image of individual fibers, which indicates the size of an individual fiber—7.5 μm. Accelerating voltage is 15 kV, magnification is 3000, scale bar corresponds to 5 μm

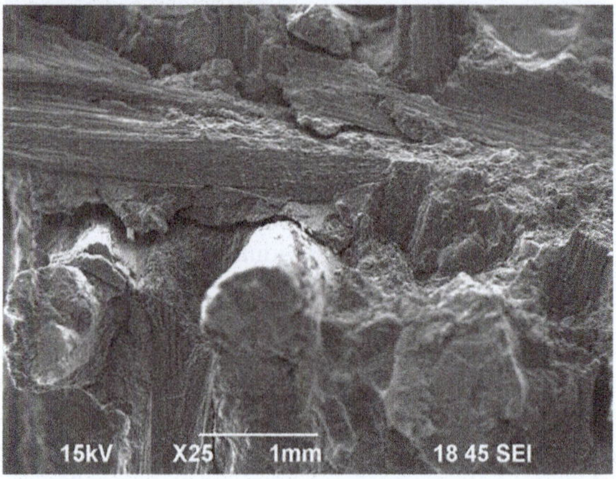

Fig. 3.165 Secondary electron image of fiber bundle ends on the specimen slice. Accelerating voltage is 15 kV, magnification ×25, scale bar corresponds to 1 mm

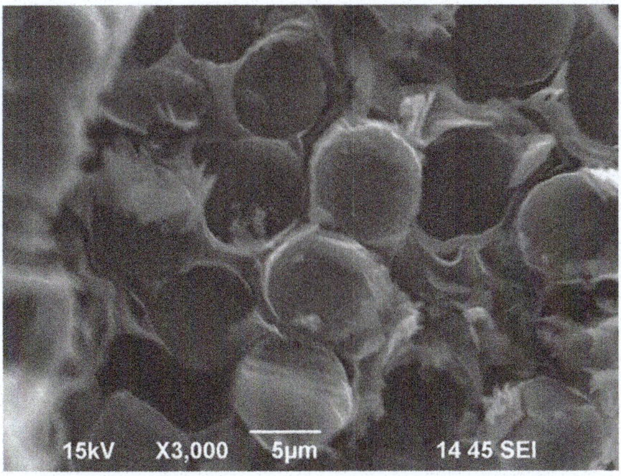

Fig. 3.166 Secondary electron image of individual fiber ends on specimen section. Accelerating voltage is 15 kV, magnification ×3000, scale bar corresponds to 5 μm

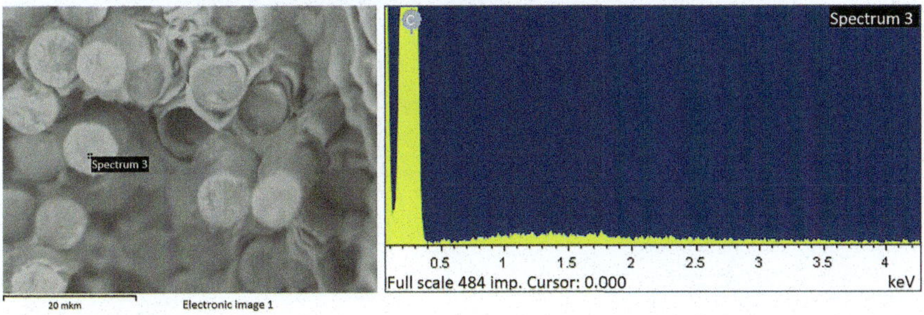

Element	Weight %
C K	100.00
Total	100.00

Fig. 3.167 Results of X-ray microanalysis of fibers on specimen 10.2. The measurement point is on fiber end

concentrations are given in mass percentages (as is seen from the table, fibers are 100% carbon). The concentration of C (carbon) was measured on K lines of characteristic X-ray emission. Figure 3.167 also presents the spectra of characteristic X-ray emission from this measurement point.

Figure 3.168 presents an example of the results of X-ray microanalysis of the fiber of specimen 10.2 section. The microanalysis was performed at the point located on fiber lateral surface, which is shown in the backscattered electron image. The concentration of chemical elements was measured on K lines of characteristic X-ray emission and is given in mass percentages. Figure 3.168 also presents the spectra of characteristic X-ray emission from this measurement point.

Figure 3.169 shows an example of the results of X-ray microanalysis of composite matrix (binder) for the section of specimen 10.2. The microanalysis was performed

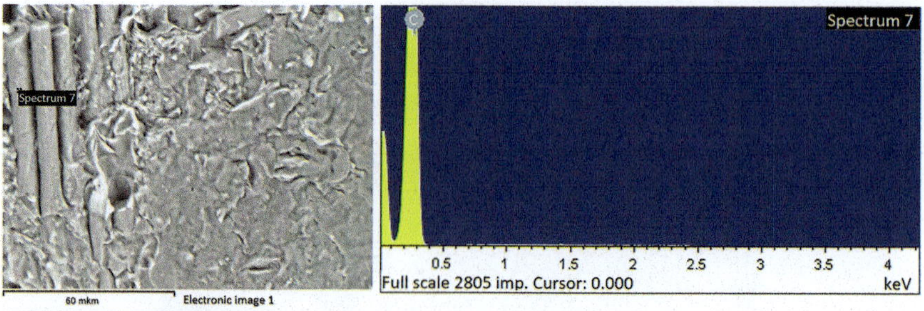

Element	Weight %
C K	100.00
Total	100.00

Fig. 3.168 Results of fiber X-ray microanalysis of the specimen section. Microanalysis was performed at the point located on fiber lateral surface

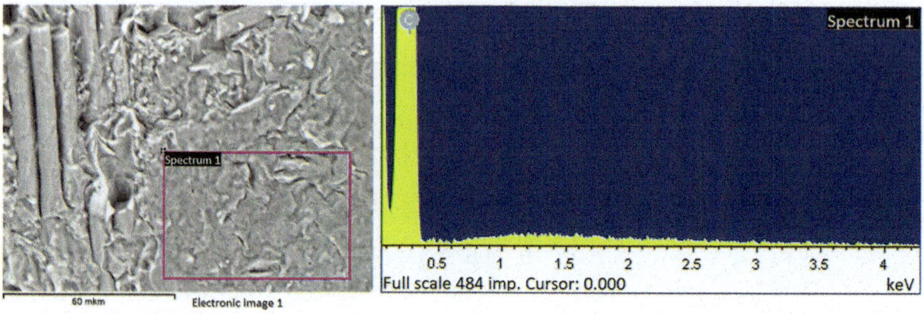

Element	Weight %
C K	100.00
Total	100.00

Fig. 3.169 Results of X-ray microanalysis of composite matrix (binder) on the specimen section. Microanalysis was performed in the rectangular area shown in backscattered electron image

in the rectangular area, which is shown in the backscattered electron image. The concentration of chemical elements was measured on K lines of characteristic X-ray emission and is given in mass percentages. The fibers consist of !00% carbon, like those in Figs. 3.167 and 3.168. Figure 3.169 also presents the spectra of characteristic X-ray emission from this measurement area.

Single particles (approximately one particle per 30×30 μm) were observed on the specimen section in the matrix region. Figure 3.170 presents an example of X-ray microanalysis of one of these particles. Microanalysis was performed at the point shown in backscattered electron image. Chemical element concentrations are given in mass percentages. The concentrations of all chemical elements were measured on K lines of characteristic X-ray emission. Figure 3.170 also presents the spectra of characteristic X-ray emission from this measurement point.

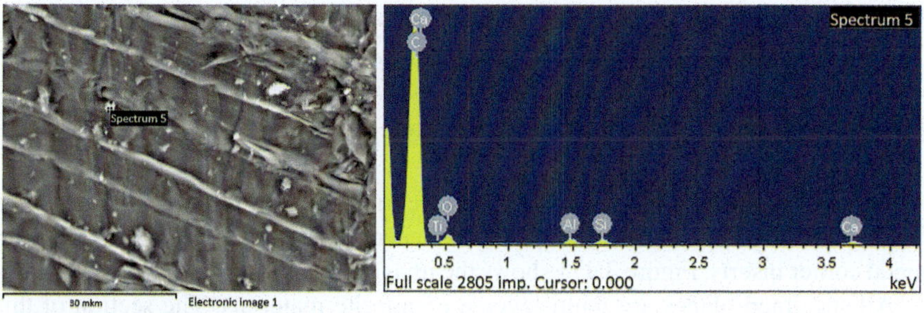

Element	Weight %
C K	84.03
O K	9.12
Al K	0.91
Si K	0.94
Ca K	1.67
Ti K	3.95
Total	100.00

Fig. 3.170 Results of X-ray microanalysis of a particle observed on the specimen section

Thus, the examined specimen of a fragment of the nozzle unit insert of *Zircon* hypersonic missile (specimen 10.2) is a composite material where carbon fibers of 7.5 μm diameter with 100% carbon concentration are used as filler. The fibers form bundles crossing at an angle, and their transverse dimension is 550 μm.

The matrix, like the fibers, consists of 100% carbon. Single particles observed on the specimen section in the matrix region have the following chemical composition: 84.03% C (carbon), 9.12% O (oxygen), 3.95% Ti (titanium), 1.67% Ca (calcium), 0.94% Si (silicon), and 0.91% Al (aluminum). However, during the measurement of matrix chemical composition in the rectangular area (Fig. 3.169), the chemical elements present in the particles were not detected in the matrix due to a very low density of these particles. Hence, the matrix, like the fibers, practically consists of 100% carbon.

Summing up the above, the specimen studied can be considered as belonging to relatively new composites—carbon–carbon materials in which both the filler and the matrix consist of practically 100% carbon.

3.3.4 *Results of Studying Fragments of Loitering Munition (LM) and Unmanned Aerial Vehicles (UAV)*

3.3.4.1 Fragment of *Shahed-136* LM Body

A fragment of *Shahed-136* LM body with a corner insert (further—specimen) is a T-shaped structure whose vertical plane is connected with the horizontal plane with metal corner inserts. Figure 3.171 shows the images of this fragment from both sides.

All specimen planes are multi-layered composite materials. The section of this material is schematically shown in Fig. 3.172.

The vertical specimen plane has the structure of layers identical to that of its horizontal plane, but it has a wider layer of honeycomb porous structure.

Figure 3.173 presents a secondary electron image of the surface of black polymer layer, which is present on both sides of each specimen plane.

Figures 3.174 and 3.175 present a secondary electron image of the surface of artificial fabric layer at different magnifications.

Figures 3.176, 3.177, 3.178, and 3.179 show secondary and backscattered (Fig. 3.179) electron images of the material used to glue metal inserts around their perimeters.

Figure 3.180 presents a secondary electron image of porous honeycomb layer (honeycombs are empty, without filler).

Fig. 3.171 Image of a fragment of *Shahed-136* LM body

№	Symbol	Name of material layers in section
1		Black polymer layer of the coating
2		Polymer layer reinforced with artificial fabric
3		Polymer layer
4		Layer with honeycomb porous structure (honeycombs are empty, without filler)
5		Polymer layer
6		Polymer layer reinforced with artificial fabric
7		Black polymer layer of the coating

Fig. 3.172 Schematic presentation of specimen section. Metal inserts are attached to the planes of fragment structure with screws, and glue is applied around the perimeter

Fig. 3.173 Secondary electron image of the surface of black polymer layer, which is present on both sides of each specimen plane. Accelerating voltage is 15 kV, magnification ×500, scale mark—50 μm

Fig. 3.174 Secondary electron image of the surface of artificial fabric layer. Accelerating voltage is 15 kV, magnification ×500, scale mark—50 μm

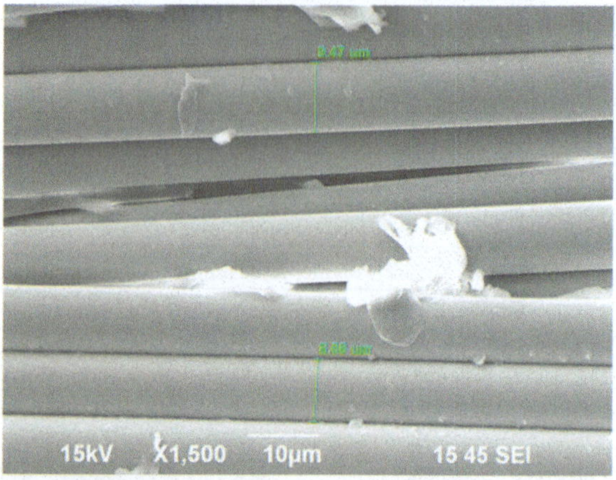

Fig. 3.175 Secondary electron image of artificial fabric layer. Accelerating voltage is 15 kV, magnification ×1,500, scale mark—10 μm. The image shows the sizes of artificial fabric fibers (8.68–9.47 μm)

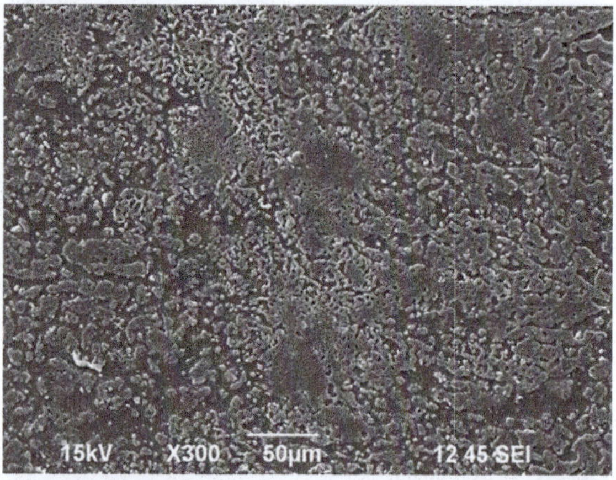

Fig. 3.176 Secondary electron image of the surface of the material used to glue metal inserts around their perimeters. Accelerating voltage is 15 kV, magnification ×300, scale mark—50 μm

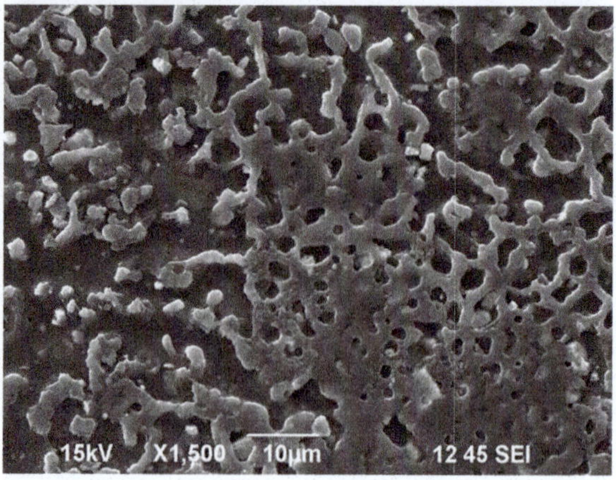

Fig. 3.177 Secondary electron images of the surface of the material used to glue metal inserts around their perimeters. Accelerating voltage is 15 kV, magnification ×1,500, scale mark—10 μm

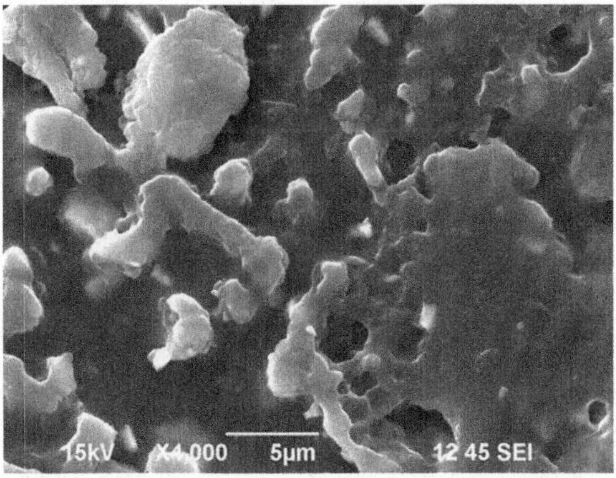

Fig. 3.178 Secondary electron images of the surface of the material used to glue metal inserts around their perimeters. Accelerating voltage is 15 kV, magnification ×4,000, scale mark—5 μm

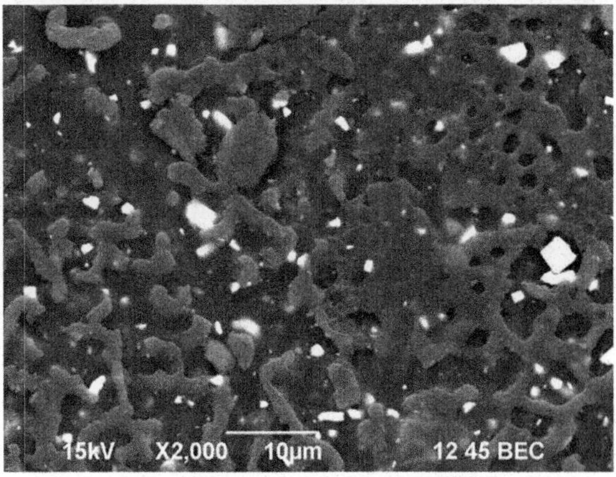

Fig. 3.179 Backscattered electron image of the surface of the material used to glue metal inserts around their perimeters. Accelerating voltage is 15 kV, magnification ×4,000, scale mark—5 μm

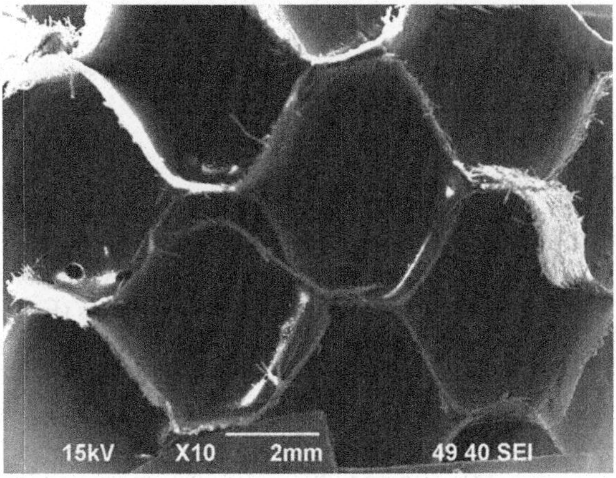

Fig. 3.180 Secondary electron image of porous honeycomb layer (honeycombs are empty, without filler). Accelerating voltage is 15 kV, magnification ×10, scale mark—2 mm

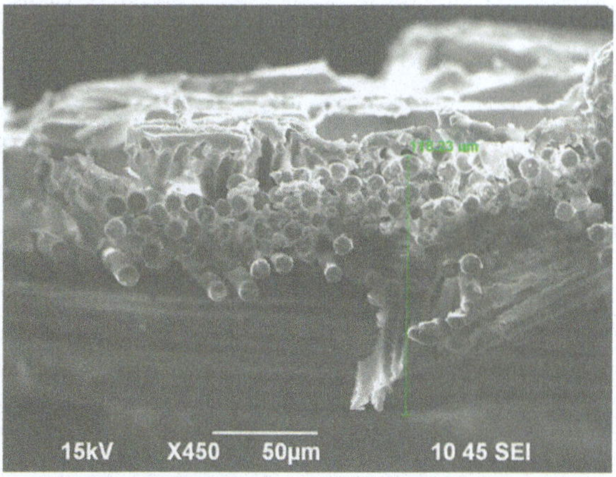

Fig. 3.181 Secondary electron image of the slice of artificial fabric layer. Accelerating voltage is 15 kV, magnification ×450, scale mark—50 μm

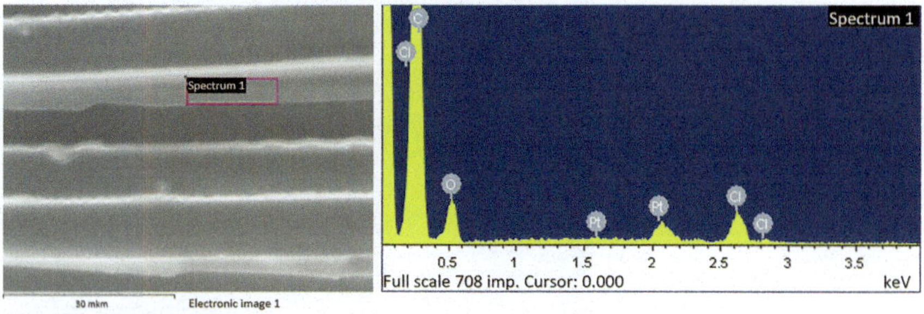

Element	Weight %
C K	80.30
O K	17.19
Cl K	2.51
Total	100.00

Fig. 3.182 Results of X-ray microanalysis on the surface of black polymer layer covering all specimen planes

Figure 3.181 shows a secondary electron image of the slice of artificial fabric layer.

Figure 3.182 presents the results of X-ray microanalysis on the surface of black polymer layer which cover all specimen planes (Fig. 3.171). The microanalysis was performed in a definite rectangular area. The measured concentrations of chemical elements were averaged within this area, which is shown in backscattered electron image. Chemical element concentrations are given in mass percentages. The concentrations of all chemical elements were measured on K lines of characteristic X-ray emission. Figure 3.182 also presents the spectrum of characteristic X-ray emission of this measurement area.

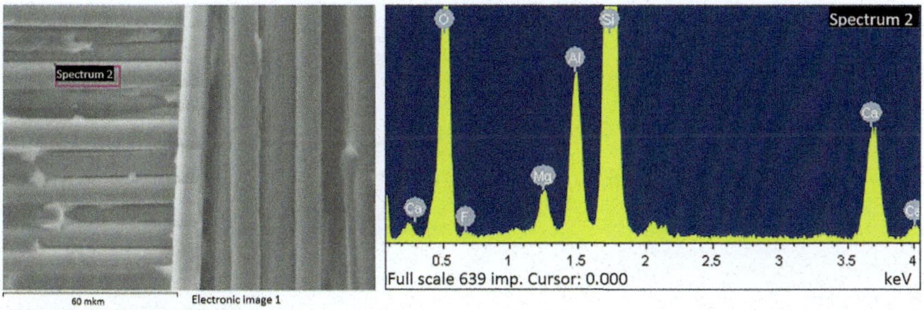

Element	Weight %
O K	51.82
Mg K	2.38
Al K	7.20
Si K	26.05
Ca K	12.55
Total	100.00

Fig. 3.183 Results of X-ray microanalysis of the fibers of artificial fabric layer

Figure 3.183 shows the results of X-ray microanalysis of the fibers of artificial fabric layer. The microanalysis was performed in a definite rectangular area. The measured concentrations of chemical elements were averaged within this area, which is shown in backscattered electron image. Chemical element concentrations are given in mass percentages. The concentrations of all chemical elements were measured on K lines of characteristic X-ray emission. Figure 3.183 also presents the spectrum of characteristic X-ray emission of this measurement area.

Figure 3.184 demonstrates the results of X-ray microanalysis of the polymer situated below and above the porous honeycomb layer (see Fig. 3.172). The microanalysis was performed in a definite rectangular area. The measured concentrations of chemical elements were averaged within this area, which is shown in backscattered electron image. Chemical element concentrations are given in mass percentages. The concentrations of all chemical elements were measured on K lines of characteristic X-ray emission of this measurement area.

Figure 3.185 shows the results of X-ray microanalysis of the substance of matrix (binder) of polymer layer reinforced with artificial fabric. The microanalysis was performed in a definite rectangular area. The measured concentrations of chemical elements were averaged within this area, which is shown in backscattered electron image of the fragment edge cut. Chemical element concentrations are given in mass percentages. The concentrations of all chemical elements were measured on K lines of characteristic X-ray emission. Figure 3.185 also presents the spectrum of characteristic X-ray emission of this measurement area.

Figure 3.186 presents the results of X-ray microanalysis of the material in the porous honeycomb layer (honeycombs are empty, without filler). The microanalysis was performed in a definite rectangular area. The measured concentrations of chemical elements were averaged within this area, which is shown in the backscattered

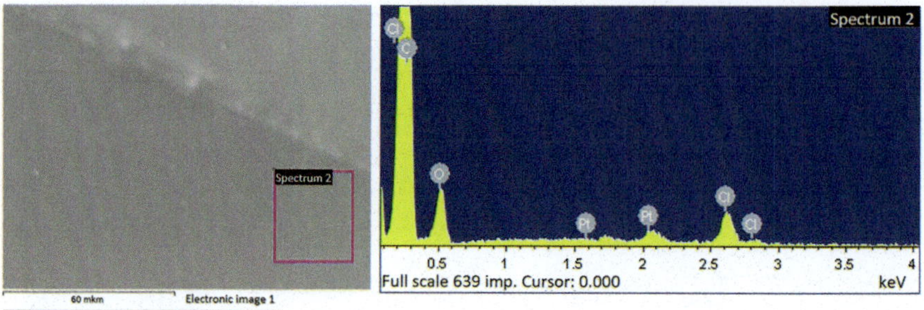

Element	Weight %
C K	79.63
O K	18.25
Cl K	2.12
Total	100.00

Fig. 3.184 Results of the analysis of the polymer located above and below the porous honeycomb layer (see Fig. 3.172

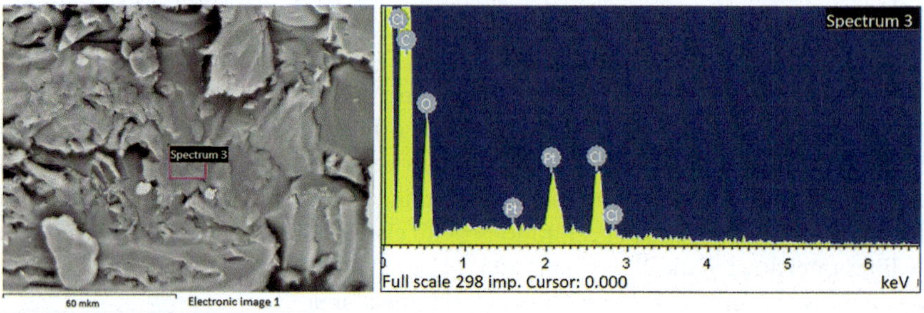

Element	Weight %
C K	80.29
O K	18.83
Cl K	2.43
Total	100.00

Fig. 3.185 Results of X-ray microanalysis of the matrix (binder) of polymer layer reinforced with artificial fabric

electron image. Chemical element concentrations are given in mass percentages. The concentrations of all chemical elements were measured on K lines of characteristic X-ray emission. Figure 3.186 also presents the spectrum of characteristic X-ray emission of this measurement area.

Figure 3.187 shows the results of X-ray microanalysis of the material used to glue metal inserts around their perimeters. The microanalysis was performed in a definite rectangular area. The measured concentrations of chemical elements were averaged within this area, which is shown in backscattered electron image. Chemical element concentrations are given in mass percentages. The concentrations of all chemical elements were measured on K lines of characteristic X-ray emission, except that

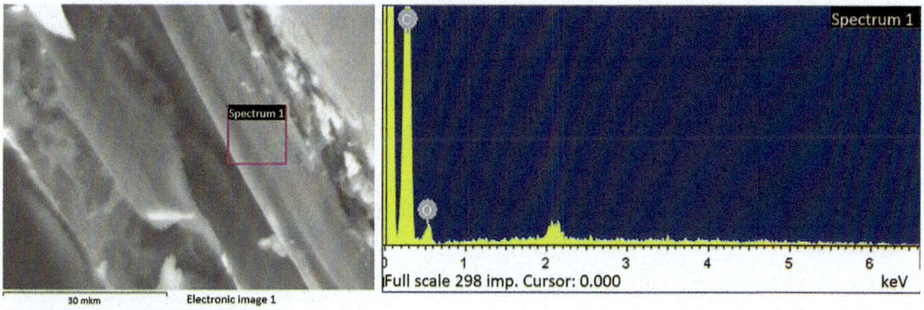

Element	Weight %
C K	83.05
O K	16.95
Total	100.00

Fig. 3.186 Results of X-ray microanalysis of the material of porous honeycomb layer (honeycombs are empty, without filler)

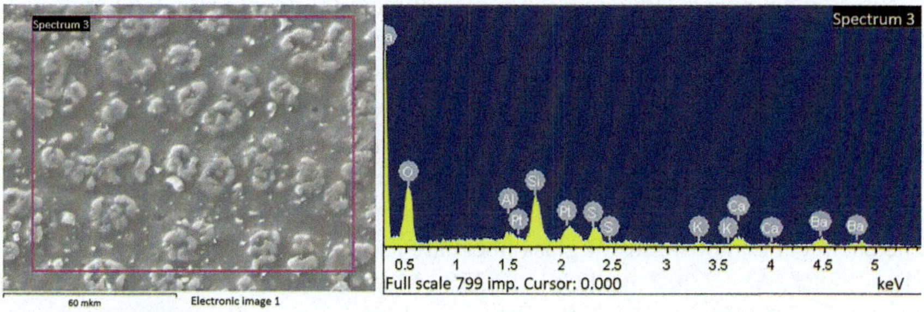

Element	Weight %
C K	67.55
O K	22.28
Al K	0.44
Si K	2.50
S K	1.33
K K	0.34
Ca K	0.99
Ba L	4.57
Total	100.00

Fig. 3.187 Results of X-ray microanalysis of the material used to glue metal inserts around their perimeters

of Ba (barium), whose concentration was measured on L lines. Figure 3.187 also presents the spectrum of characteristic X-ray emission of this measurement area

Conclusions

All specimen planes are multi-layered composite materials consisting of seven layers. Vertical plane thickness is 12,000 μm and horizontal plane thickness is 5,840 μm.

Table 3.3 Results of studying the structure of *Shahed-136* LM body fragment

Name of specimen outer coating layer (the main part)	Layer thickness, μm
Black polymer layer	50
Polymer layer reinforced with artificial fabric	300
Polymer layer	70
Porous honeycomb layer (honeycombs are empty, without any filler)	5000
Polymer layer	70
Polymer layer reinforced with artificial fabric	300
Black polymer layer	50

In the plane bulk, under black polymer layer, polymer layer reinforced with artificial fabric and polymer layer, a porous honeycomb layer is observed, with empty honeycombs without filler, and of 4.2 μm in diameter.

Situated below is a table with the names of main fragment parts, with their layer thicknesses indicated. The layer order in the table corresponds to actual layer placement in the fragment planes (Table 3.3).

Black polymer layer has the following chemical composition and chemical element concentration: C—80.30%, O—17.19%, and Cl—2.51%.

Polymer layers reinforced with artificial fabric are composites consisting of a glass fabric with 8—10 μm glass fibers, and their matrices (binders) are polymers. The chemical compositions of glass fibers are as follows: Si—6.05%, O—51.82%, Mg—2.38%, Al—7.20%, and Ca—12.55%. Matrix (binder) chemical composition is such: C—80.29%, O—18.83%, and Cl—2.43%.

Polymer layers placed below and above porous honeycomb layer (the thickness of these layers is 70 μm) have such chemical compositions: C—79.63%, O—18.25%, and Cl—2.12%.

The material of porous honeycomb layer is a polymer and consists of such chemical elements in respective chemical concentrations: C—83.05% and O—6.95%.

The specimen vertical plane has chemical composition, chemical element concentrations, structure, and all layer sizes identical to all those of the horizontal plane except one, viz., the thickness of porous honeycomb layer in the vertical plane is 1,1160 μm, due to which the whole vertical plane is 12,000 μm thick.

The material used to glue metal inserts around their perimeters has the following chemical composition: C—67.55%, O—22.28%, Al—0.44%, Si—2.50%, S—1.33%, K—0.34%, Ca—0.99%, and Ba—4.57%.

3.3.4.2 Fragment of *Shahed-136* LM Warhead Mount Element

Figures 3.188, 3.189, and 3.190 present secondary electron images (SEI) of a surface fragment of *Shahed-136* LM warhead mount element at different magnifications.

To study the structure and measure chemical element concentrations in the bulk of a fragment of *Shahed-136* warhead mount element (specimen), a slice of this fragment was prepared. Its secondary electron images are shown in Figs. 3.191, 3.192, 3.193, and 3.194.

Figure 3.195 presents the results of X-ray microanalysis on the specimen surface. The microanalysis was performed in a definite rectangular area. The measured concentrations of chemical elements were averaged within this area, which is shown in secondary electron image. Chemical element concentrations are given in mass percentages. The concentrations of all chemical elements were measured on K lines

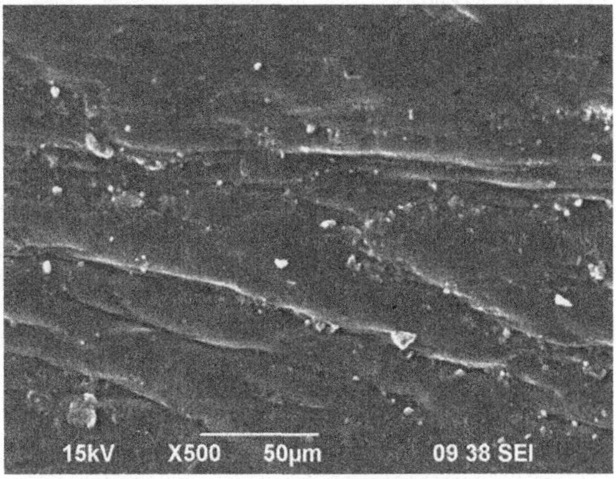

Fig. 3.188 Secondary electron image of a surface fragment of *Shahed-136* LM warhead mount element. Accelerating voltage is 15 kV, magnification ×500, scale mark—50 μm

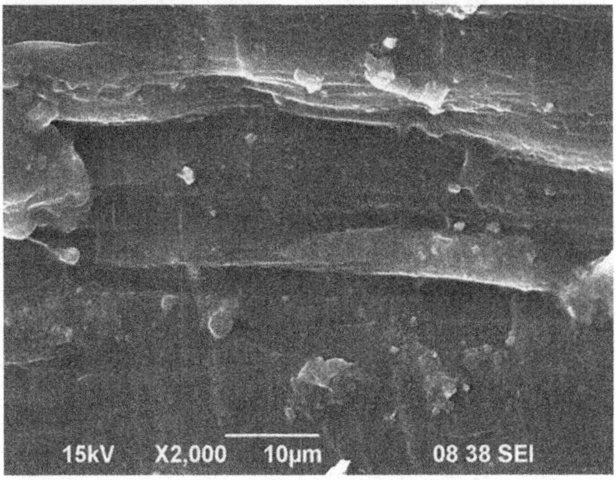

Fig. 3.189 Secondary electron image of a surface fragment of *Shahed-136* warhead mount element. Accelerating voltage is 15 kV, magnification ×2,000, scale mark—10 μm

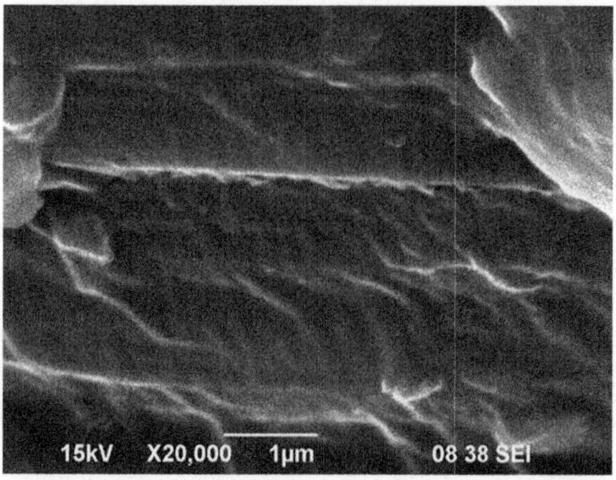

Fig. 3.190 Secondary electron image of a surface fragment of *Shahed-136* warhead mount element. Accelerating voltage is 15 kV, magnification ×20,000, scale mark—1 μm

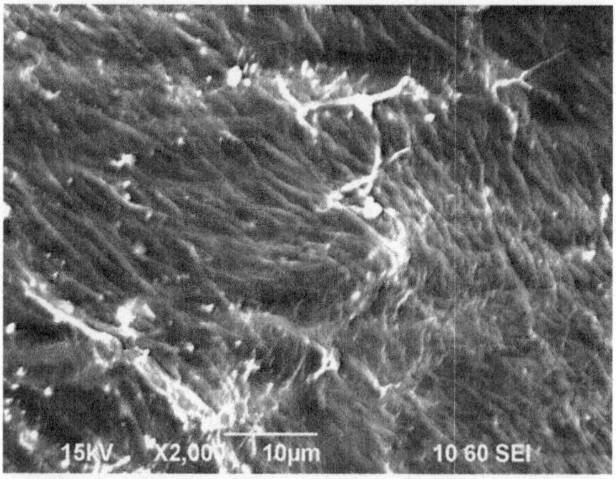

Fig. 3.191 Secondary electron image of a specimen cut. Accelerating voltage is 15 kV, magnification ×2,000, scale mark—10 μm

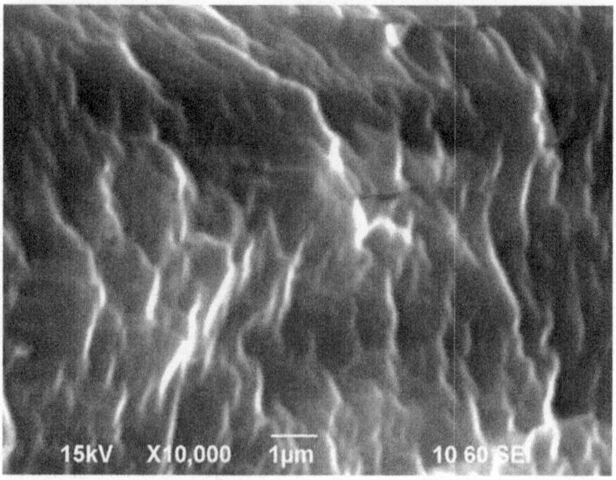

Fig. 3.192 Secondary electron image of the specimen slice. Accelerating voltage is 15 kV, magnification ×10,000, scale mark—10 μm

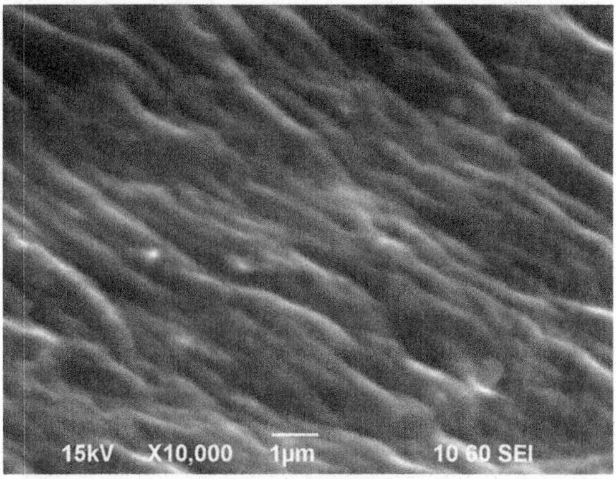

Fig. 3.193 Secondary electron image of the specimen slice. Accelerating voltage is 15 kV, magnification ×10,000, scale mark—10 μm (the images in Fig. 3.192 and that in Fig. 3.193 show different parts of the surface at the same magnification)

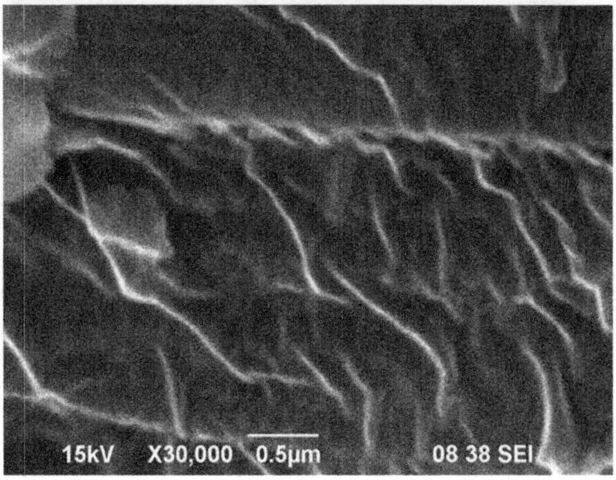

Fig. 3.194 Secondary electron image of the specimen slice. Accelerating voltage is 15 kV, magnification ×30,000, scale mark—0.5 μm

of characteristic X-ray emission. Figure 3.195 also presents the spectrum of characteristic X-ray emission of this measurement area.

Figure 3.196 shows the results of X-ray microanalysis on the surface of the specimen slice. The microanalysis was performed in a definite rectangular area. The measured concentrations of chemical elements were averaged within this area, which is shown in a secondary electron image. Chemical element concentrations are given in mass percentages. The concentrations of all chemical elements were measured on K lines of characteristic X-ray emission. Figure 3.196 also shows the spectrum of characteristic X-ray emission of this measurement area.

To perform the microanalysis, the specimen was coated with a thin (10–20 nm) platinum layer to provide the drainage of specimen electrical charge that appears when electrons irradiate non-conductive specimens. Therefore, the spectra of characteristic X-ray emission demonstrate platinum peaks (Figs. 3.195 and 3.196), but

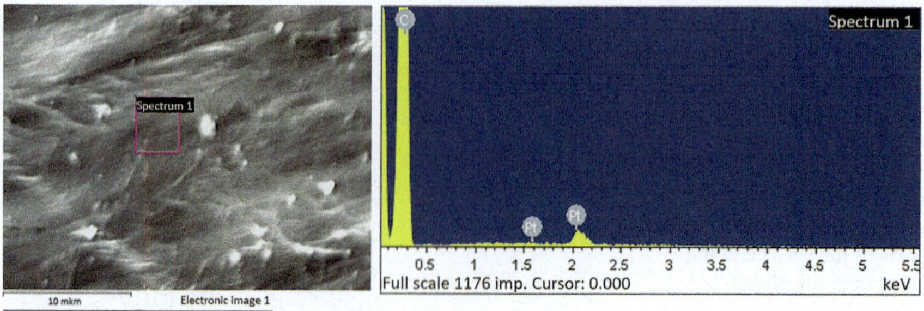

Element	Weight %
C K	100.00
Total	100.00

Fig. 3.195 Results of X-ray microanalysis on the specimen surface

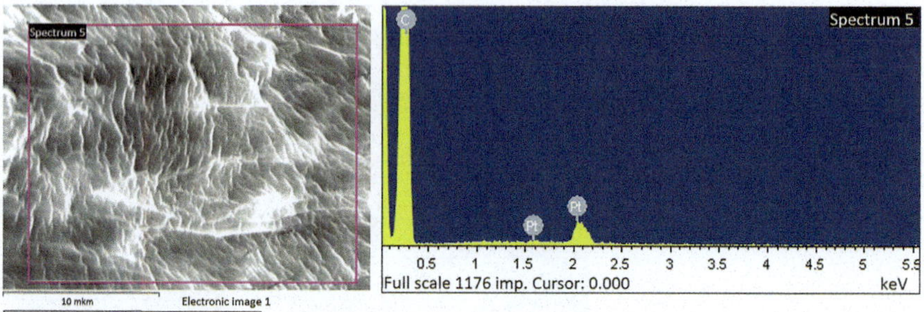

Element	Weight %
C K	100.00
Total	100.00

Fig. 3.196 Results of X-ray microanalysis on the surface of specimen slice

platinum concentration is not taken into account in the microanalysis results since platinum is not actually present in the specimen itself.

The examination of the structure and chemical composition of the fragment of *Shahed-136* warhead mount give reasons to consider that the material of this fragment is a polymer thermoplastic material whose molecules only consist of two elements—carbon and hydrogen.

3.3.4.3 Fragment of the Outer Coating of *Shahed-136* Loitering Munition Body

The examined fragment of the outer coating of *Shahed-136* loitering munition (LM) body is a flat multi-layer composite material (Fig. 3.197).

The main middle part of the sample is thicker than the grey margin along the fragment edge. In the main part, under the polymer layer reinforced with artificial

Fig. 3.197 Fragment of the outer coating of *Shahed-136* LM body (further—specimen) (both sides are shown)

№	Sign	Names of specimen material layers in section (the main part)
1		Black polymer layer
2		Polymer layer reinforced with artificial fabric
3		Polymer layer
4		Layer of porous honeycomb structure (honeycombs are empty, without filler)
5		Polymer layer
6		Polymer layer reinforced with artificial fabric
7		Grey coating layer

Fig. 3.198 Schematic presentation of the material of specimen's main part

fabric and polymer layer, a layer of porous honeycomb structure is observed (the honeycombs are empty, without filler). On the back side of the fragment, a continuous grey coating is observed.

The section of this material is schematically shown in Fig. 3.198.

Besides, a margin is situated along the specimen's edge (which is thinner than the main part of the fragment), in which porous honeycomb layer is absent. The section of this margin is schematically shown in Fig. 3.199.

Figures 3.200, 3.201, 3.202, and 3.203 present secondary and backscattered electron images of grey layer surface of the coating (covering one side of the examined specimen, see Fig. 3.197) at different magnifications.

№	Sign	Name of specimen material layers in section
1		Black polymer layer
2		Polymer layer reinforced with artificial fabric
3		Polymer layer
4		Polymer layer reinforced with artificial fabric
5		Grey polymer layer

Fig. 3.199 Schematic image of specimen margin section

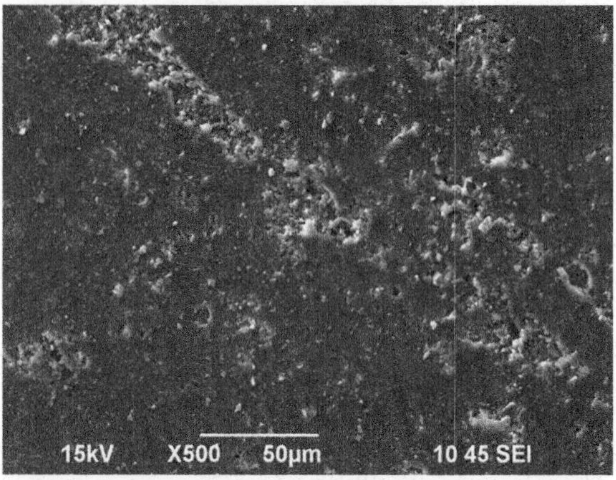

Fig. 3.200 Secondary electron image of grey coating layer surface. Accelerating voltage is 15 kV, magnification ×500, scale mark—50 μm

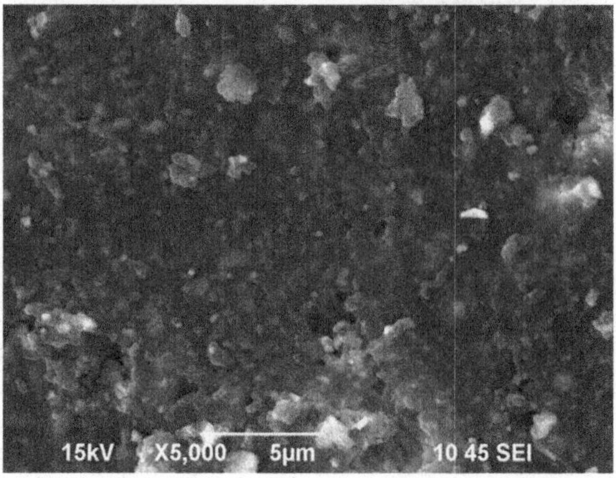

Fig. 3.201 Secondary electron image of grey surface coating layer. Accelerating voltage is 15 kV, magnification ×5,000, scale mark—5 μm

Fig. 3.202 Secondary electron image of grey coating layer surface. Accelerating voltage is 15 kV, magnification ×10,000, scale mark—1 μm

Fig. 3.203 Backscattered electron image of grey coating layer surface. Accelerating voltage is 15 kV, magnification ×5.000, scale mark—5 μm

Figures 3.204 and 3.205 present secondary electron images of artificial fabric layer surface at different magnifications. Figure 3.205 shows the sizes of artificial fabric fibers, which are of 8.2–8.62 μm size.

Figures 3.206 and 3.207 present secondary and backscattered (Fig. 3.207) electron images of matrix (binder) layer of the polymer reinforced with artificial fabric.

Figure 3.208 presents a secondary electron image of porous honeycomb layer (honeycombs are empty, without filler).

Figure 3.209 presents a secondary electron image of the surface of black polymer layer covering one of the sides of the studied specimen (Fig. 3.197). Surface relief is affected by the artificial fabric layer situated under it.

To study the structure and measure chemical element concentrations in the specimen bulk, as well as measure the thicknesses of specimen layers, a slice of the specimen mentioned was prepared. Backscattered electron images of the specimen margin slice at different magnifications are shown in Figs. 3.210 and 3.211.

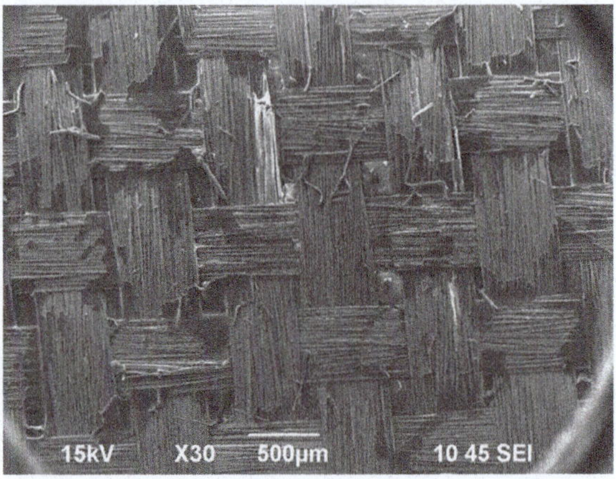

Fig. 3.204 Secondary electron image of artificial fabric layer. Accelerating voltage is 15 kV, magnification ×30, scale mark—500 μm

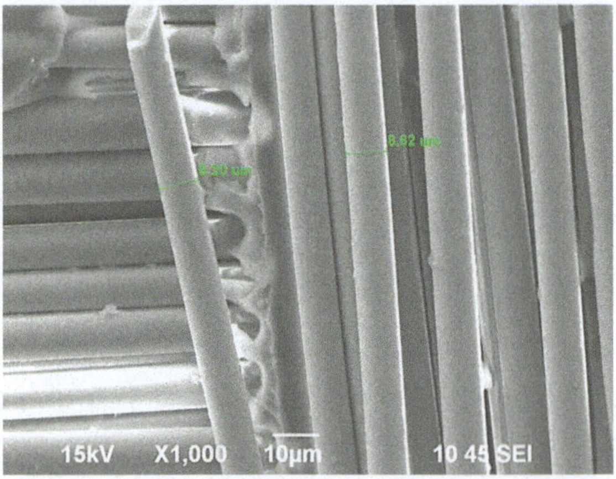

Fig. 3.205 Secondary electron image of artificial fabric layer. Accelerating voltage is 15 kV, magnification ×1,000, scale mark—10 μm. The image shows the sizes of artificial fabric fibers (8.2–8.62 μm)

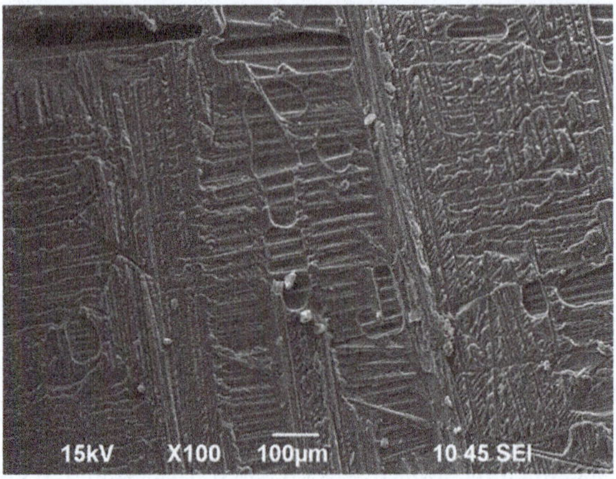

Fig. 3.206 Secondary electron image of the surface of matrix (binder) of the polymer layer reinforced with artificial fabric. Accelerating voltage is 15 kV, magnification ×100, scale mark—100 μm

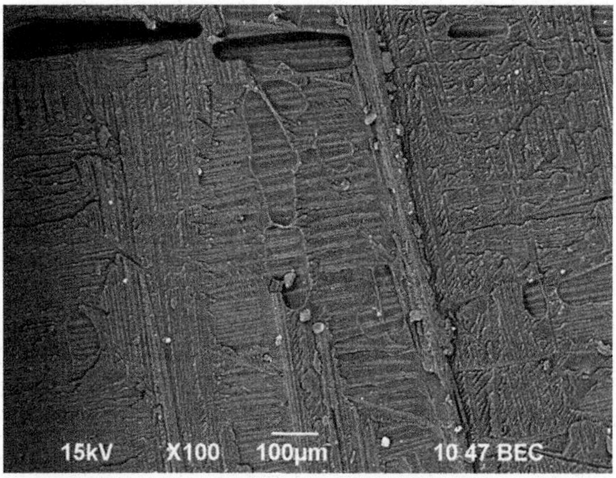

Fig. 3.207 Backscattered electron image of the surface of matrix (binder) of the polymer layer reinforced with artificial fabric. Accelerating voltage is 15 kV, magnification ×100, scale mark—100 μm

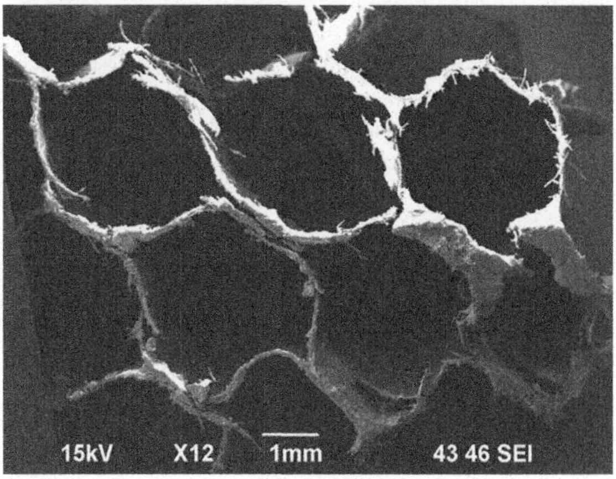

Fig. 3.208 Secondary electron image of honeycomb porous structure layer (honeycombs are empty, without filler). Accelerating voltage is 15 kV, magnification ×12, scale mark—1 mm

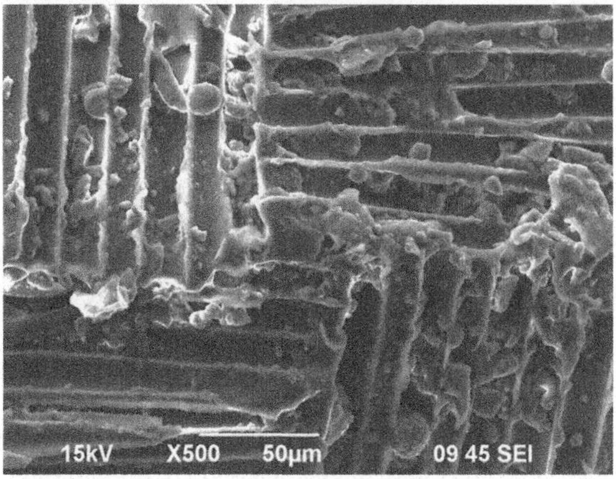

Fig. 3.209 Secondary electron image of black surface layer of the polymer coating one side of the examined specimen (Fig. 3.1//this figure?!). Accelerating voltage is 15 kV, magnification ×500, scale mark—50 μm

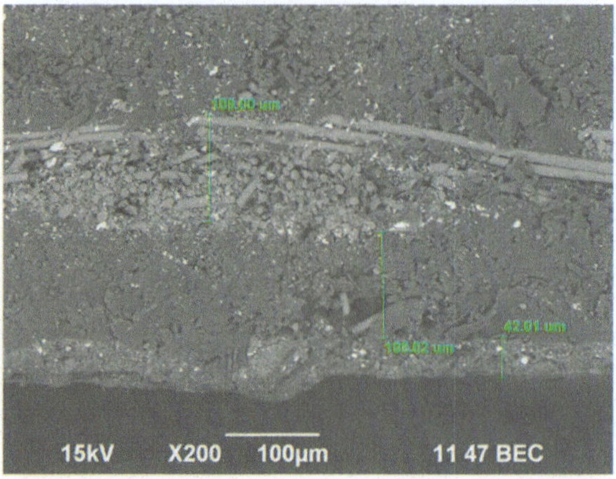

Fig. 3.210 Backscattered electron image of the specimen margin slice (at the bottom is the grey layer image). Magnification ×200, scale mark—100 μm

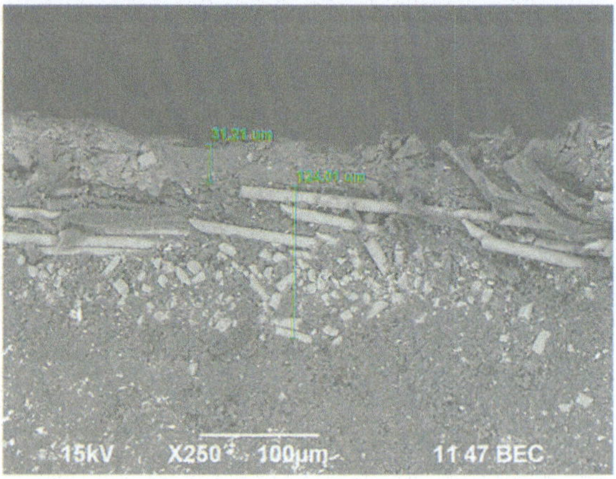

Fig. 3.211 Backscattered electron image of the specimen margin slice (at the top is the black layer image). Magnification ×250, scale mark—100 μm

Figure 3.212 shows the results of X-ray microanalysis on the surface of grey layer of the specimen. The microanalysis was performed in a definite rectangular area. The measured concentrations of chemical elements were averaged within this area, which is shown in the backscattered electron image. Chemical element concentrations are given in mass percentages. The concentrations of all chemical elements were measured on K lines of characteristic X-ray emission. Figure 3.212 also presents the spectrum of characteristic X-ray emission of this measurement area.

Figure 3.213 presents the results of X-ray microanalysis of the fibers of artificial fabric layer. The microanalysis was performed in a definite rectangular area. The measured concentrations of chemical elements were averaged within this area, which is shown in the backscattered electron image. Chemical element concentrations are given in mass percentages. The concentrations of all chemical elements were measured on K lines of characteristic X-ray emission. Figure 3.213 also presents the spectrum of characteristic X-ray emission of this measurement area.

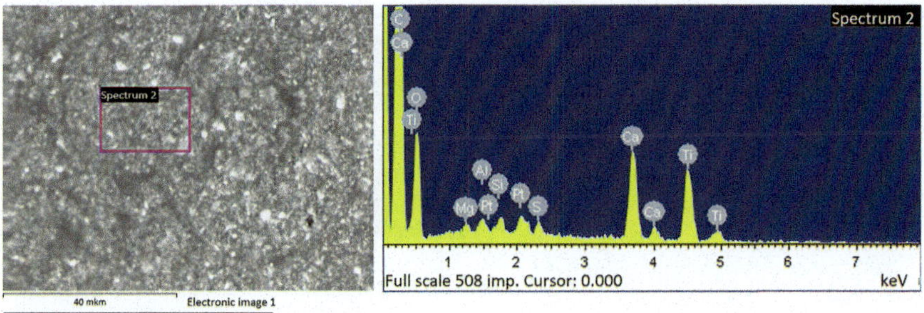

Element	Weight %
C K	60.81
O K	23.50
Mg K	0.25
Al K	0.43
Si K	0.33
S K	0.51
Ca K	5.92
Ti K	8.25
Total	100.00

Fig. 3.212 Results of X-ray microanalysis on the surface of grey specimen layer

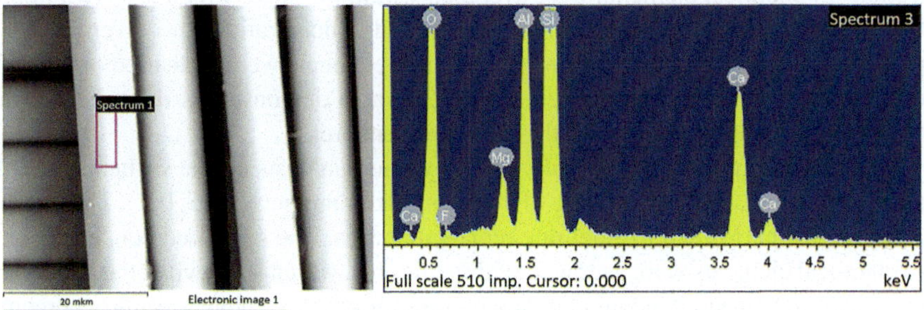

Element	Weight %
O K	49.82
Mg K	2.38
Al K	7.92
Si K	26.25
Ca K	13.63
Total	100.00

Fig. 3.213 Results of X-ray microanalysis of the fibers of artificial fabric layer

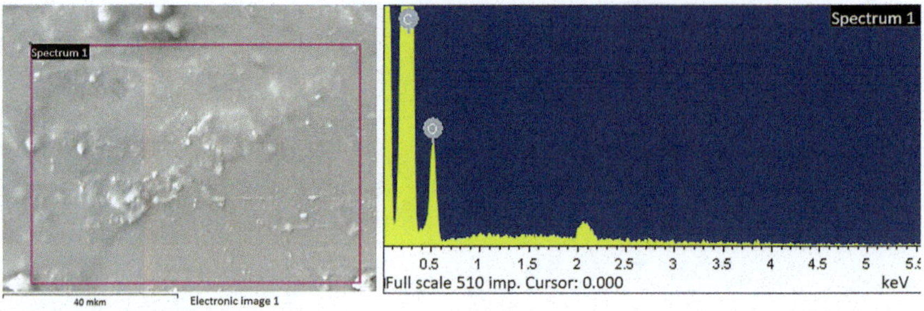

Element	Weight %
C K	83.00
O K	17.00
Total	100.00

Fig. 3.214 Results of X-ray microanalysis of porous honeycomb material

Figure 3.214 demonstrates the results of X-ray microanalysis of the material of honeycomb porous structure layer (honeycombs are empty, without filler). The microanalysis was performed in a definite rectangular area. The measured concentrations of chemical elements were averaged within this area, which is shown in the backscattered electron image. Chemical element concentrations are given in mass percentages. The concentrations of all chemical elements were measured on K lines of characteristic X-ray emission. Figure 3.214 also demonstrates the spectrum of characteristic X-ray emission of this measurement area.

Figure 3.215 presents the results of X-ray microanalysis on the surface of black polymer layer that covers one side of the examined fragment (Fig. 3.197). The microanalysis was performed in a definite rectangular area. The measured concentrations of chemical elements were averaged within this area, which is shown in the backscattered electron image. Chemical element concentrations are given in mass percentages. The concentrations of all chemical elements were measured on K lines of characteristic X-ray emission. Figure 3.215 also presents the spectrum of characteristic X-ray emission of this measurement area.

Figure 3.216 presents the results of X-ray microanalysis of the polymer placed below and above the layer of porous honeycomb structure (the main part of the fragment, see Fig. 3.198). The microanalysis was performed in a definite rectangular area. The measured concentrations of chemical elements were averaged within this area, which is shown in backscattered electrons. Chemical element concentrations are given in mass percentages. The concentrations of all chemical elements were measured on K lines of characteristic X-ray emission. Figure 3.216 also presents the spectrum of characteristic X-ray emission of this measurement area.

Figure 3.217 shows the results of X-ray microanalysis of matrix (binder) polymer layer reinforced with artificial fabric. The microanalysis was performed in a definite rectangular area. The measured concentrations of chemical elements were averaged within this area, which is shown in the backscattered electrons image of the fragment

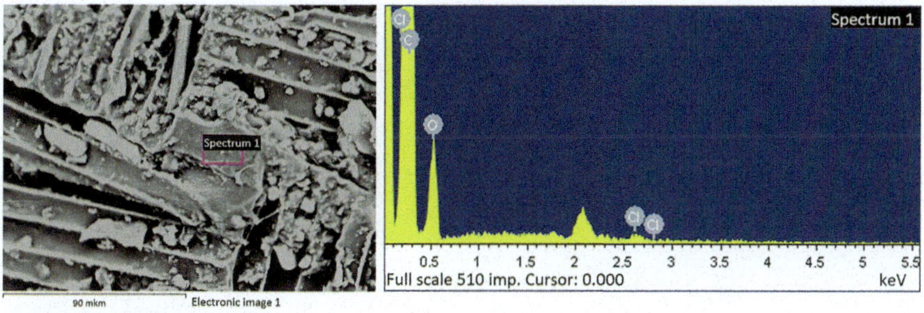

Element	Weight %
C K	81.84
O K	17.80
Cl K	0.36
Total	100.00

Fig. 3.215 Results of X-ray microanalysis on the surface of black polymer layer covering one of the sides of the examined fragment (Fig. 3.1)

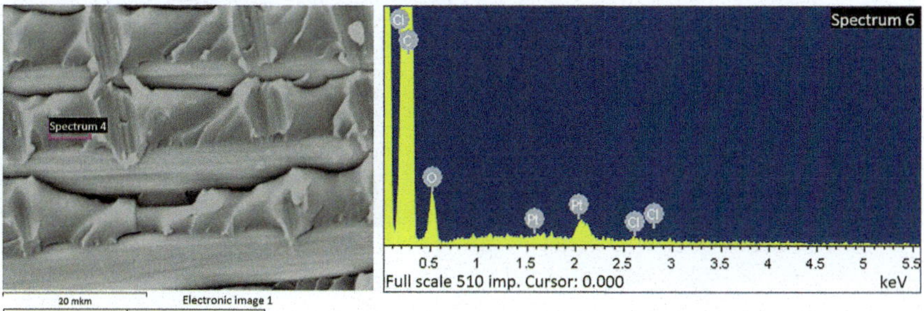

Element	Weight %
C K	86.08
O K	13.74
Cl K	0.18
Total	100.00

Fig. 3.216 Results of X-ray microanalysis of the polymer placed below and above porous honeycomb layer (the main part of the fragment, see Fig. 3.198)

margin slice. Chemical element concentrations are given in mass percentages. The concentrations of all chemical elements were measured on K lines of characteristic X-ray emission. Figure 3.217 also presents the spectrum of characteristic X-ray emission of this measurement area.

Figure 3.218 presents the results of X-ray microanalysis of the polymer layer situated between polymer layers reinforced with artificial fabric (Fig. 3.199). The microanalysis was performed in a definite rectangular area. The measured concentrations of chemical elements were averaged within this area, which is shown in the backscattered electron image of the fragment margin slice. Chemical element concentrations are given in mass percentages. The concentrations of all chemical

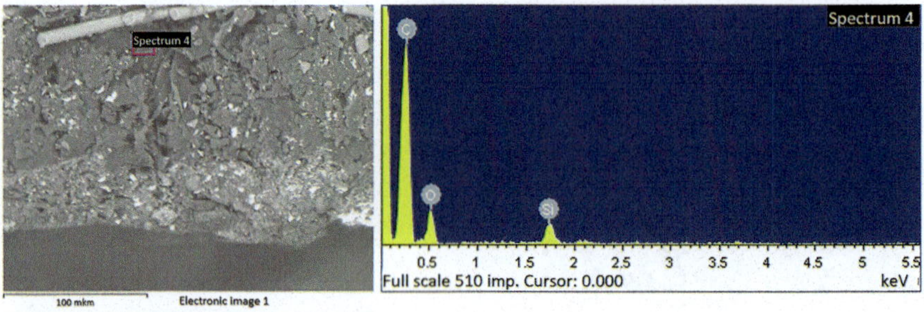

Element	Weight %
C K	88.38
O K	11.32
Si K	0.30
Total	100.00

Fig. 3.217 Results of X-ray microanalysis of the matrix (binder) of the polymer layer reinforced with artificial fabric

Element	Weight %
C K	87.47
O K	12.53
Total	100.00

Fig. 3.218 Results of X-ray microanalysis of the polymer layer situated between the polymer layers reinforced with artificial fabric (Fig. 3.199)

elements were measured on K lines of characteristic X-ray emission. Figure 3.218 also presents the spectrum of characteristic X-ray emission of this measurement area.

Conclusions

Thus, the fragment of the outer coating of *Shahed-136* LM body is a flat multi-layer composite material (Fig. 3.197). The main middle part of the fragment consists of seven layers and is thicker (6,020 μm) than the margin on the fragment edge, which consists of five layers and is 1,390 μm thick. In the bulk of the main part, below black polymer layer, that of the polymer reinforced with artificial fabric and the polymer layer, a layer of porous honeycomb material with empty honeycombs (of 3.2 mm

Table 3.4 Results of studying the structure of the specimen's main part

Name of the fragment of *Shahed-136* body outer coating (the main part)	Layer thickness, μm
Black polymer layer	50
Polymer layer reinforced with artificial fabric	300
Polymer layer	70
Layer of porous honeycomb layer (honeycombs are empty, without any filler)	5155
Polymer layer	70
Polymer layer reinforced with artificial fabric	300
Grey coating layer	75

diameter) is observed. In the margin bulk, the porous honeycomb layer is absent. A continuous grey coating can be seen on the fragment back.

A table with the names of the main part layers, with layer thicknesses shown, is given below. The order of layers in the table corresponds to the actual layer placement in the main part of the fragment.

Grey coating layer has the following chemical composition and chemical element concentrations: C—60.81%, O—23.50%, Mg—0.25%, Al—.43%, Si—0.33%, S—0.51%, Ca—5.92%, and Ti—8.25%.

The polymer layers reinforced with artificial fabric consist of a composite whose filler is glass fabric with 8—10 μm glass fibers, and its matrix (binder) is a polymer. The chemical composition of glass fibers is the following: Si—26.25%, O—49.82%, Mg—2.38%, Al—7.92%, and Ca—13.63%. The chemical composition of the matrix (binder) is: C—88.38%, O—11.32%, and Si—0.30% (Table 3.4).

Polymer coatings situated above and below the porous honeycomb layer (their thickness is 70 μm) have such chemical composition: C—86.08%, O—13.74%, and Cl—0.18%.

The material of porous honeycomb layer is a polymer and consists of such chemical elements with respective concentrations: C—83.00% and O—17.00%.

Black polymer layer covering one of the examined fragment sides (Fig. 3.197) has such chemical composition: C—81.84%, O—17.80%, and Cl—0.36.

A table showing the names of fragment margins and layer thicknesses is presented below. The layer placement order in the table corresponds to the actual layer placement in the fragment (Table 3.5).

Grey coating layer is common for the fragment main part and its margin (in Fig. 3.197, a continuous grey coating can be seen on the fragment back) and, naturally, it has the same chemical composition and the same chemical element concentrations for both measurement areas.

The layers reinforced with artificial fabric are also common for the fragment main part and its margin; they consist of a composite whose filler is glass fabric with 8–10 μm glass fibers and its matrix (binder)—a polymer. The chemical composition of glass fibers is as follows: Si—26.25%, O—49.82%, Mg—2.38%, Al—7.92%, and

Table 3.5 Results of studying the structure of the specimen's main part

Names of margin layers in the outer coating fragment of *Shahed-136* body	Layer thickness, μm
Black polymer layer	50
Polymer layer reinforced with artificial fabric	300
Polymer layer	665
Polymer layer reinforced with artificial fabric	300
Grey coating layer	75

Ca—13.63%. Matrix (binder) chemical composition is: C—88.38%, O—11.32%, and Si—0.30%.

The polymer layer located between the polymer layers reinforced with artificial fabric has such chemical composition: C—87.47% and O—12.53%.

Black coating layer is common for the main part of the fragment and its margin (Fig. 3.197 shows a continuous black coating); it has the same chemical composition and similar concentrations of chemical elements: C—81.84%, O—17.80%, and Cl—0.36%.

3.3.4.4 Fragment of *Shahed-136* Loitering Munition (LM) Body Hatch

A fragment of the body hatch (A-720) of *Shahed-136* LM is a flat triangular structural element manufactured from multi-layered composite material. The images of its both sides are shown in Fig. 3.219.

Fig. 3.219 Images of both sides of the fragment of Shahed-136 LM body hatch (further—specimen)

№	Material symbol	Names of specimen material layers in section
1		Polymer layer reinforced with coated artificial fabric
2		Polymer layer
3		Porous honeycomb layer (honeycombs are empty, without filler)
4		Polymer layer
5		Polymer layer reinforced with artificial fabric
6		Grey coating layer

Fig. 3.220 Schematic presentation of the section of the main part of specimen material

№	Material symbol	Name of specimen layer materials in section
1		Black polymer layer
2		Layer reinforced with artificial fabric
3		Polymer layer
6		Layer reinforced with artificial fabric
7		Grey coating layer

Fig. 3.221 Schematic presentation of specimen edge material in section

The main middle part of the specimen is thicker than the dark-colored edge around the fragment perimeter. In the main part, under polymer layer reinforced with an artificial fabric with a coating and a polymer layer, a porous honeycomb layer (with empty honeycombs, without filler) is observed. A continuous grey coating can be seen on the fragment's back (Fig. 3.219).

A section of this material is schematically shown in Fig. 3.220.

Besides, around the specimen perimeter there is an edge (thinner than the fragment main part) in whose composition porous honeycomb layer is absent. The section of this edge is schematically shown in Fig. 3.221.

Figures 3.222, 3.223, and 3.224 present secondary electron images of grey coating layer surface, and Fig. 3.225—backscattered electron image at different magnifications. On the right-hand side of Fig. 3.225, the next (to grey coating layer) artificial fabric layer can be seen.

Figure 3.226 demonstrates secondary electron image of artificial fabric layer, and Figs. 3.227 and 3.228 show its backscattered electron images at different magnifications. Figure 3.220 shows this layer above grey coating layer.

Figures 3.229 and 3.230 present secondary electron images of the outer surface of coated artificial fabric at different magnifications. In Fig. 3.220, this layer is topmost.

Figure 3.231 presents secondary electron image of porous honeycomb structure (honeycombs are empty, without filler).

To study the structure and determine the concentrations of chemical elements in the specimen bulk, as well as measure material layer thicknesses, a slice of the

Fig. 3.222 Secondary electron image of grey coating layer surface. Accelerating voltage is 15 kV, magnification ×500, scale mark—50 μm

Fig. 3.223 Secondary electron image of grey coating layer surface. Accelerating voltage is 15 kV, magnification ×500, scale mark—5 μm

Fig. 3.224 Secondary electron image of grey coating layer surface. Accelerating voltage is 15 kV, magnification ×10,000, scale mark—1 μm

Fig. 3.225 Backscattered electron image of grey coating layer surface. Accelerating voltage is 15 kV, magnification ×1,000, scale mark—10 μm

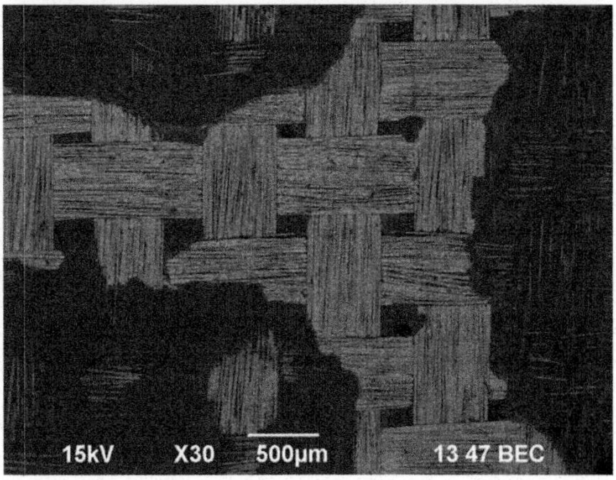

Fig. 3.226 Backscattered electron image of artificial fabric layer. Accelerating voltage is 15 kV, magnification ×30, scale mark—500 μm. Black polymer residues can be seen in the image

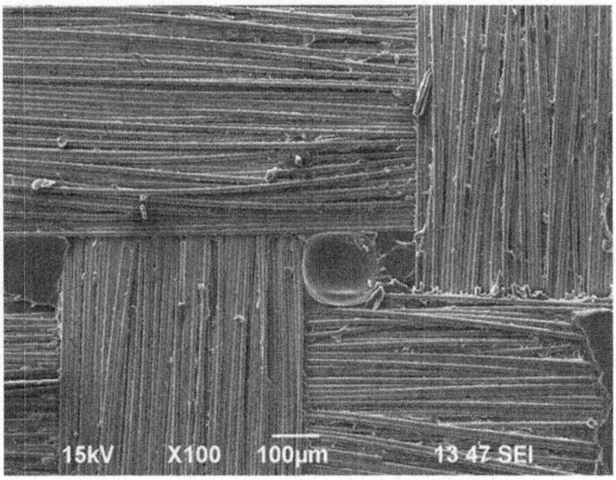

Fig. 3.227 Secondary electron image of artificial fabric layer. Accelerating voltage is 15 kV, magnification ×100, scale mark—100μm

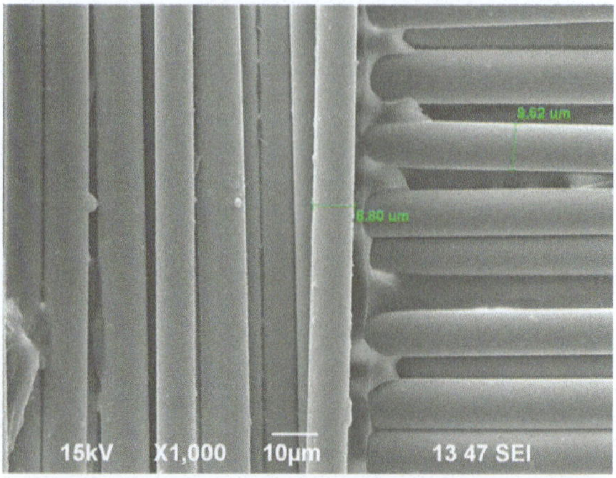

Fig. 3.228 Secondary electron image of artificial fabric. Accelerating voltage is 15 kV, magnification ×1,000, scale mark—10 μm. The size of artificial fabric fibers is shown (8.8–9.6 μm)

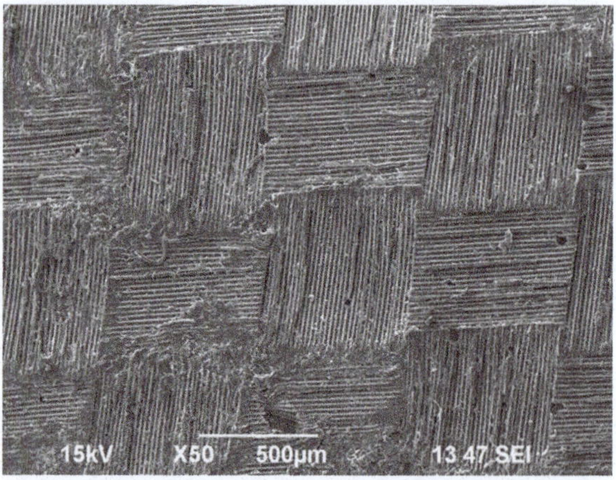

Fig. 3.229 Secondary electron image of coated artificial fabric (the topmost layer in Fig. 3.220). Accelerating voltage is 15 15 kV, magnification ×50, scale mark—500 μm

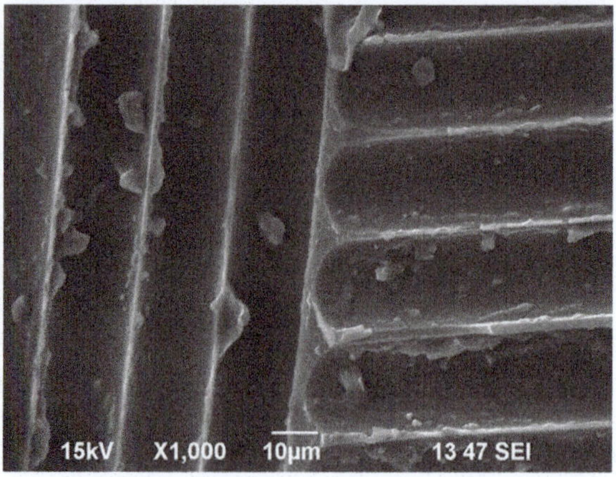

Fig. 3.230 Secondary electron image of the outer surface of artificial fabric layer (the topmost layer in Fig. 3.220). Accelerating voltage is 15 kV, magnification ×1,000, scale mark—10 μm

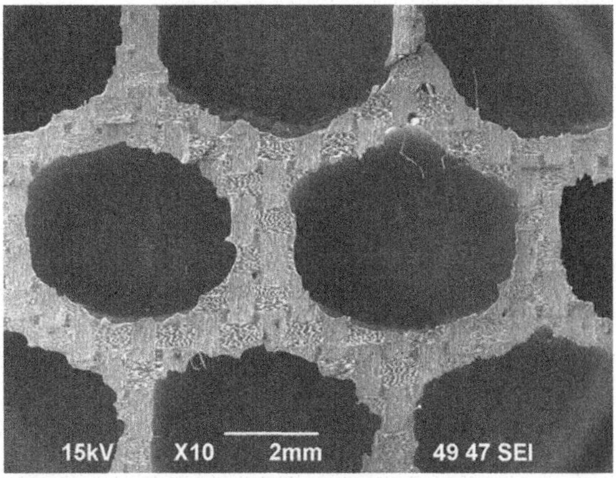

Fig. 3.231 Secondary electron image of the layer of porous honeycomb structure. Honeycombs are empty, without filler. Accelerating voltage is 15 kV, magnification ×10, scale mark—2 mm

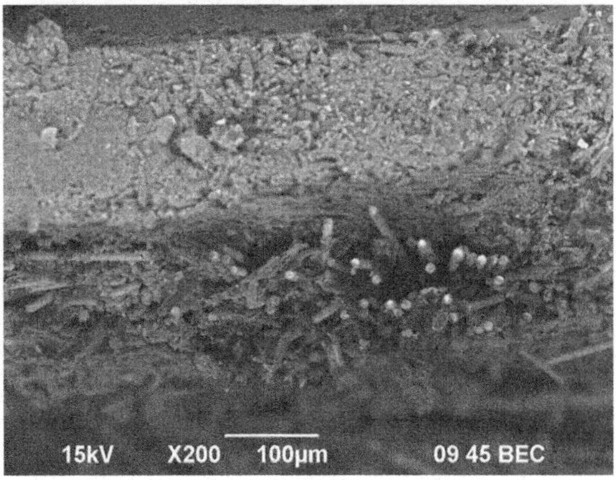

Fig. 3.232 Backscattered electron image of specimen's main part. Accelerating voltage is 15 kV, magnification ×200, scale mark—100 µm

specimen mentioned was prepared. Backscattered electron images of the slice of specimen's main part at different magnifications are shown in Figs. 3.232 and 3.233.

Figure 3.234 demonstrates backscattered electron image of the specimen edge (above grey coating in the image).

Figure 3.235 presents backscattered electron image of the specimen edge slice (black polymer layer at the image bottom).

Figure 3.236 presents the results of X-ray microanalysis on the surface of grey coating surface. The microanalysis was performed in a definite rectangular area. The measured concentrations of chemical elements were averaged within this area, which is shown in backscattered electron image. Chemical element concentrations are given in mass percentages. The concentrations of all chemical elements were measured on K lines of characteristic X-ray emission. Figure 3.236 also shows the spectrum of characteristic X-ray emission of this measurement area.

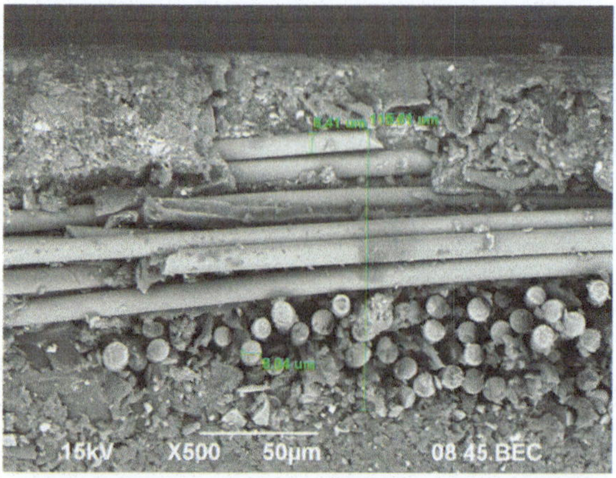

Fig. 3.233 Backscattered electron image of specimen's main part. Accelerating voltage is 15 kV, magnification ×500, scale mark—50 μm

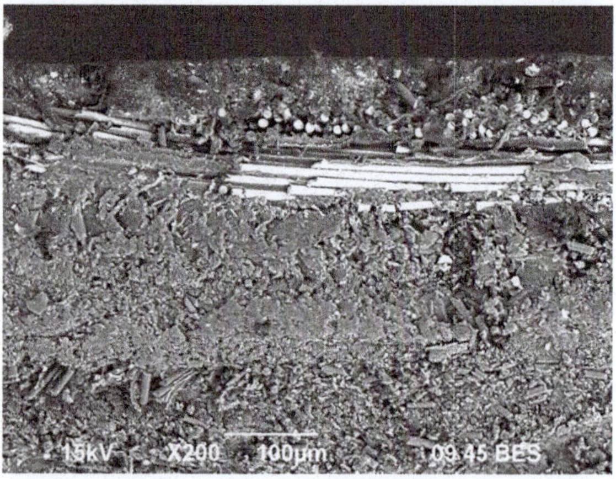

Fig. 3.234 Backscattered electron image of specimen edge slice (shown above grey coating in the image). Accelerating voltage is 15 kV, magnification ×200, scale mark—100 μm

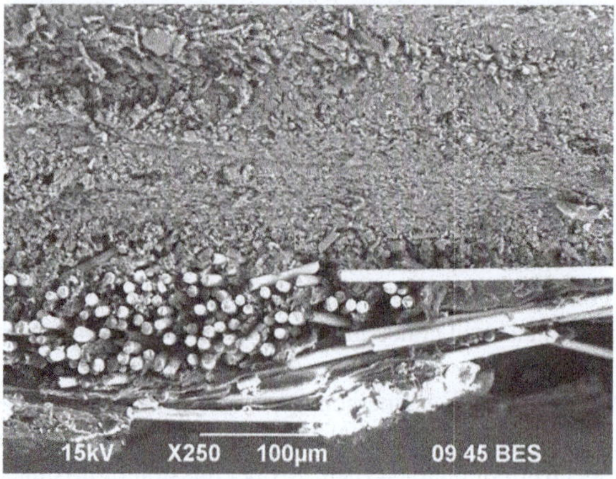

Fig. 3.235 Backscattered electron image of the edge fragment slice (black polymer layer at the image bottom). Accelerating voltage is 15 kV, magnification ×250, scale mark—100 μm

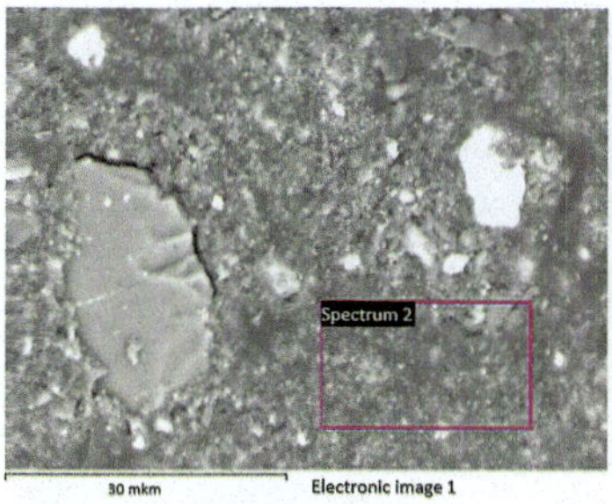

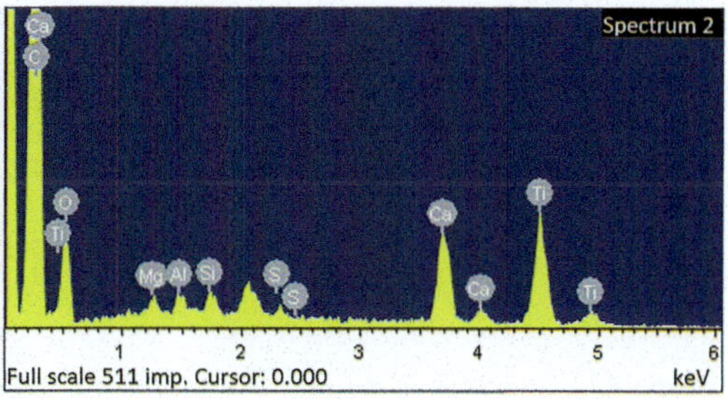

Element	Weight %
C K	67.02
O K	15.31
Mg K	0.45
Al K	0.48
Si K	0.53
S K	0.20
Ca K	5.69
Ti K	10.32
Total	100.00

Fig. 3.236 Results of X-ray microanalysis on the surface of grey coating layer

Light-colored particles are observed in backscattered electron image (Fig. 3.236) of grey coating surface. Figure 3.237 presents the results of X-ray microanalysis on the surface of grey coating layer in the region of light-colored particles. The microanalysis was performed in a definite rectangular area. The measured concentrations of chemical elements were averaged within this area, which is shown in backscattered electron image. Chemical element concentrations are given in mass percentages. The concentrations of chemical elements were measured on K lines of characteristic X-ray emission (except that of Ba (barium), whose concentration was measured on K line). Figure 3.236 also presents the spectrum of characteristic X-ray emission of this measurement area.

Figure 3.238 presents the results of microanalysis on the surface of black coating layer observed on specimen edge. The microanalysis was performed in a definite rectangular area. The measured concentrations of chemical elements were averaged within this area, which is shown in backscattered electron image. Chemical element concentrations are given in mass percentages. The concentrations of all chemical elements were measured on K lines of characteristic X-ray emission. Figure 3.238 also presents the spectrum of characteristic X-ray emission of this measurement area.

Figure 3.239 shows the results of X-ray microanalysis of the fibers of artificial fabric layer. The microanalysis was performed in a definite rectangular area. The measured concentrations of chemical elements were averaged within this area, which is shown in backscattered electron image. Chemical element concentrations are given in mass percentages. The concentrations of all chemical elements were measured on

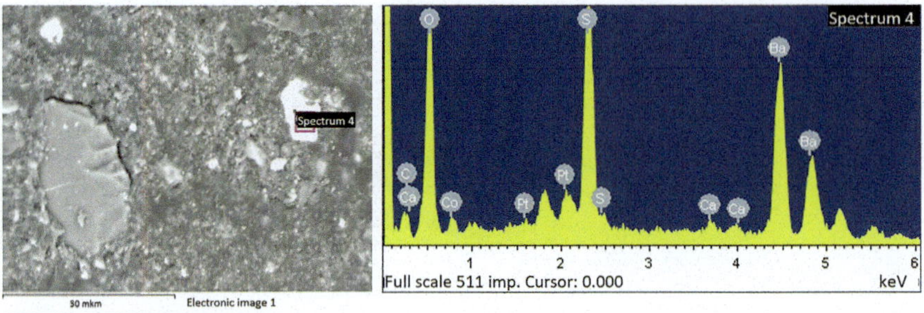

Element	Weight %
C K	9.23
O K	21.69
S K	13.05
Ca K	0.74
Co K	0.01
Ba L	59.86
Total	100.00

Fig. 3.237 Results of X-ray microanalysis on the surface of grey coating layer in the region of light-colored particles. It can be seen that light-colored particles have high Ba concentration (about 60 %)

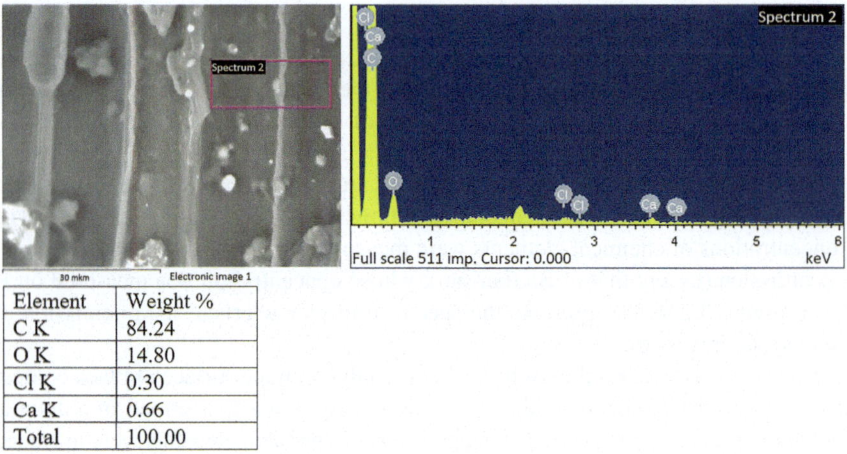

Element	Weight %
C K	84.24
O K	14.80
Cl K	0.30
Ca K	0.66
Total	100.00

Fig. 3.238 Results of X-ray microanalysis on the surface of black coating layer observed on the fragment edge of *Shahed-136* body hatch

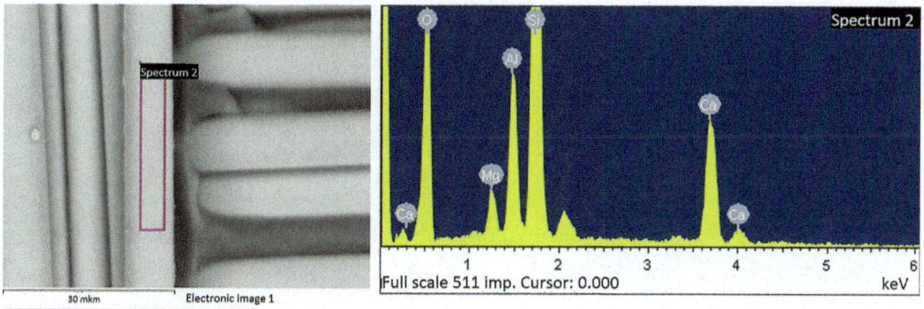

Element	Weight %
O K	49.48
Mg K	2.40
Al K	7.73
Si K	26.02
Ca K	14.36
Total	100.00

Fig. 3.239 Results of X-ray microanalysis of the fibers of artificial fabric layer

K lines of characteristic X-ray emission. Figure 3.239 also shows the spectrum of characteristic X-ray emission of this emission area.

Figure 3.240 presents the results of X-ray microanalysis of the polymer located below and above porous honeycomb layer (the main part of the fragment, see Fig. 3.220). The microanalysis was performed in a definite rectangular area. The measured concentrations of chemical elements were averaged within this area, which is shown in backscattered electron image. Chemical element concentrations are given in mass percentages. The concentrations of all chemical elements were measured on K lines of characteristic X-ray emission. Figure 3.240 also presents the spectrum of characteristic X-ray emission of this measurement area.

Figure 3.241 presents the results of X-ray microanalysis of the outer surface of artificial fabric layer (the topmost layer in Fig. 3.220). The microanalysis was performed in a definite rectangular area. The measured concentrations of chemical elements were averaged within this area, which is shown in backscattered electron image. Chemical element concentrations are given in mass percentages. The concentrations of all chemical elements were measured on K lines of characteristic X-ray emission. Figure 3.241 also presents the spectrum of characteristic X-ray emission of this measurement area.

Figure 3.242 illustrates the results of X-ray microanalysis of porous honeycomb layer material (honeycombs are empty, without filler). The microanalysis was performed in a definite rectangular area. The measured concentrations of chemical elements were measured within this area, which is shown in backscattered electron image. Chemical element concentrations are given in mass percentages. The concentrations of chemical elements were measured on K lines of characteristic X-ray emission. Figure 3.242 also shows the spectrum of characteristic X-ray emission of this measurement area.

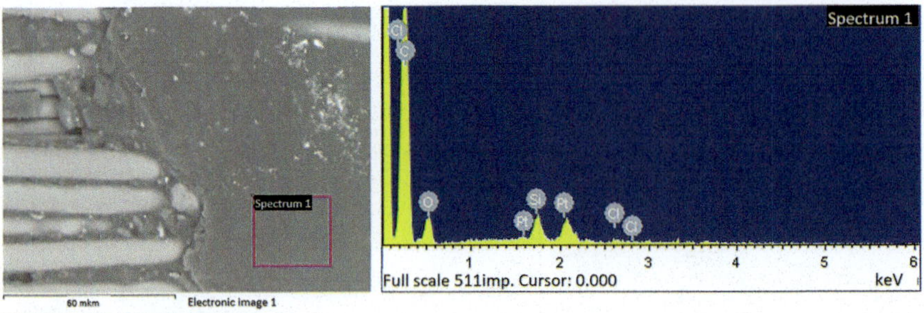

Element	Weight %
C K	80.66
O K	17.03
Si K	1.86
Cl K	0.46
Total	100.00

Fig. 3.240 Results of X-ray analysis of the polymer located below and above porous honeycomb layer

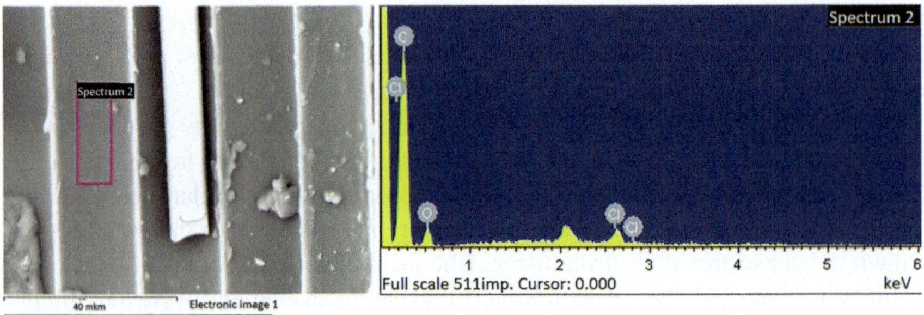

Element	Weight %
C K	83.01
O K	14.59
Cl K	2.41
Total	100.00

Fig. 3.241 Results of microanalysis of the outer surface of artificial fabric layer (the topmost layer in Fig. 3.220)

Figure 3.243 presents backscattered electron image of fragment edge slice with 80 μm scale mark, as well as the image of the same area in characteristic X-ray emission (the spatial distribution of chemical elements over the examined specimen slice). White dots correspond to the presence of the chemical element denoted below each image.

Figure 3.244 demonstrates the results of X-ray microanalysis of the polymer layer located between polymer layers reinforced with artificial fabric (Fig. 3.221). The microanalysis was performed in a definite rectangular area. The measured concentrations of chemical elements were averaged within this area, which is shown

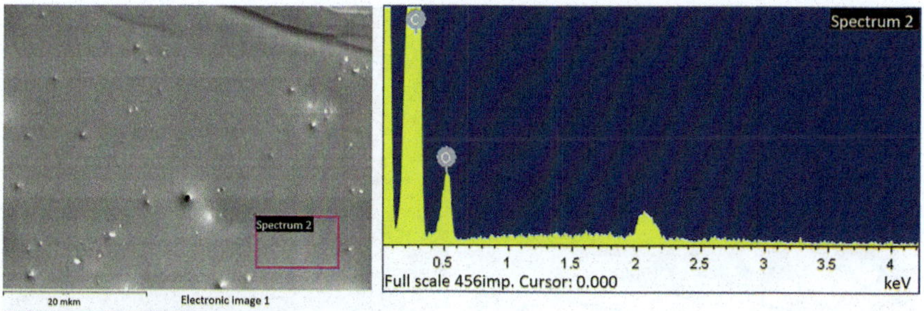

Element	Weight %
C K	82.60
O K	17.40
Total	100.00

Fig. 3.242 Results of X-ray microanalysis of the material of porous honeycomb material

in backscattered electron image of the fragment edge slice. Chemical element concentrations are given in mass percentages. The concentrations of chemical elements were measured on K lines of characteristic X-ray emission. Figure 3.244 also shows the spectrum of characteristic X-ray emission of this measurement area.

Figure 3.245 presents the results of X-ray microanalysis of the matrix (binder) substance layer reinforced with artificial fabric. The microanalysis was performed in a definite rectangular area. The measured concentrations of chemical elements were averaged within this area, which is shown in backscattered electron image of the fragment edge slice. Chemical element concentrations are given in mass percentages. The concentrations of all chemical elements were measured on K lines of characteristic X-ray emission. Figure 3.245 also demonstrates the spectrum of characteristic X-ray emission of this measurement area.

Conclusions

The fragment of *Shahed-136* body hatch is manufactured from a multi-layered composite material (Fig. 3.220). The dark-colored edge middle part of the fragment consists of six layers and is thicker (3620 μm) than the dark-colored edge around its perimeter, which consists of five layers and is 850 μm thick. In the bulk of the main part, under the layer of coated artificial fabric and polymer layer, a porous honeycomb layer is observed (honeycombs are empty, without filler, their diameter is 4.7 mm). In the bulk of dark-colored edge, there is no porous honeycomb layer. A continuous grey coating is present at the fragment's back.

The table showing the names of the main part layers and their thicknesses is presented below. The order of layers in the table corresponds to their actual places in the main part of the fragment (Table 3.6).

Grey coating layer has the following chemical composition and such chemical element concentrations: C—67.02%, O—15.31%, Mg—0.45%, Al—0.48%, Si—0.53%, S—0.20%, Ca—5.69%, and Ti—10.32%. Besides, particles of about 5 μm size and having high Ba (barium) concentration are noticed on grey coating surface (Fig. 3.237).

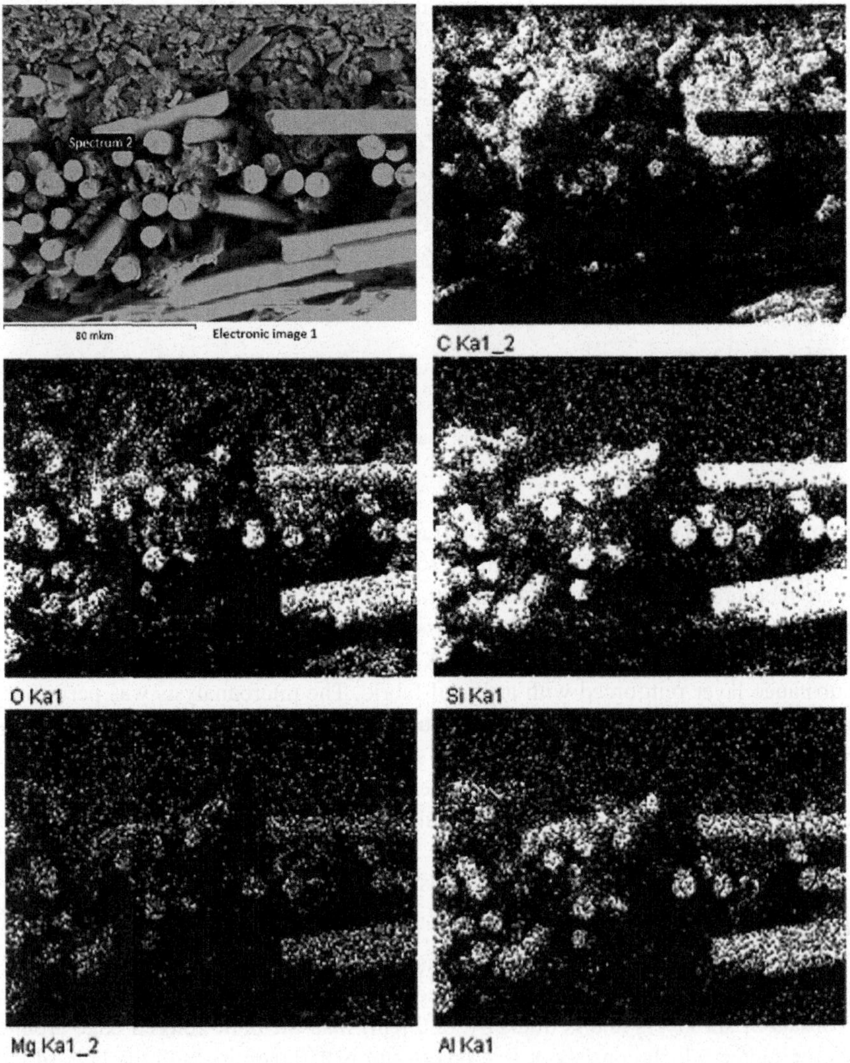

Fig. 3.243 Spatial distribution of chemical elements over fragment edge slice. White dots correspond to the presence of the chemical element denoted below each image. It is seen from the figure that Si, Al, O, Mg have the same spatial distribution

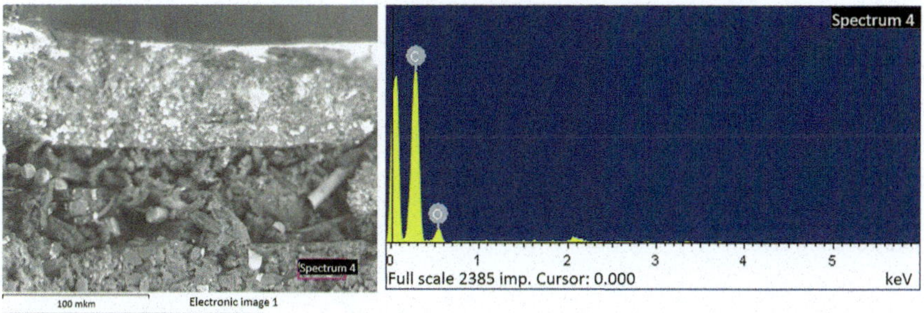

Element	Weight %
C K	83.85
O K	16.15
Total	100.00

Fig. 3.244 Results of X-ray microanalysis of the layer of polymer located between artificial fabric layers (Fig. 3.221). The topmost layer in backscattered electron image is grey coating layer, situated under it is artificial fabric layer

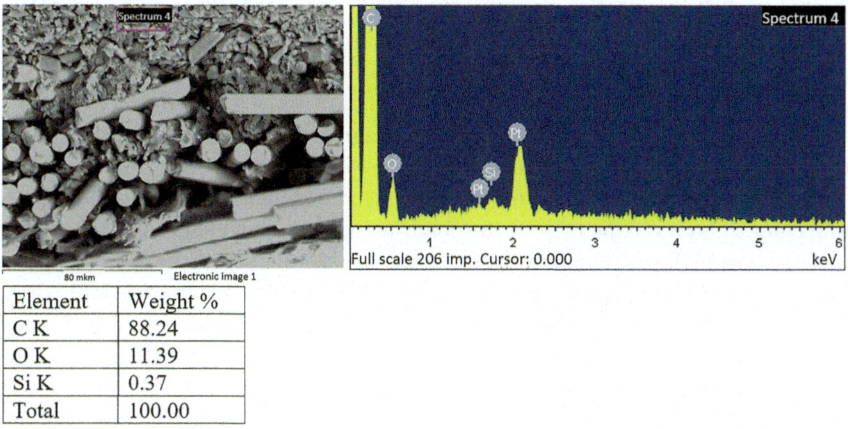

Element	Weight %
C K	88.24
O K	11.39
Si K	0.37
Total	100.00

Fig. 3.245 Results of X-ray microanalysis of matrix (binder) material of the layer reinforced with artificial fabric

Table 3.6 Results of studying the structure of the material of specimen's main part

Layer names	Layer thickness, μm
Polymer layer reinforced with coated artificial fabric	350
Polymer layer	70
Porous honeycomb layer (honeycombs are empty, without filler)	125
Polymer layer	70
Polymer layer reinforced with artificial fabric	300
Grey coating layer	75

Table 3.7 Results of studying the structure of the material of specimen's main part

Name of fragment edge layer	Layer thickness, μm
Black polymer layer	50
Polymer layer reinforced with artificial fabric	300
Polymer layer	483
Polymer layer reinforced with artificial fabric	300
Grey coating layer	75

Polymer layers reinforced with artificial fabric are a composite whose filler is a glass fabric with 8–10 μm glass fibers, and its matrix (binder) is a polymer. The chemical composition of glass fibers: Si—26.02%, O—49.48%, Mg—2.40%, Al—7.73%, and Ca—14.36%. Chemical composition of the polymer: C—88.24%; O—11.39%; Si—0.37%.

The outer coating of polymer layer reinforced with glass fabric (the layer is 145 μm thick) has such chemical composition: C—83.01%, O—14.59%, and Cl—2.41%.

Polymer layers located below and above porous honeycomb layer (they are 70 μm thick) have the following chemical composition: C—80.66%, O—17.03%, Si—1.86%, and Cl—0.46%.

The material of porous honeycomb layer is a polymer and consists of such chemical elements in respective concentrations: C—82.60% and O—17.40%.

Table 3.7 presents the names of fragment edge layers and shows their thicknesses. The order of layers in the table corresponds to their actual location in the fragment.

Grey coating layer is common for the main part of the fragment and its edge (in Fig. 3.219 a continuous grey coating can be noticed on the fragment back) which, naturally, has the same chemical composition and chemical element concentrations.

The layers reinforced with artificial fabric are also identical for the fragment's main part and fragment's edge. They are a composite whose filler is glass fabric with glass fibers 9–10 μm in size, and its matrix is a polymer. Chemical composition of fiberglass fibers is : Si—26.02%, O—49.48%, Mg—2.40%, Al—7.73%, and Ca—14.36%. Chemical composition of the polymer: C—88.24%, O—11.39%, and Si—0.37%.

The polymer layer located between artificial fabric layers has such chemical composition: C—83.85% and O—16.15%.

Black coating layer observed on the fragment of the edge of *Shahed-136* body hatch is a polymer of the following chemical composition: C—84.24%, O—14.80%, Cl—0.30%, and Ca—0.66%.

Fig. 3.246 Convex surface of the fragment. The surface is painted black

3.3.4.5 Fragment of Upgraded *Shahed-136* LM Body

The fragment of upgraded *Shahed-136* LM body is an arched specimen that has convex (Fig. 3.246) and concave surfaces (Fig. 3.247, left). The convex surface is covered with black paint (Fig. 3.246). In the fragment bulk, there are two composite layers whose fillers consist of artificial fabric. A porous honeycomb layer with empty honeycombs is located between those layers (Fig. 3.247, right). Hence, the above-mentioned fragment is a multi-layered structure schematically shown in section in Fig. 3.248.

On the arched surface of the LM body fragment, there are areas where black paint layer is damaged and where the presence of a composite layer with artificial fabric filler, as well as fragments of fabric-forming fibers, can be seen under the paint layer.

Figure 3.249 presents secondary electron image of the surface of black paint layer.

Figure 3.250 demonstrates secondary electron image of an area of composite with artificial fabric filler, which is located under paint layer. Figure 3.251 shows backscattered electron image of the fragments of fabric-forming fibers.

Figure 3.252 presents secondary electron image of an area of the composite with artificial fabric filler, which is located below porous honeycomb layer. Figure 3.253 shows secondary electron image of the fragments of fibers that form this artificial fabric.

Figure 3.254 presents secondary electron image of an area of porous honeycomb layer.

Fig. 3.247 Concave surface and a part of porous honeycomb layer (right)

№	Material symbol	Names of specimen layers in section
1		Black paint layer (convex surface)
2		Polymer layer reinforced with artificial fabric
4		Porous honeycomb layer (honeycombs are empty, without filler)
6		Polymer layer reinforced with artificial fabric

Fig. 3.248 Schematic presentation of the section of specimen material

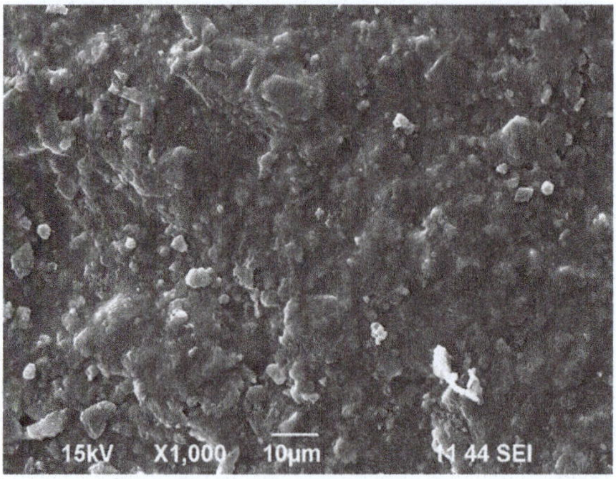

Fig. 3.249 Secondary electron image of the surface of black paint layer. Accelerating voltage is 15 kV, magnification ×1000, scale mark—10 μm

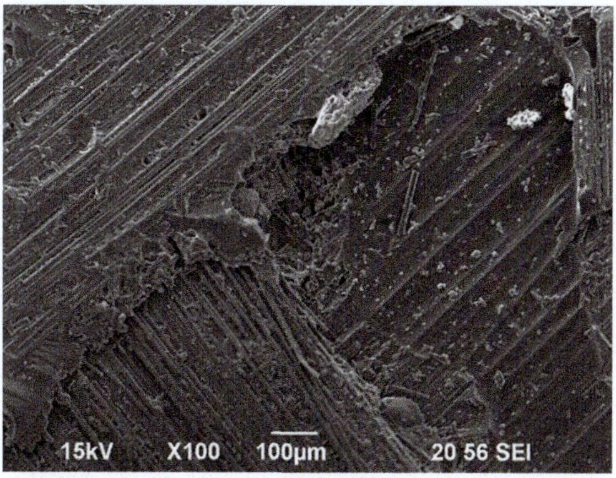

Fig. 3.250 Secondary electron image of an area of composite with artificial fabric filler. Accelerating voltage is 15 kV, magnification ×100, scale mark—100 μm

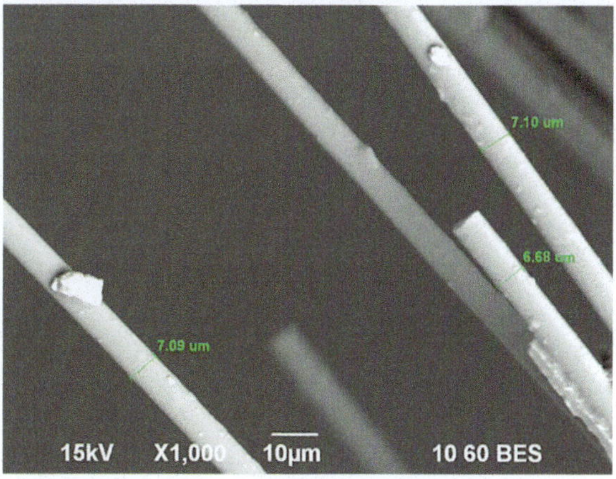

Fig. 3.251 Backscattered electron image of fragments of the fibers forming artificial fabric. Acceleration voltage is 15 kV, magnification ×1000, scale mark—10 μm. The image shows fiber thicknesses

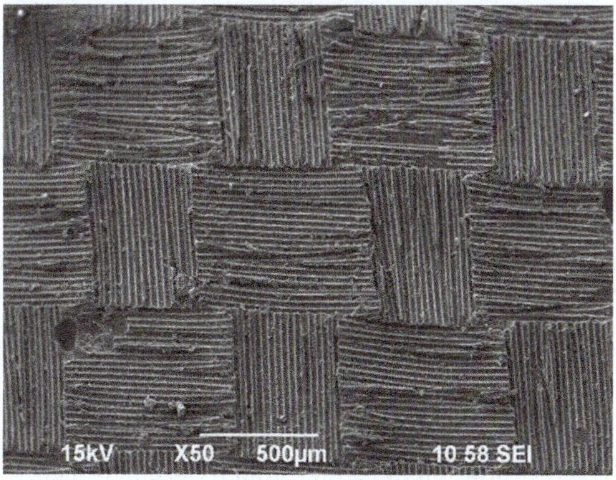

Fig. 3.252 Secondary electron image of an area of the composite with artificial fabric filler. The composite is located under porous honeycomb layer. Accelerating voltage is 15 kV, magnification ×50, scale mark—500 μm

Fig. 3.253 Secondary electron image of the fragments of fibers forming the artificial fabric. Accelerating voltage is 15 kV, magnification ×1500, scale mark—10 μm. The figure shows fiber thicknesses

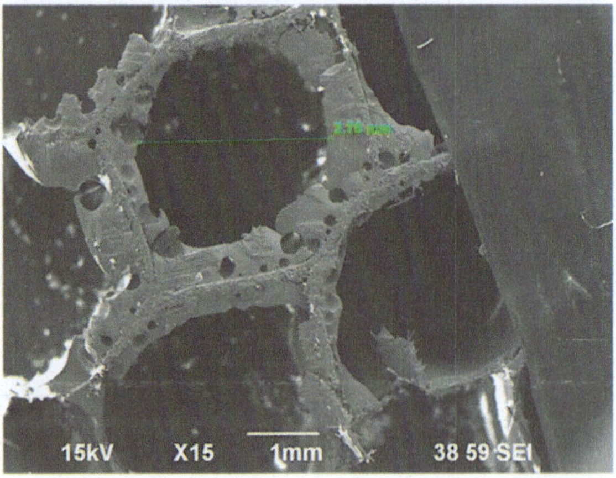

Fig. 3.254 Secondary electron image of an area of porous honeycomb layer. Accelerating voltage is 15 kV, magnification ×15, scale mark—1 mm. The figure shows the size of honeycombs

To study the structure and determine the chemical element concentrations in the fragment bulk, as well as measure the thicknesses of various layers, slices of LM body fragment were prepared.

Figure 3.255 presents secondary electron image of an area of the slice of the composite with artificial fabric filler. The layer is located under that of paint. This layer is shown full width, with its size indicated.

Figures 3.256 and 3.257 demonstrate backscattered electron images of the area of LM body fragment slice at different magnifications, where fibers forming artificial fabric can be seen, along with black paint layer (at the bottom of the images). Figure 3.257 shows the width of black paint layer.

Figure 3.258 presents secondary electron images of an area of slice of composite layer with artificial fabric filler, which is located under porous honeycomb layer. This layer is shown in full width, with its size indicated.

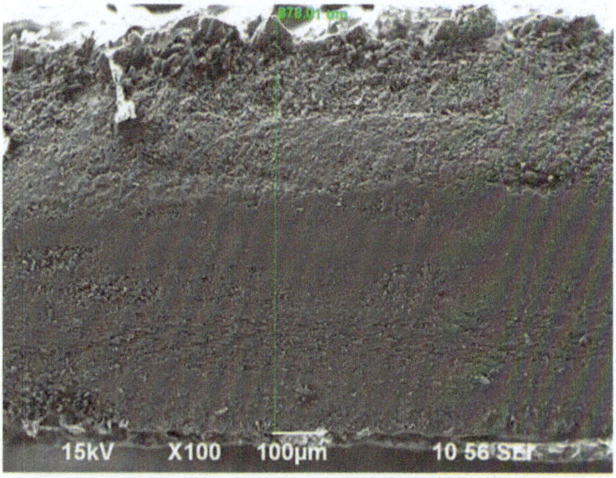

Fig. 3.255 Secondary electron image of a slice layer area of the composite with artificial fabric filler. The layer is located below that of paint. Accelerating voltage is 15 kV, magnification ×100, scale mark—100 μm. Layer width is denoted

Fig. 3.256 Backscattered electron image of an area of LM body fragment slice, which shows fibers forming artificial fabric, along with black paint layer (at the bottom of the image). Accelerating voltage is 15 kV, magnification ×350, scale mark—50 μm

Fig. 3.257 Backscattered electron image of an area of LM body fragment slice, which shows fibers forming artificial fabric, alongside with black paint layer (at the bottom of images). Accelerating voltage is 15 kV, magnification ×700, scale mark—20 μm. The image shows the width of black paint layer

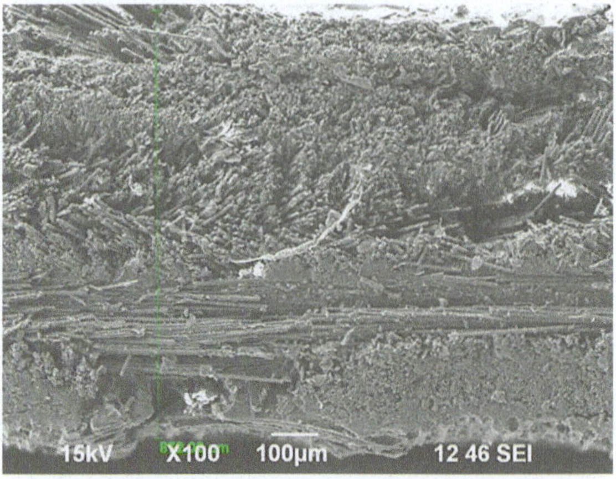

Fig. 3.258 Secondary electron image of an area of layer slice of the composite with artificial fabric filler. The composite layer is located below that with porous honeycomb structure. Accelerating voltage is 15 kV, magnification ×100, scale mark—μm. The width of the whole layer is 872 μm

Fig. 3.259 Secondary electron image of an area of slice of composite layer located under porous honeycomb layer. The image shows fibers forming the artificial fabric. Accelerating voltage is 15 kV, magnification ×200, scale mark—100 μm

Figures 3.259 and 3.260 demonstrate secondary electron images of an area of the slice of composite layer located under porous honeycomb layer at different magnifications. The images show fibers forming the artificial fabric. Figure 3.260 denotes fiber thickness.

Figure 3.261 presents the results of the X-ray microanalysis of the artificial fabric in the composite layer located under the layer of black paint. The microanalysis was performed at a definite point, which is shown in backscattered electron image. Chemical element concentrations are given in mass percentages. The concentration of C (carbon) was measured on K line of characteristic X-ray emission. Figure 3.261 also shows the spectrum of characteristic X-ray emission of this measurement point.

Figure 3.262 shows backscattered electron image (with 60 μm scale mark) of an area of composite layer located under the layer of black paint, which demonstrates fibers forming the artificial fabric. It also shows the image of this area produced with characteristic X-ray emission (spatial distribution of chemical elements over

Fig. 3.260 Secondary electron image of an area of the slice of composite layer located under porous honeycomb layer. The image shows fibers forming the artificial fabric. Accelerating voltage is 15 kV, magnification ×2500, scale mark—10 μm. Fiber thicknesses are denoted

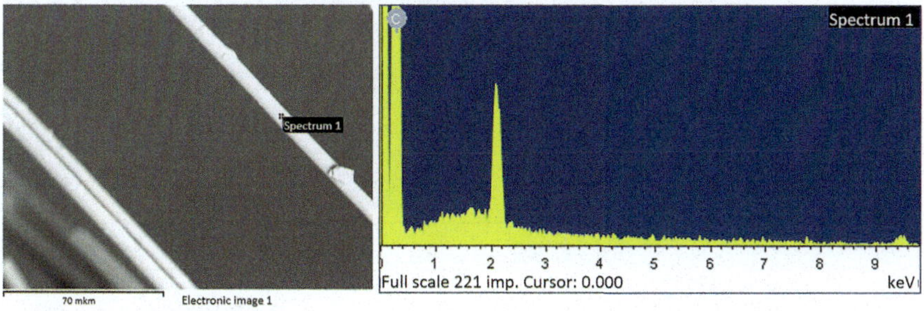

Element	Weight %
C K	100.00
Total	100.00

Fig. 3.261 Results of X-ray microanalysis of artificial fabric fibers of the composite layer located under that of black paint

the studied area). White dots correspond to the presence of the chemical element denoted below the image.

The results of examinations presented in Figs. 3.261 and 3.262 confirm that fibers forming the artificial fabric of the composite layer that is under the layer of black paint are carbon-containing ones with carbon concentration of up to 100%.

Figure 3.263 presents the results of X-ray microanalysis of artificial fabric fibers of the composite layer located under the porous honeycomb one. The microanalysis was performed at a definite point, which is shown in backscattered electron image. Chemical element concentrations are given in mass percentages. The concentrations of chemical elements were measured on K lines of characteristic X-ray emission.

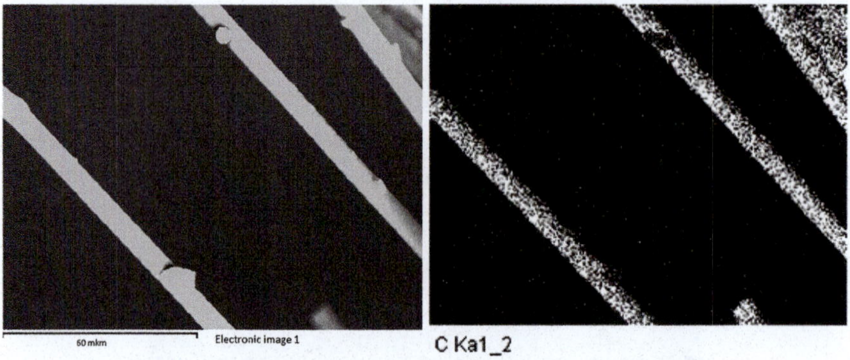

Fig. 3.262 Backscattered electron image (with 60 μm scale mark) of an area of composite layer located under the layer of clack paint, which shows fibers that form artificial fabric, as well as the image of this area produced by characteristic X-ray emission (spatial distribution of chemical elements over the studied area). White dots correspond to the presence of the chemical element denoted below the image

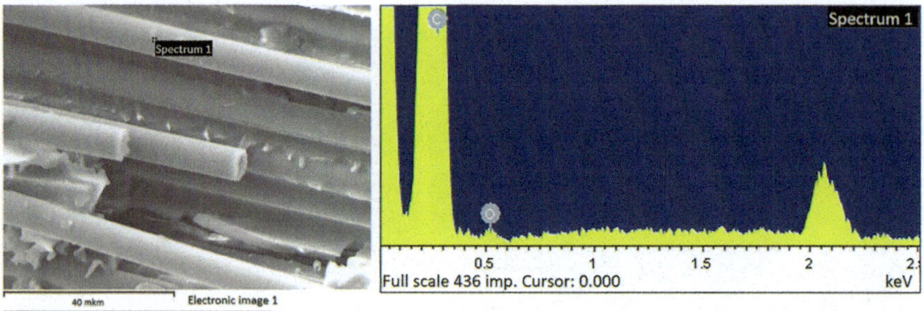

Element	Weight %
C K	91.01
O K	0.99
Total	100.00

Fig. 3.263 Results of X-ray microanalysis of artificial fabric fibers of the composite layer located under porous honeycomb one

Figure 3.263 also presents the spectrum of characteristic X-ray emission of this measurement point.

Figure 3.264 shows the results of X-ray microanalysis of artificial fabric fibers of the composite layer located under porous honeycomb layer. The microanalysis was performed at a definite point, which is shown in backscattered electron image. Chemical element concentrations are given in mass percentages. The concentrations of chemical elements were measured on K lines of characteristic X-ray emission of this measurement point. Figure 3.264 also presents the spectrum of characteristic X-ray emission of this measurement point. Unlike the concentration measurements, whose results are presented in Fig. 3.263, this measurement was performed at the point located at the fiber end, not on its lateral surface.

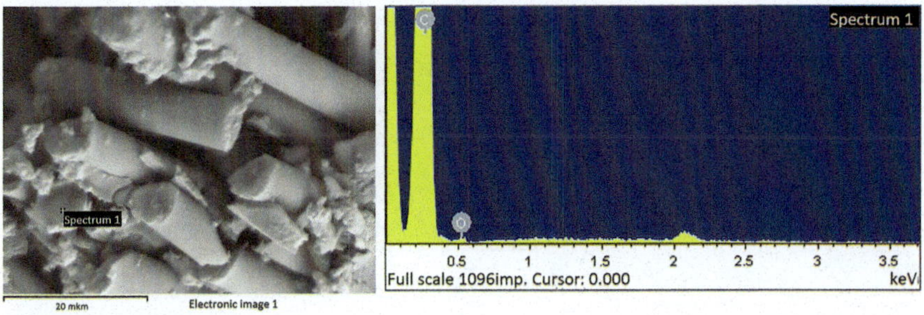

Element	Weight %
C K	99.31
O K	0.69
Total	100.00

Fig. 3.264 Results of X-ray microanalysis of artificial fabric fibers in the composite layer located under porous honeycomb layer

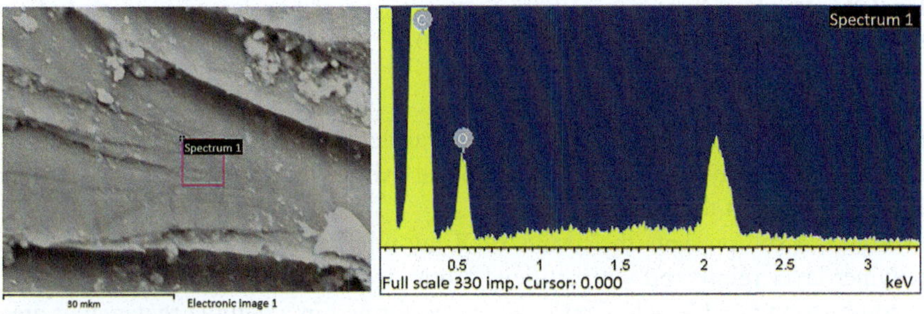

Element	Weight %
C K	82.42
O K	17.58
Total	100.00

Fig. 3.265 Results of X-ray microanalysis of porous honeycomb layer material

Figure 3.265 presents the results of X-ray microanalysis of the porous honeycomb layer material. The microanalysis was performed in a definite rectangular area. The measured concentrations of chemical elements were averaged within this area, which is shown in backscattered electron image. Chemical element concentrations are given in mass percentages. The concentrations of chemical elements were measured on K lines of characteristic X-ray emission. Figure 3.265 also presents the spectrum of characteristic X-ray emission of this measurement area.

Figure 3.266 illustrates the results of X-ray microanalysis of matrix (binder) of polymer layer reinforced with artificial fabric, which is located under black paint layer. The microanalysis was performed in a definite rectangular area. The measured concentrations of chemical elements were averaged within this area, which is shown in backscattered electron image. Chemical element concentrations are given in mass

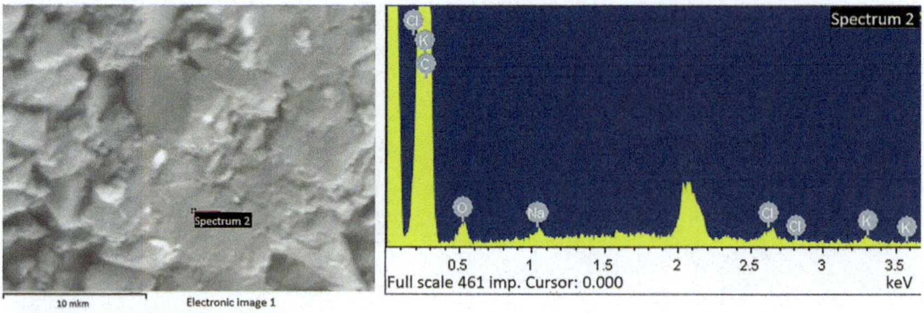

Element	Weight %
C K	78.72
O K	18.62
Na K	1.16
Cl K	0.65
K K	0.85
Total	100.00

Fig. 3.266 Results of X-ray microanalysis of matrix (binder) of the polymer layer reinforced with artificial fabric, which is located under black paint layer

percentages. The concentrations of all chemical elements were measured on K lines of characteristic X-ray emission. Figure 3.266 also shows the spectrum of characteristic X-ray emission of this measurement area.

Figure 3.267 presents the results of X-ray microanalysis of matrix (binder) of polymer layer reinforced with artificial fabric, which is located under porous honeycomb structure. The microanalysis was performed in a definite rectangular area. The measured concentrations of chemical elements were averaged within this area, which is shown in backscattered electron image. Chemical element concentrations are given in mass percentages. The concentrations of all chemical elements were measured on K lines of characteristic X-ray emission. Figure 3.267 also presents the spectrum of characteristic X-ray emission of this measurement area.

Figure 3.268 shows the results of X-ray microanalysis of the layer of black paint applied onto the convex surface of LM body fragment. The microanalysis was performed in a definite rectangular area. The measured concentrations of chemical elements were averaged within this area, which is shown in backscattered electron image. Chemical element concentrations are given in mass percentages. The concentrations of all chemical elements were measured on K lines of characteristic X-ray emission. Figure 3.268 also presents the spectrum of characteristic X-ray emission of this measurement area.

Conclusions

The studied fragment of *Shahed-136* LM (upgraded) body is manufactured from multi-layered composite material. Presented below is a table with the names of fragment layers, with their thicknesses denoted (Table 3.8).

The fragment has four layers two of those are 870-μm-thick composite consisting of polymer matrix and filler in the form of artificial fabric composed of carbon fibers. In the fragment bulk, between polymer layers reinforced with artificial fabric made

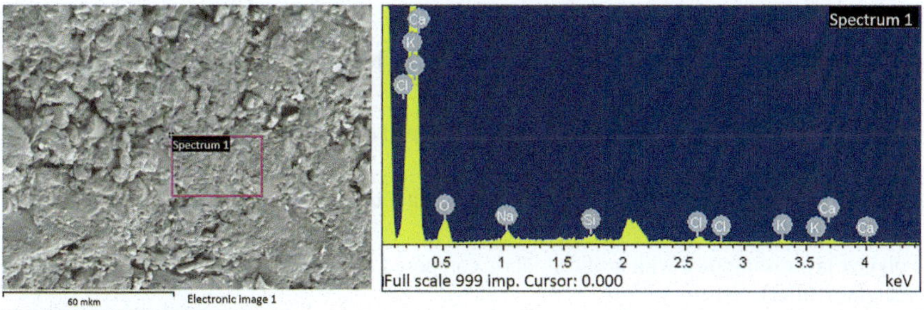

Element	Weight %
C K	78.35
O K	18.80
Na K	1.12
Si K	0.13
Cl K	0.60
K K	0.65
Ca K	0.35
Total	100.00

Fig. 3.267 Results of X-ray microanalysis of matrix (binder) of polymer layer reinforced with artificial fabric, which is located under porous honeycomb layer

Element	Weight %
C K	51.14
O K	32.48
Na K	0.36
Mg K	1.48
Al K	0.58
Si K	1.44
Ca K	12.53
Total	100.00

Fig. 3.268 Results of X-ray microanalysis of black paint layer, which is applied onto the convex surface of LM body fragment

Table 3.8 Results of studying the material structure of *Shahed-136* LM (upgraded) body fragment

Layer names of *Shahed-136* LM body fragment (BLACK)	Layer thickness, μm
Layer of black paint applied onto the convex surface	40
Polymer layer reinforced with artificial fabric composed of carbon fiberglass (filler)	870
Porous honeycomb layer (honeycombs are empty, without any filler)	5000
Polymer layer reinforced with artificial fabric composed of carbon fiberglass (filler)	870

Table 3.9 Results of studying chemical element concentrations in polymer matrices of composite material layers of *Shahed-136* LM (upgraded) body fragment

Chemical element	Chemical element concentrations (expressed in mass %) of the matrix of polymer layer located under black paint layer	Chemical element concentrations (expressed in mass %) of the matrix of polymer layer located under porous honeycomb layer
C, carbon	78.72	78.35
O, oxygen	18.62	18.80
Cl, chlorine	0.65	0.60
Na, sodium	1.16	1.12
Ca, calcium	–	0.35
K, potassium	0.85	0.65
Si, silicon	–	0.13

of carbon fibers (filler), there is a porous honeycomb layer of 5.000 μm thickness (the honeycombs are empty, without filler, and are about 3 mm in diameter). Convex surface is painted black, paint layer thickness is 40 μm.

Carbon fibers of 7 μm thickness that form artificial fabric (the filler in the two composite layers mentioned above) practically consist of a single chemical element—carbon. Carbon concentrations determined with X-ray analysis amount to 100%.

The polymer (binder of the two composite layers mentioned) has the following chemical composition, which is identical for both layers (except that Ca and Si in small concentrations are present in the layer located under porous honeycomb structure) (Table 3.9).

The material of porous honeycomb layer is a polymer and consists of the following chemical elements in respective concentrations: C—82.42% and O—17.58%. It should be stressed that honeycomb chemical composition, diameter (3μm) of honeycombs and thickness (5,000 μm) of porous honeycomb layer of *Shahed-136* LM (upgraded) fragment and the fragment of *Shahed-136* body outer coating, which had been examined earlier, are practically the same.

Table 3.10 Results of studying chemical element concentrations in polymer matrices of *Shahed-136* LM (upgraded) body fragment

Chemical element	Chemical element concentrations expressed in mass % in black paint
C, carbon	51.14
O, oxygen	32.48
Mg, magnesium	1.48
Al, aluminum	0.58
Si, silicon	1.44
Ca, calcium	12.53
Na, sodium	0.36

It should be noted that in the investigation of polymer chemical composition the concentration of hydrogen was not measured since the equipment used permits the measurement of the concentrations of chemical elements starting from Z = 6. So, hydrogen could also be present in polymer chemical compositions.

Presented below is the table with concentrations of chemical elements present in black paint layer applied onto the body of *Shahed-136* LM (upgraded) (Table 3.10).

Thus, the fragment of Shahed-136 UAV is manufactured from composite material of the total thickness about 7 mm. It contains two polymer layers reinforced with artificial fabric which are identical in chemical composition and width (870 μm). Unlike the fragments of *Shakhed-136* LM studied earlier, which contained composite layers with polymer matrix and glass fabric filler, artificial fabric filler made of carbon fibers is used in this fragment. Hence, these two layers perform two functions: strengthening the structure and ensuring the absorption of electromagnetic emission. The RF absorption is due to electrical conductivity of carbon fibers.

To produce black color of the paint, finely dispersed carbon black with carbon concentration of 85–98% in its particles and particle sizes 200 μm–25 nm is added as the pigment. When studying black fiber, no individual particles, fibers, or other formations containing high carbon concentrations (higher than average carbon concentration on the paint surface) were detected. Most likely, this results from the fact that pigment particles are not determined because of high average carbon concentration in the paint and small size of pigment particles. Pigment particles could contribute to the absorption of electromagnetic emission, but because of small paint layer thickness (40 μm) this contribution is not significant. For example, for the well-known RF absorbent paint ZIPSIL 750 RPM-K, backscattering coefficient smaller than 10 Db is achieved for the paint layer at least 900 μm thick. Therefore, RF absorption in our case occurs in the layers of polymer reinforced with artificial fabric made of carbon fibers.

Taking into account the above, the following conclusion can be made: the examined fragment of Shahed-136 (upgraded) UAV is an RF absorbent dielectric (non-magnetic) material consisting of two polymer layers with the filler in the form of artificial fabric made of carbon fibers.

3.3.4.6 Glider Fragment of *LANCET* UAV and Loitering Munition

Figures 3.269 and 3.270 present, respectively, the images of the external and internal parts of a fragment of *LANCET* UAV and Loitering Munition glider (further—specimen), which was chosen for studying.

Figure 3.271 shows the secondary electron image (SEI) of an area of specimen's external surface.

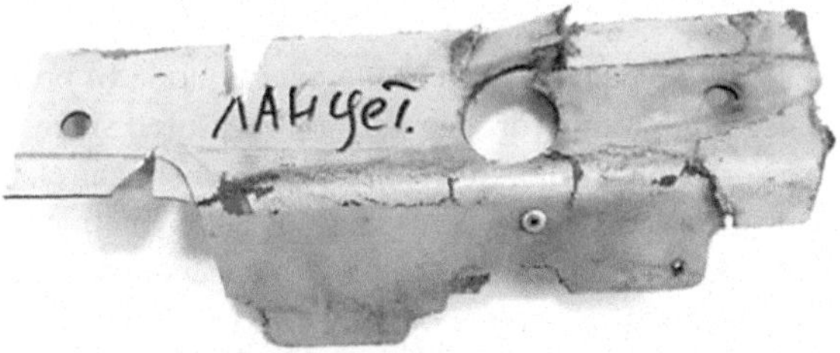

Fig. 3.269 Image of the external side of *LANCET* glider fragment

Fig. 3.270 Image of the internal side of *LANCET* glider fragment

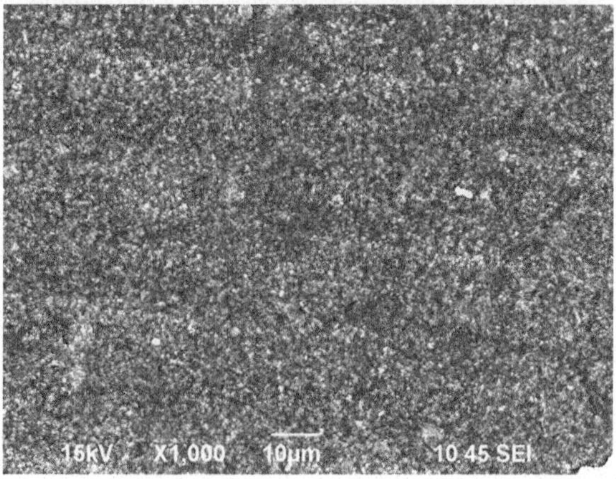

Fig. 3.271 Secondary electron image (SEI) of an area of specimen's external surface. Accelerating voltage is 15 kV, magnification ×1000, scale mark—10 μm

Fig. 3.272 Secondary electron image (SEI) of an area of specimen's internal surface (the dark-colored region). Accelerating voltage is 15 kV, magnification ×50, scale mark—500 μm

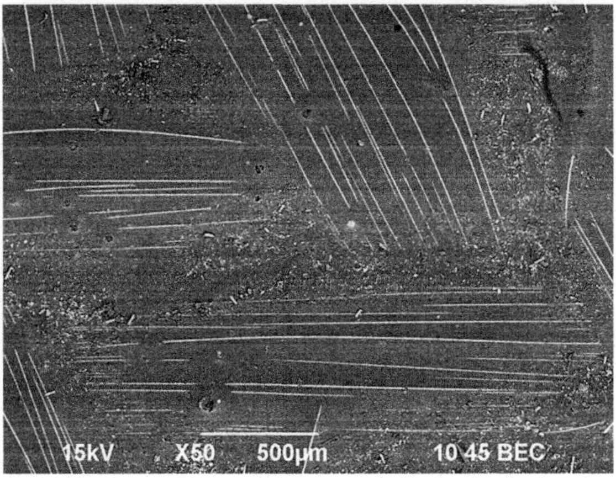

Fig. 3.273 Backscattered electron (BEC) image of an area of the specimen's internal surface (dark-colored region). Accelerating voltage is 15 kV, magnification ×50, scale mark—500 μm

Figures 3.272 and 3.273 present, respectively, the secondary electron (SEI) and backscattered electron (BEC) images of an area of *LANCET* internal surface fragment. In Figs. 3.272 and 3.273, artificial fabric partially covered by a composite can be seen. The studied area is located in the dark-colored area of the internal surface, where the artificial fabric is covered by a thicker layer of the composite.

Figures 3.274 and 3.275 present, respectively, the secondary electron (SEI) and backscattered electron (BEC) images of the area of specimen's internal surface. Figures 3.274 and 3.275 also show the artificial fabric partially covered by the composite. The studied area is situated in the light-colored region of the internal surface, where the artificial fabric is covered by the composite layer that is thinner as compared to the dark-colored region of the internal surface. There composite-free fibers are also present, which permits their diameters to be measured (see Fig. 3.276).

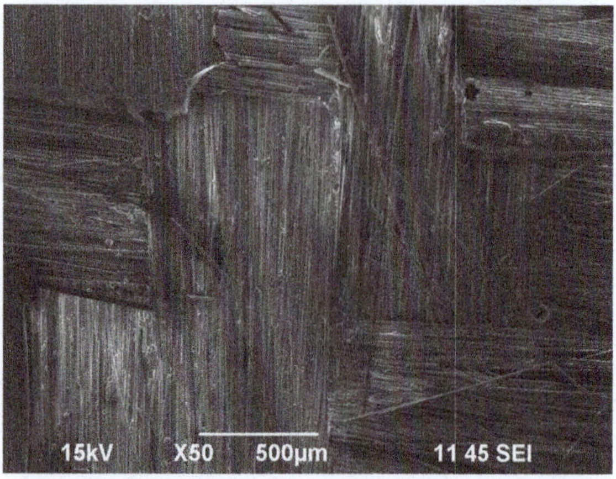

Fig. 3.274 Secondary electron image (SEI) of the specimen's internal surface (light-colored region). Accelerating voltage is 15 kV, magnification ×50, scale mark—500 μm

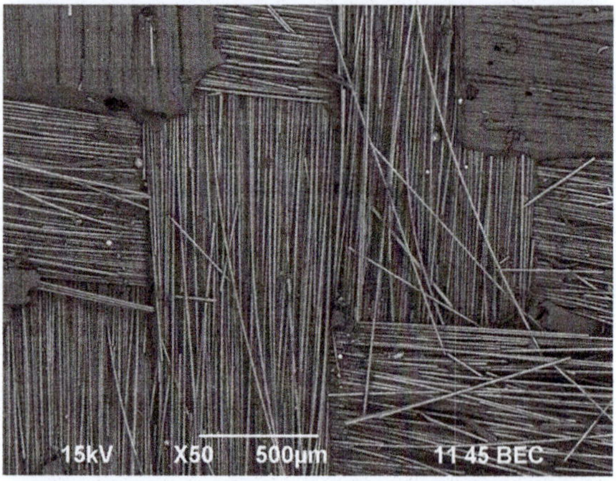

Fig. 3.275 Backscattered electron image (BEC) of the specimen's inner surface (light-colored region). Accelerating voltage is 15 kV, magnification ×50, scale mark—500 μm

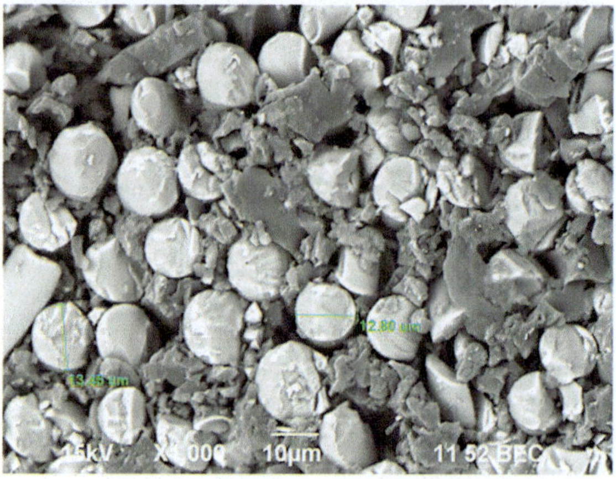

Fig. 3.276 Backscattered electron image (BEC) of artificial fabric fibers. Accelerating voltage is 15 kV, magnification ×1000, scale mark—10 μm. The image shows fiber diameter

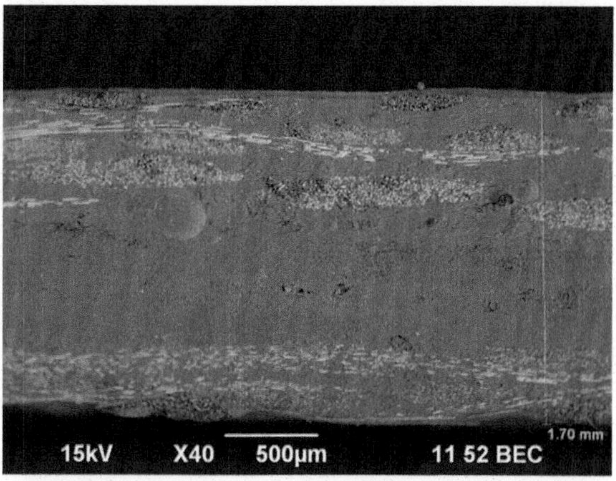

Fig. 3.277 Backscattered electron image of the specimen's slice. Accelerating voltage is 15 kV, magnification ×40, scale mark—500 μm

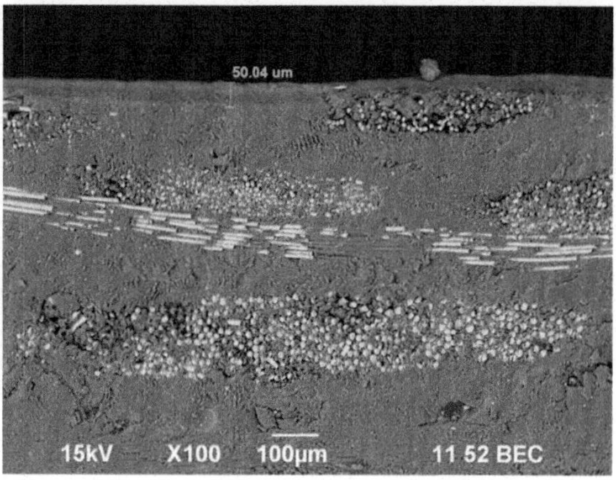

Fig. 3.278 Backscattered electron image of the top part (the nearest to the external surface) of the glider specimen's slice. Accelerating voltage is 15 kV, magnification is ×100, scale mark—100 μm

To study the specimen's structure and determine chemical element concentrations in the specimen bulk, as well as measure the thicknesses of different layers of the specimen, its slice was produced.

Figure 3.277 presents the backscattered electron image of the slice of glider fragment specimen. Its thickness is indicated in the image.

Figure 3.278 presents backscattered electron image of the top part (the nearest to grey external surface) of the glider specimen's slice; it also shows the thickness of external coating layer.

Figure 3.279 presents backscattered electron image of the slice in the glider specimen's region where artificial fabric fibers are located. Light-colored longitudinal structures in Figs. 3.278 and 3.279 correspond to the presence of artificial fabric fibers parallel to the upper margin of the specimen's slice, while dot structures in the images correspond to the fibers that are perpendicular to the longitudinal ones.

Fig. 3.279 Backscattered electron image of the slice in the region where fibers of artificial fabric layer are located. Accelerating voltage is 15 kV, magnification ×1000, scale mark—10 μm

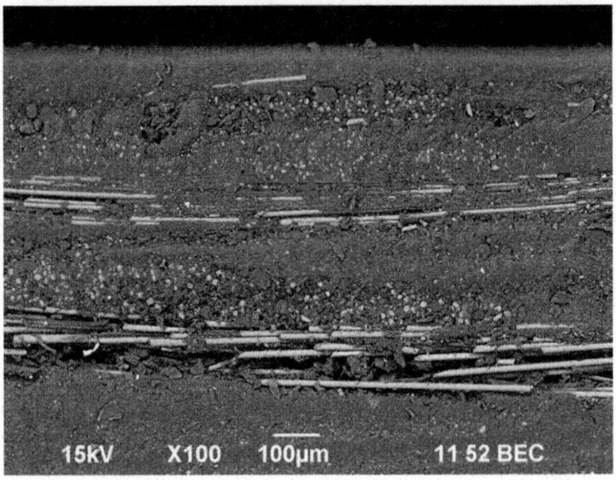

Fig. 3.280 Backscattered electron image of the specimen's slice in the region of the edge. Accelerating voltage is 15 kV, magnification ×100, scale mark—100 μm

It can be seen from Figs. 3.278 and 3.279 that artificial fabric layers are located throughout the entire bulk of the examined *LANCET* specimen and that artificial fabric layers are parallel to the external surface of the fragment.

It can be seen from Fig. 3.269 that along the margin of the fragment there is an edge somewhat thinner than its main part. The examination of the fragment slice in the region of the edge has shown that the specimen thickness in the region of the edge is 1.4 μm, and the thickness of the external coating layer, like that in the fragment main part, is 50 μm. Figure 3.280 shows that there are artificial fabric layers in the entire bulk of the examined part of the glider specimen.

Figure 3.281 demonstrates an example of the results of X-ray microanalysis on the area of *LANCET* glider external surface. The microanalysis was performed in the rectangular area shown in the backscattered electron image. Chemical element concentrations are given in mass percentages. The concentrations of all chemical

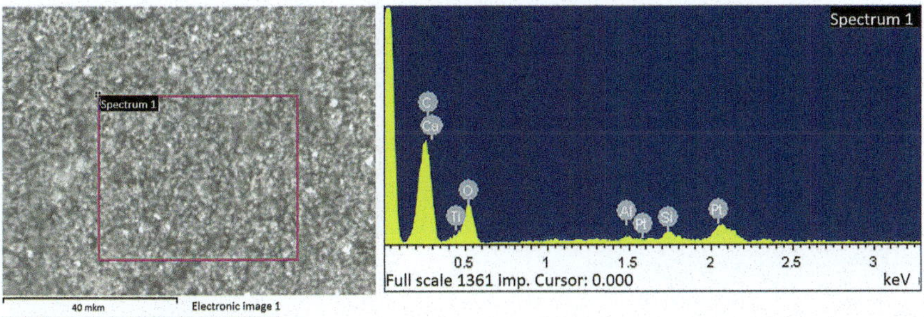

Element	Weight %
C K	57.44
O K	28.53
Al K	0.49
Ti K	13.54
Total	100.00

Fig. 3.281 Results of X-ray microanalysis on the shown area of the external surface of *LANCET* fragment specimen

elements were measured on K lines of characteristic X-ray emission. Figure 3.281 also shows the spectra of characteristic X-ray emission in this measurement region.

Figure 3.282 presents an example of the results of X-ray microanalysis of artificial fabric fibers in the rectangular area of *LANCET* fragment specimen. The microanalysis was performed in the rectangular region shown in the backscattered electron image. Chemical element concentrations are given in mass percentages. The concentrations of all chemical elements were measured on K lines of characteristic X-ray emission. Figure 3.282 also shows the spectra of characteristic X-ray emission at this measurement point.

Figure 3.283 demonstrates the results of X-ray microanalysis of the composite matrix (binder) on an area of the internal surface of *LANCET* fragment specimen. The microanalysis was performed in the rectangular region shown in the backscattered electron image. Chemical element concentrations are given in mass percentages. The concentrations of all chemical elements were measured on K lines of characteristic X-ray emission. Figure 3.283 also presents the spectra of characteristic X-ray emission in this measurement region.

Figure 3.284 shows the backscattered electron image (with 30 μm scale mark) of the internal surface of *LANCET* fragment specimen, along with the image of the same area in X-ray characteristic emission (spatial distribution of chemical elements over the examined area). White dots correspond to the presence of the chemical element specified under each image.

It can be seen from Fig. 3.284 that the X-ray emission images of the distribution of the chemical elements present in artificial fabric fibers (Si, Mg, Ca, O, Al) clearly correlate with backscattered electron images of the fibers, and X-ray emission images of the distribution of chlorine (Cl), which is present in the chemical composition of

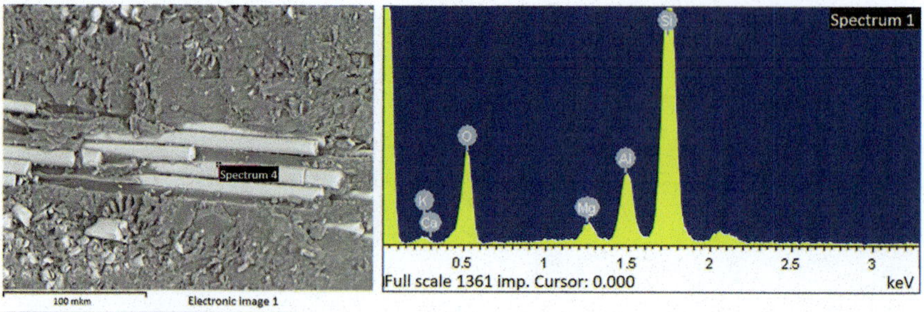

Element	Weight %
O K	39.62
Mg K	2.88
Al K	8.33
Si K	30.95
Ca K	17.59
K K	0.63
Total	100.00

Fig. 3.282 Results of X-ray microanalysis of artificial fabric fiber on the rectangular area of the slice of *LANCET* fragment specimen

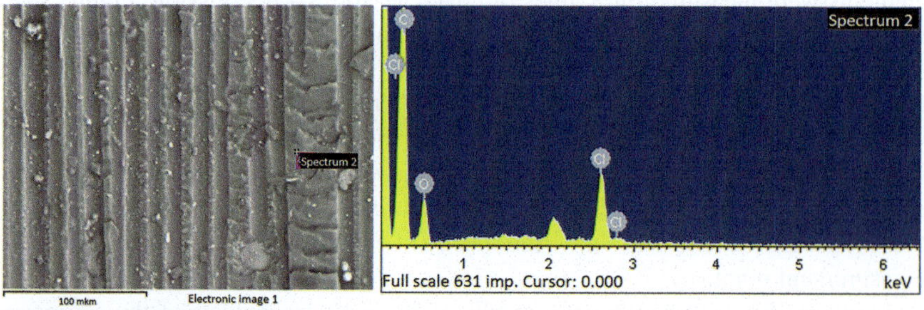

Element	Weight %
C K	76.25
O K	18.09
Cl K	5.66
Total	100.00

Fig. 3.283 Results of X-ray microanalysis of the composite matrix (binder) on the shown area of specimen internal surface

the composite matrix, clearly correlate with the regions of matrix location on the backscattered electron image (dark-colored regions).

Figure 3.285 presents the results of X-ray microanalysis of the composite matrix (binder) on the area of specimen slice. The microanalysis was performed in the rectangular region shown in the backscattered electron image. Chemical element concentrations are given in mass percentages. The concentrations of all chemical

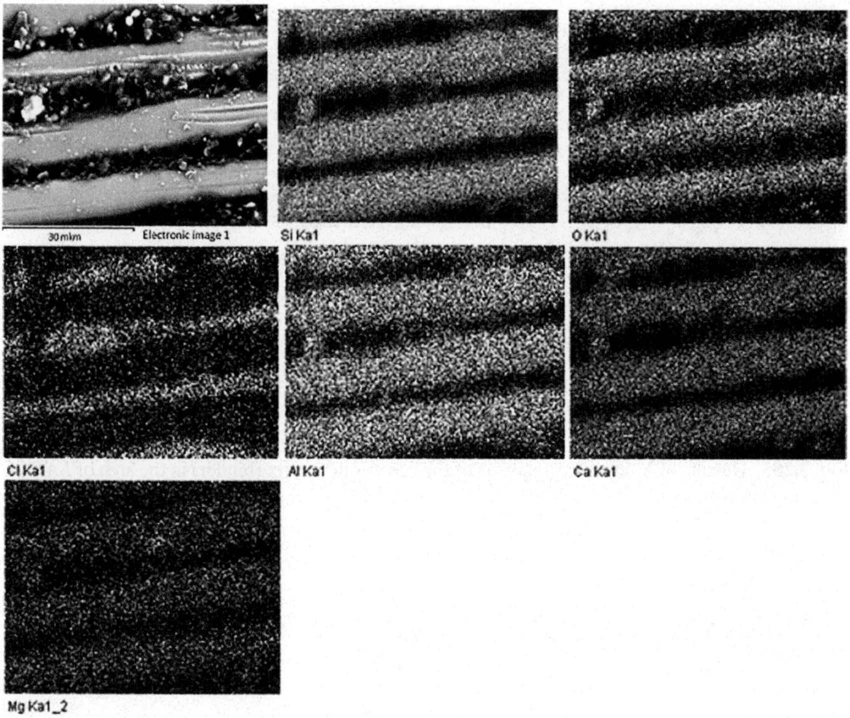

Fig. 3.284 Backscattered electron image (with 30 μm scale mark) of specimen internal surface, along with the image of the same area in X-ray characteristic emission (the spatial distribution of chemical elements over the studied area). White dots correspond to the presence of the chemical element specified under each image

elements were measured on the lines of characteristic X-ray emission. Figure 3.285 also shows the spectra of characteristic X-ray emission in this measurement region.

If the results of X-ray microanalysis which are presented in Figs. 3.283 and 3.285 are compared, it can be seen that the measured concentrations of the chemical elements of the composite matrix which were obtained in these two measurements are practically the same.

Thus, the examined material of *LANCET* glider fragment is a composite with a polymer matrix and with inclusions of layers of artificial fiberglass fabric with 1.7 mm total thickness and nearly 50-μm-thick coating layer applied onto its external surface. The layers of fiberglass artificial fabric (composite inclusions) are situated throughout the entire composite bulk, including the edge. Studies of the specimen slice in the region of its edge show that the fragment thickness in this area is 1.4 μm, and the external coating layer, like that of the fragment bulk, is 50 μm thick.

The diameters of glass fibers are about 13 μm, and their chemical compositions are shown in Table 3.11.

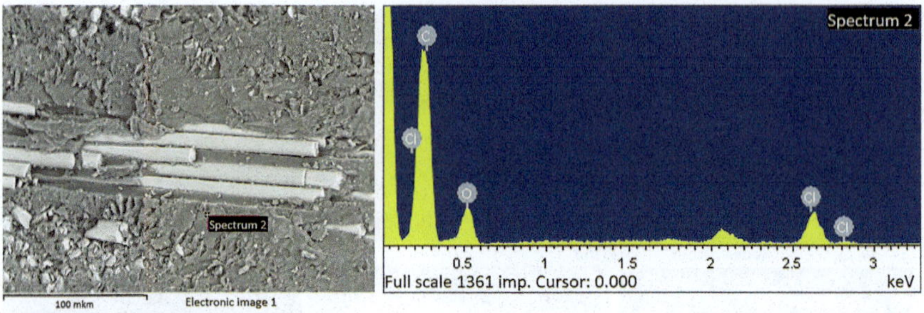

Element	Weight %
C K	76.10
O K	18.22
Cl K	5.68
Total	100.00

Fig. 3.285 Results of X-ray microanalysis of the composite matrix (binder) in the area of *LANCET* fragment specimen slice

Table 3.11 Results of studying the chemical element concentrations in glass fibers of the specimen composite material

Chemical elements	Mass concentrations, %
Si, silicon	30.95
O, oxygen	39.62
Ca, calcium	17.59
K, potassium	0.63
Al, aluminum	8.33
Mg, manganese	2.88

Table 3.12 Results of studying chemical element concentrations in the matrix of specimen composite material

Chemical elements	Mass concentrations, %
C, carbon	76.10
O, oxygen	18.22
Cl, chlorine	5.68

The results of studying chemical element concentrations in the matrix of specimen composite material are presented in Table 3.12.

The layer about 50 μm thickness, which is applied onto the specimen's external surface, is a paint and its chemical composition is shown in Table 3.13.

Table 3.13 Results of studying chemical element concentrations in the layer of paint applied onto specimen's external surface

Chemical elements	Mass concentrations, %
C, carbon	57.44
O, oxygen	28.53
Ti, titanium	13.54
Al, aluminum	0.49

Fig. 3.286 Image of a fragment of ORLAN UAV glider

3.3.4.7 Glider Fragment of *ORLAN* UAV Complex

Figure 3.286 demonstrates the image of the studied glider fragment of *ORLAN* unmanned aerial vehicle (UAV) complex (further—the specimen), which is a composite material on whose surface a grey layer can be seen.

Figure 3.287 presents secondary electron image (SEI) of an area of grey external surface of the specimen's fragment.

Figures 3.288 and 3.289 present, respectively, the secondary electron image (SEI) and the backscattered electron one (BEC) of an area of specimen internal surface. Artificial fabric is observed in Figs. 3.288 and 3.289.

Figure 3.290 presents a backscattered electron image (BEC) of artificial fabric fibers present on the area of specimen's internal surface. The image shows fiber size (diameter).

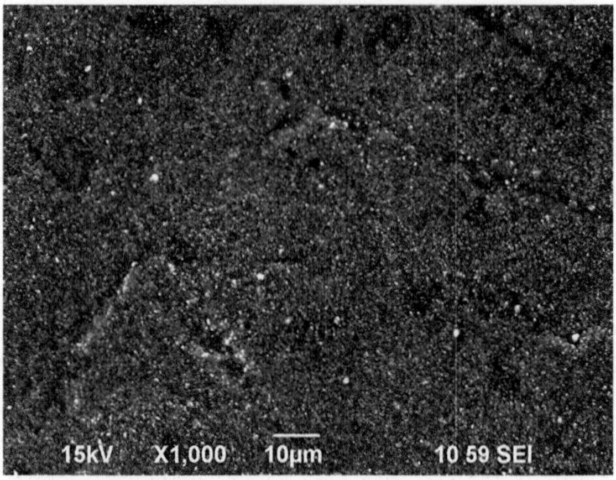

Fig. 3.287 Demonstrates secondary electron image (SEI) of an area of grey external surface of the specimen's fragment. Accelerating voltage is 15 kV, magnification ×1000, scale mark—10 μm

Fig. 3.288 Secondary electron image (SEI) of the specimen's internal surface. Accelerating voltage is 15 kV, magnification ×50, scale mark—500 μm

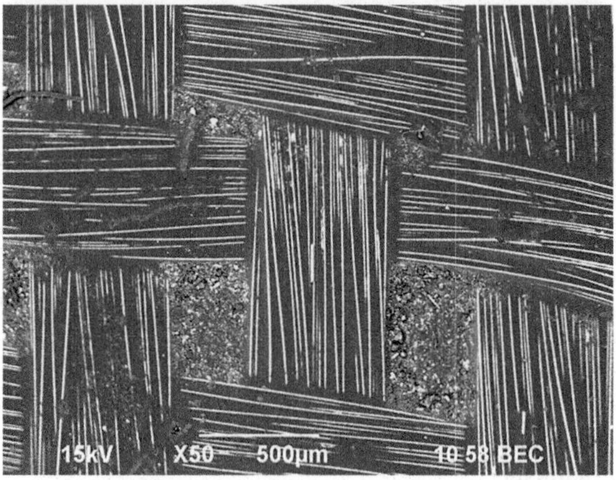

Fig. 3.289 Backscattered electron image (BEC) of an area of specimen's internal surface. Accelerating voltage is 15 kV, magnification ×50, scale mark—500 μm

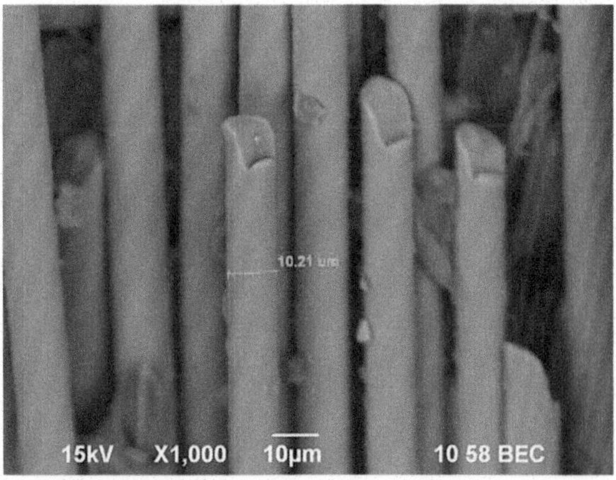

Fig. 3.290 Backscattered electron image (BEC) of artificial fabric fibers present on the area of specimen internal surface. Accelerating voltage is 15 kV, magnification ×1000, scale mark—10 μm

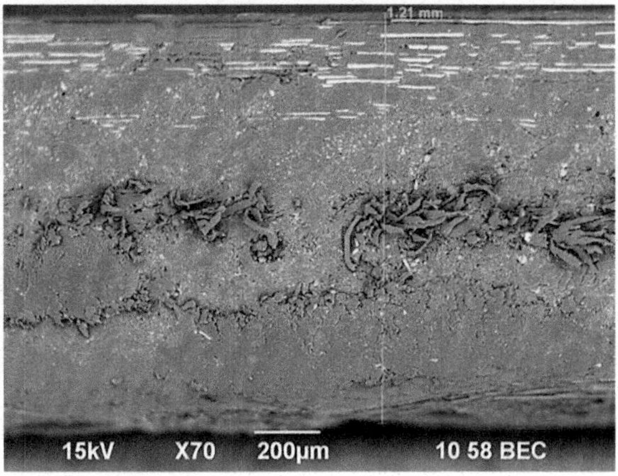

Fig. 3.291 Backscattered electron image of the specimen's slice. Accelerating voltage is 15 kV, magnification ×70, scale mark—200 μm

As noted above, Figs. 3.289 and 3.290 present backscattered electron images. In Sect. 1.7, it was shown that in this mode of electron microscope functioning, contrast depends on atomic number's difference: the larger the atomic number the lighter are respective regions on backscattered electron images. Hence, the average atomic number of artificial fabric fibers is higher than that of the composite matrix (binder). Further X-ray microanalysis confirmed this conclusion.

To study the structure and determine chemical element concentrations in the specimen's bulk as well as measure the thickness of various specimen layers, a slice of this specimen was produced.

Figure 3.291 presents the backscattered electron image of the specimen's slice; it also shows specimen's thickness.

Figure 3.292 demonstrates the backscattered electron image of the upper part (the nearest to grey external surface) of the specimen's slice. The image specifies the thickness of grey external coating layer.

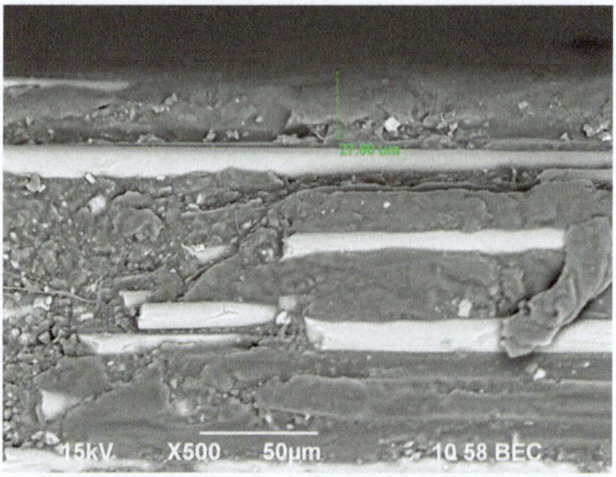

Fig. 3.292 Backscattered electron image of grey upper part (adjacent to grey surface) of the specimen's slice. Accelerating voltage is 15kV, magnification ×500, scale mark—50 µm

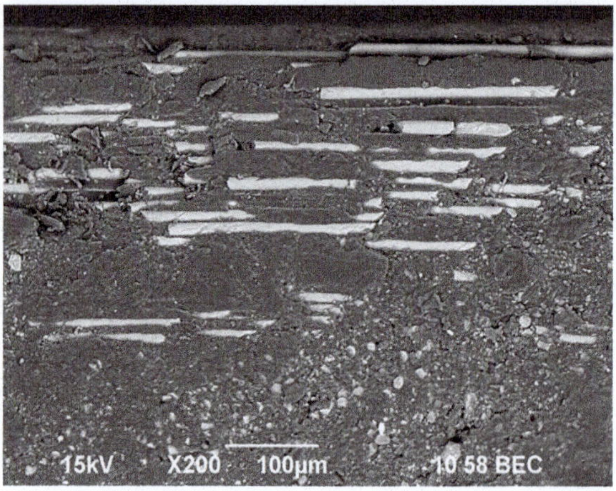

Fig. 3.293 Backscattered electron image of the upper and middle parts of the specimen's slice. Accelerating voltage is 15 kV, magnification ×200, scale mark—100 µm

Figure 3.293 shows the backscattered electron image of the upper and middle parts of the specimen's slice, and Fig. 3.294 presents that of the lower part of the specimen's slice. Light-colored longitudinal structures in the images correspond to the presence of artificial fabric fibers that are parallel to the upper margin of the specimen's slice, while light-colored dot structures in the images correspond to the fibers that are perpendicular to longitudinal ones.

It can be seen from Figs. 3.293 and 3.294 that artificial fabric layers are present throughout the entire bulk of the examined specimen and also that artificial fabric layers are located in parallel to the fragment's external surface.

Figure 3.295 presents the results of X-ray microanalysis on an area of grey specimen's grey surface. The microanalysis was performed in the rectangular region shown in the backscattered electron image. Chemical element concentrations are given in mass percentages. The concentrations of all chemical elements were mea-

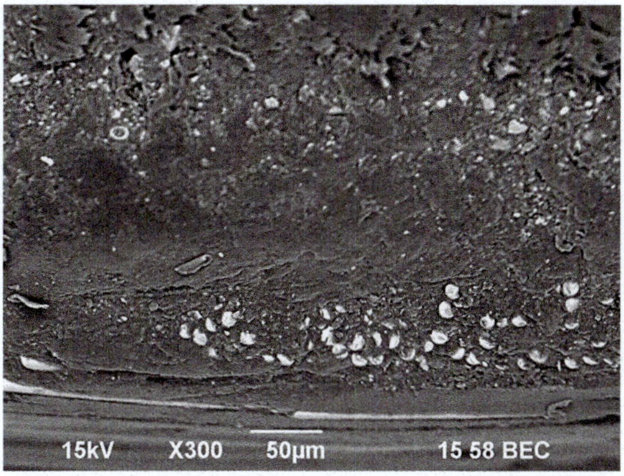

Fig. 3.294 Backscattered electron image of the lower part of the specimen's slice. Accelerating voltage is 15 kV, magnification ×300, scale mark—50 µm

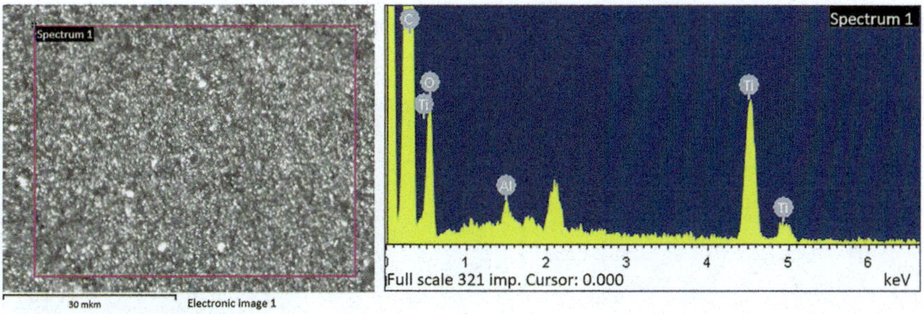

Element	Weight %
C K	65.59
O K	22.14
Al K	0.47
Ti K	11.80
Total	100.00

Fig. 3.295 Results of X-ray microanalysis on an area of grey external surface of the specimen

sured on K lines of characteristic X-ray emission. Figure 3.295 also shows the spectra of characteristic X-ray emission in this measurement region.

Figure 3.296 presents the results of X-ray microanalysis of artificial fabric fibers on an area of specimen's internal surface. The microanalysis was performed at the point shown in the backscattered electron image. Chemical element concentrations are given in mass percentages. The concentrations of all chemical elements were measured on K lines of characteristic X-ray emission. Figure 3.296 also presents the spectra of characteristic X-ray emission at this measurement point.

Figure 3.297 demonstrates the results of X-ray microanalysis of the composite matrix (binder) on the area of the specimen's internal surface. The microanalysis

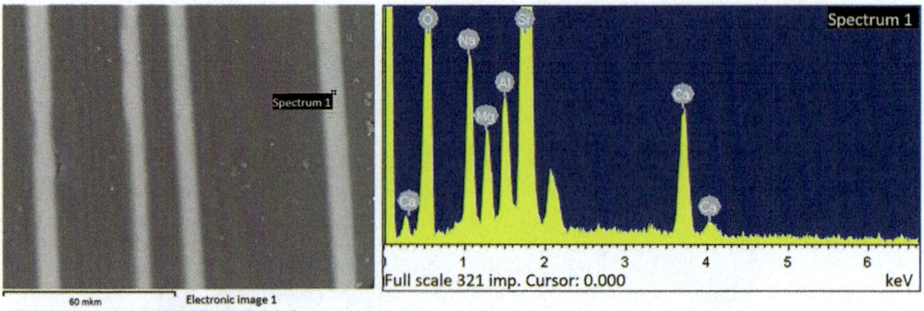

Element	Weight %
O K	39.71
Na K	7.26
Mg K	3.20
Al K	3.70
Si K	37.37
Ca K	8.76
Total	100.00

Fig. 3.296 Results of X-ray microanalysis of artificial fabric fiber on the area of specimen's internal surface

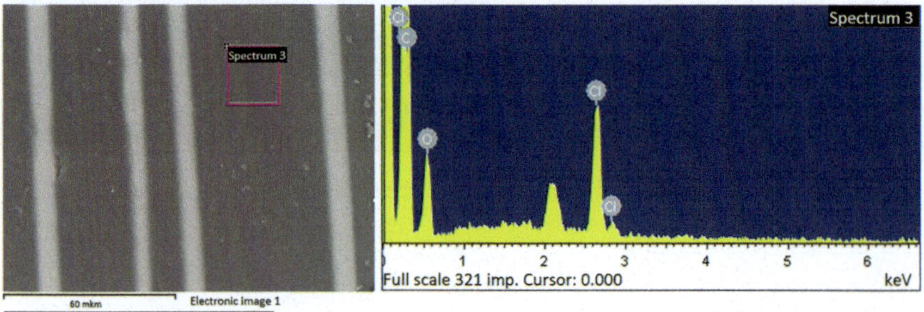

Element	Weight %
C K	75.06
O K	18.42
Cl K	6.52
Total	100.00

Fig. 3.297 Results of X-ray microanalysis of the composite matrix on specimen's internal surface

was performed in the rectangular region shown in the backscattered electron image. Chemical element concentrations are given in mass percentages. The concentrations of all chemical elements were measured on K lines of characteristic X-ray emission. Figure 3.297 also shows the spectra of characteristic X-ray emission in this measurement region.

Figure 3.298 presents the backscattered electron image of an area of the specimen's internal surface (with scale mark 60 μm), as well as the image of the same area in the X-ray characteristic emission (the spatial distribution of chemical elements over

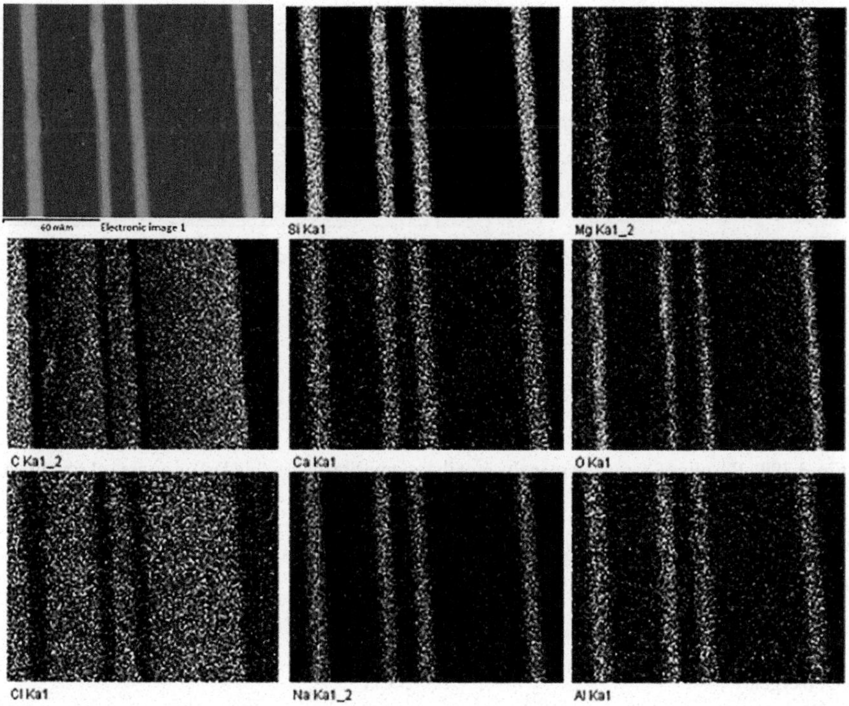

Fig. 3.298 Backscattered electron image of the internal surface area of *ORLAN* UAV fragment specimen (with 60 μm scale mark), alongside with the image of the same area in X-ray characteristic emission (the spatial distribution of chemical elements over the area studied). White dots correspond to the presence of the chemical element specified under each image

the examined area). White dots correspond to the presence of the chemical element specified under each image.

It can be seen from Fig. 3.298 that the images of the X-ray distribution of the chemical elements present in artificial fabric fibers (Si, Mg, Ca, O, Na, Al) clearly correlate with backscattered electron images, and X-ray images of the distribution of chemical elements in the composite matrix clearly correlate with the regions of matrix location in the backscattered electron image (the dark-colored regions).

Figure 3.299 shows the results of X-ray microanalysis of the composite matrix (binder) in an area of the specimen's slice. The microanalysis was performed in the rectangular region shown in the backscattered electron image. Chemical element concentrations are given in mass percentages. The concentrations of all chemical elements were measured on K lines of characteristic X-ray emission in this measurement region.

If the results of X-ray microanalysis presented in Figs. 3.297 and 3.299 are compared, it can be concluded that the measured concentrations of chemical elements in the composite matrix are practically the same in the two measurements.

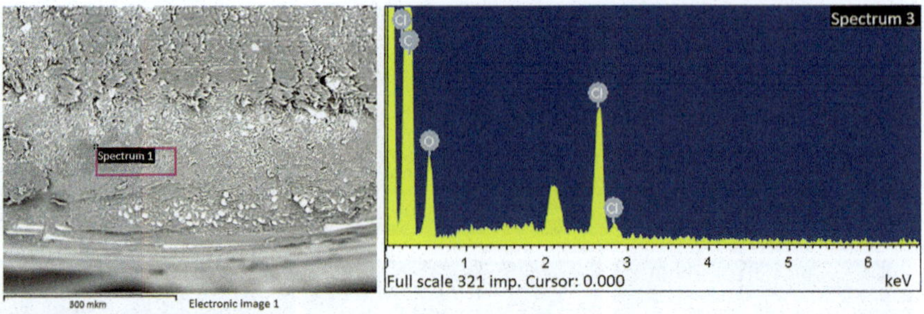

Element	Weight %
C K	75.54
O K	18.50
Cl K	5.96
Total	100.00

Fig. 3.299 Results of X-ray microanalysis of the composite matrix (binder) on an area of specimen's slice

Table 3.14 Results of studying chemical element concentrations in glass fibers of specimen's composite material

Chemical elements	Mass concentrations, %
Si, silicon	37.37
O, oxygen	39.71
Ca, calcium	8.76
Na, sodium	7.26
Al, aluminum	3.70
Mg, manganese	3.20

Thus, the studied material of *ORLAN* UAV fragment is a composite with a polymer matrix and with inclusions of fiberglass artificial fabric layers of 1.2 mm total thickness, and with the coating layer about 30 μm thick on its external surface. The layers of artificial fiberglass fabric (composite inclusions) are located throughout the entire bulk of the composite.

Glass fibers are about 10 μm in diameter, and their chemical composition is shown in Table 3.14.

The results of studying the concentrations of chemical elements in the matrix of specimen's composite material are shown in Table 3.15.

The layer about 30 μm thickness, which was applied onto the external surface of the specimen, is a paint, and its chemical composition is presented in Table 3.16.

Table 3.15 Results of studying chemical element concentrations in the matrix of specimen's composite material

Chemical elements	Mass concentrations, %
C, carbon	75.54
O, oxygen	18.50
Cl, chlorine	5.96

Table 3.16 Results of studying chemical element concentrations in the paint layer applied onto specimen's external surface

Chemical elements	Mass concentrations, %
C, carbon	65.59
O, oxygen	22.14
Ti, titanium	11.80
Al, aluminum	0.47

3.3.5 *Results of Studying Fragments of Multiple Launch Guided Missiles*

3.3.5.1 Fragment of GPS Antenna Cover of 9M544 *Tornado-C* Guided Missile

Figure 3.300 shows the fragment of GPS antenna cover of 9M544 *TornadospsC* guided missile; from it a specimen for research was cut. The fragment is shown from both sides: the convex external surface (left) and concave internal surface (right).

Figure 3.301 presents secondary electron image (SEI) of the convex, more lustrous (external), surface of the antenna cover fragment.

Figure 3.302 shows a secondary electron image (SEI) of the less lustrous (internal) surface of the antenna cover.

To study the structure and measure chemical element concentrations in the specimen's bulk, a slice of antenna cover fragment was produced. Secondary electron image of the slice was obtained with JSM-6490LV electron microscope and is shown in Fig. 3.303.

Figure 3.304 demonstrates the results of X-ray microanalysis in the bulk (in the slice) of the antenna cover fragment. The microanalysis was performed in a certain rectangular area (the measured chemical element concentrations were averaged within this area), which is shown in the secondary electron image (SEI). Chemical element concentrations are given in mass percentages. The concentrations of all chemical elements were measured on K lines of characteristic X-ray emission. Figure 3.304 also shows the spectrum of characteristic X-ray emission of this measurement area.

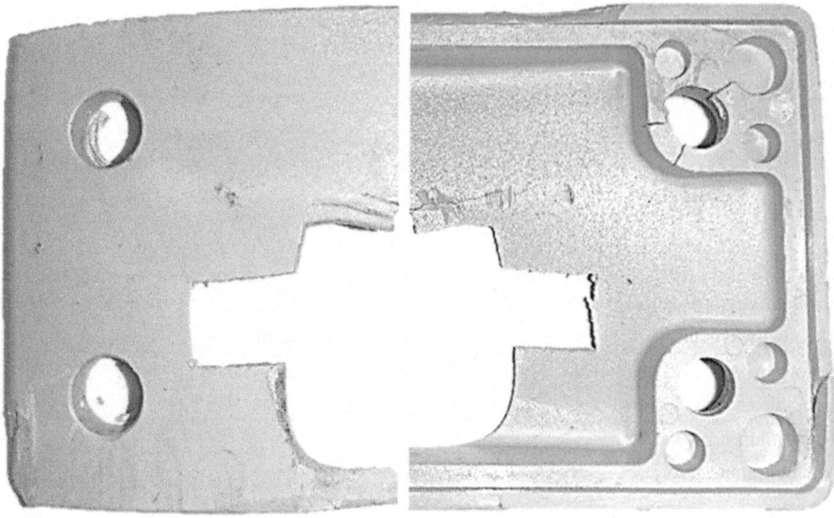

Fig. 3.300 Image of GPS antenna cover of 9M544 *Tornado-C* guided missile fragment shown from both sides: the convex external surface (left) and concave internal surface (right)

Fig. 3.301 Secondary electron image (SEI) of the convex external surface of the antenna cover fragment. Accelerating voltage is 15 kV, magnification ×1000, scale mark—10 μm

The results of X-ray microanalysis performed on the external and internal surfaces of the antenna cover fragment are identical to the results of microanalysis carried out in the bulk (in the slice) of the antenna cover fragment.

The studies of the fragment structure and chemical composition showed that this specimen, most probably, is a polymer that consists of the following chemical elements: C (carbon), O (oxygen), Ti (titanium); besides, it is most likely that H (hydrogen) is present there. It should be noted that hydrogen concentration was not measured in view of the peculiarities of scientific equipment operation.

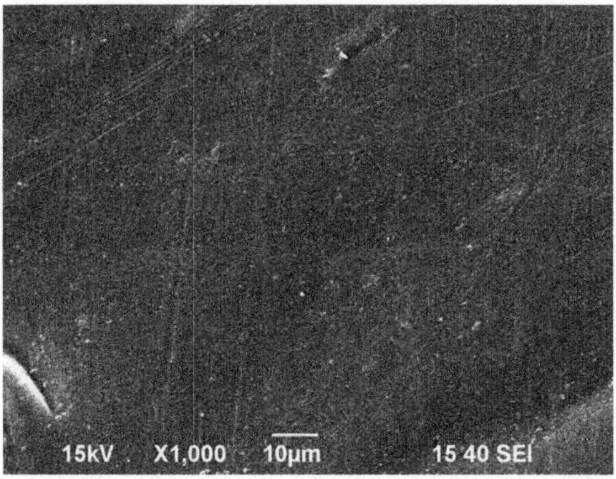

Fig. 3.302 Secondary electron image (SEI) of the concave internal surface of the antenna cover fragment. Accelerating voltage is 15 kV, magnification ×1000, scale mark—10 μm

Fig. 3.303 Secondary electron image (SEI) of the slice of antenna cover fragment. Accelerating voltage is 15kV, magnification ×1000, scale mark—10 μm

In the studies of the structure and chemical composition of the surface and bulk of the antenna cover fragment, inclusions that could lead to the absorption or scattering of electromagnetic radiation were not detected. Therefore, the studied fragment of the GPS antenna cover of 9M544 *Tornado-C* guided missile is an RF transparent material.

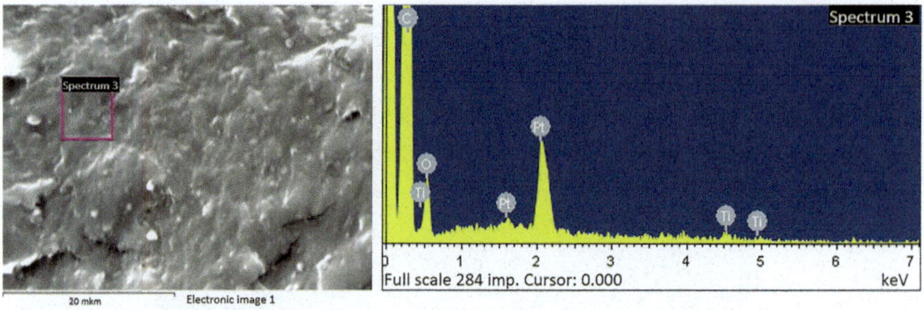

Element	Weight %
C K	86.79
O K	14.48
Ti K	0.51
Total	100.00

Fig. 3.304 Results of microanalysis in the bulk (in the slice) of the antenna cover fragment

3.3.5.2 Fragment of the Structural Element of 9M544 *Tornado-C* Guided Jet-Propelled Projectile

Figure 3.305 shows a fragment of the structural element of 9M544 *TornadospsC* guided jet-propelled projectile from which a specimen for research was cut. The fragment material has homogenous porous structure and is light colored.

Figure 3.306 demonstrates the secondary electron image (SEI) of the structural element surface.

Fig. 3.305 Image of a structural element fragment of 9M544 *Tornado-C* guided jet-propelled projectile

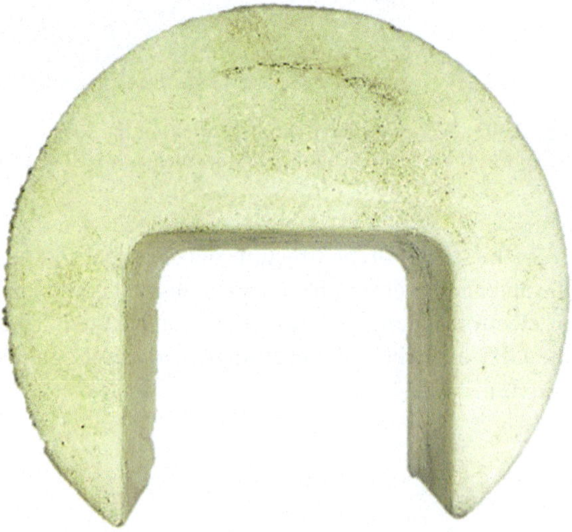

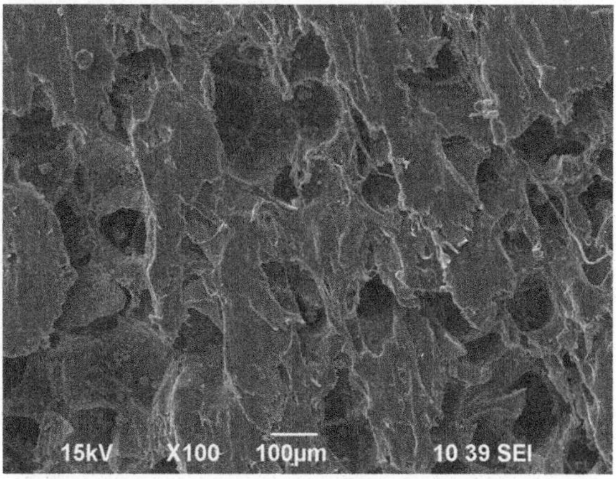

Fig. 3.306 Secondary electron image (SEI) of structural element surface. Accelerating voltage is 15kV, magnification ×100, scale mark—100 μm

Fig. 3.307 Backscattered electron image (BEC) of the structural element in the bulk (in the slice). Accelerating voltage is 15kV, magnification ×50, scale mark—500 μm

Figures 3.307 and 3.308 present backscattered electron images (BEC) in the bulk (in the slice) of the structural element at various magnifications.

It can be seen from Figs. 3.307 and 3.308 that in the bulk of the structural element there are pores, oval-shaped cavities of characteristic 50–150 μm sizes.

Figures 3.309 and 3.310 demonstrate the results of X-ray microanalysis in the bulk (in the slice) of the structural element. The microanalysis was performed in a certain rectangular area (the measured chemical element concentrations being averaged within this area), which is shown in secondary electron image. Chemical element concentrations are given in mass percentages. The concentrations of all chemical elements were measured on K lines of characteristic X-ray emission. Figure 3.309 also presents the spectrum of characteristic X-ray emission of this measurement area.

The studies of the chemical composition and structure of 9M544 *TORNADO-C* guided jet-propelled projectile showed that it consists of a foamed polymer that is uniform in structure and chemical composition. The structural element has the

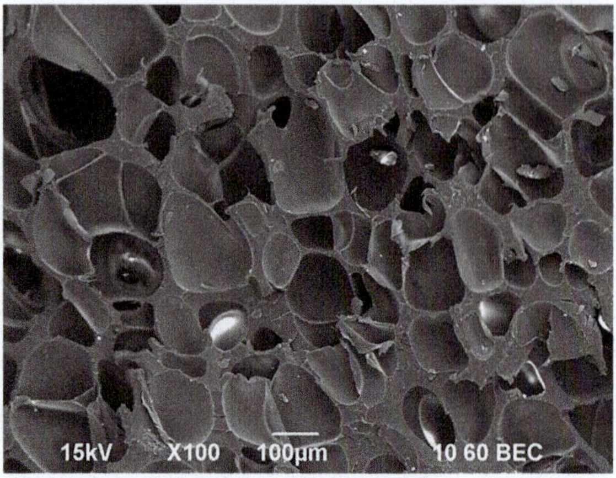

Fig. 3.308 Backscattered electron image (BEC) of the structural element in the bulk (in the slice). Accelerating voltage is 15 kV, magnification ×100, scale mark—100 μm

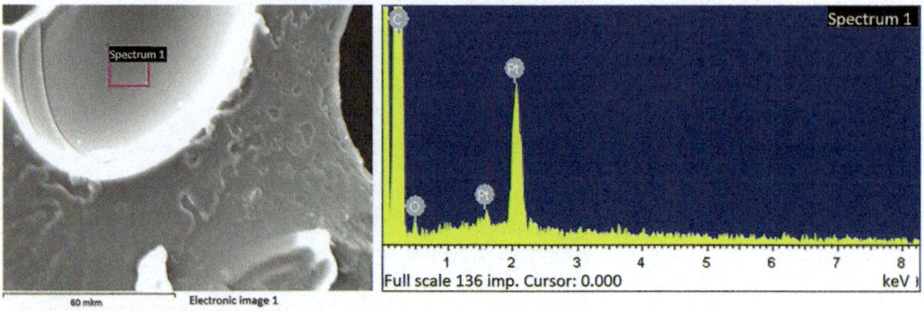

Element	Weight %
C K	97.95
O K	2.05
Total	100.00

Fig. 3.309 Results of microanalysis in the bulk (in the slice) of the structural element of 9M544 *Tornado-C* guided jet-propelled projectile. The measurements were carried out in the region of cavity cells

following chemical composition: C (carbon)—28%, O (oxygen)—1.72%, and H (hydrogen) is possibly present as well. Besides, it should be taken into consideration that the scientific equipment used does not permit the measurement of hydrogen concentration.

Taking into account the structure of the element in question (a porous foam material), the conclusion can be made that the structural element of the 9M544 *TORNADO-C* guided jet-propelled projectile has good heat insulation properties.

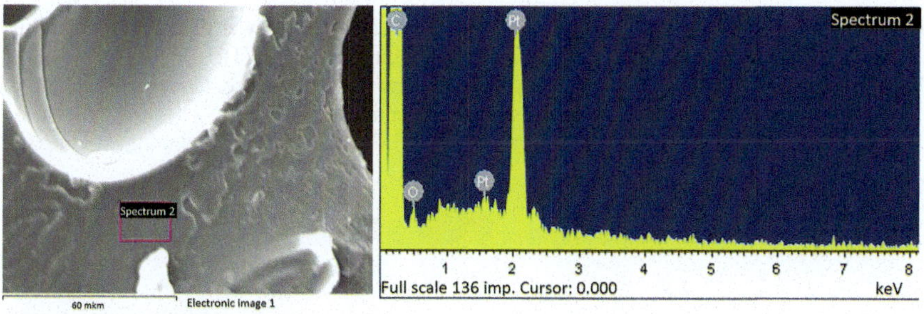

Element	Weight %
C K	98.28
O K	1.72
Total	100.00

Fig. 3.310 Results of X-ray microanalysis in the bulk (in the slice) of the structural element of 9M544 *Tornado-C* guided jet-propelled projectile. The measurements were carried out in the region outside the cavity cell

References

Ahmad Z, Jafri HZ, Basha RH (2021) Characterization techniques for composites using AI and machine learning techniques. In: Computational technologies in materials science. CRC Press, Boca Raton, London, New York

Ahmad Z, Jafri HZ, Basha RH (2022) Characterization techniques for composites using AI and machine learning techniques. In: Computational technologies in materials science. CRC Press, Boca Raton, London, New York

Baker A, Dutton S, Kelly D (2004) Composite materials for aircraft structures. American Institute of Aeronautics and Astronautics, Inc., Reston, VA

Chen J, Ding N, Li Z, Wang W (2016) Organic polymer materials in the space environment. Prog Aerosp Sci 83:37–56

Dhas JER, Arun M (2022) A review on development of hybrid composites for aerospace applications. Mater Today Proc 64:267–273

Dikin D, Kohlhaas K, Dommett G, Stankovich S, Ruoff R (2006) Scanning electron microscopy methods for analysis of polymer nanocomposites. Microsc Microanal 12(S02):674–675

Gilliam M (2014) Polymer surface treatment and coating technologies. Chem Eng Publ 11:99–124

Goldstein JI, Yakowitz H (1975) Practical scanning electron microscopy: electron and ion microprobe analysis. Plenum Press, New York, London

Herman B, Lemasters JJ (2012) Optical microscopy: emerging methods and applications. Elsewier, Amsterdam

Li Y, Xiao Y, Yu L, Ji K, Li D (2022) A review on the tooling technologies for composites manufacturing of aerospace structures: materials, structures and processes. Compos A Appl Sci Manuf 154(4):106762

Newbury DE, Yakowitz H (1975) Specimen preparation, special techniques, and applications of the scanning electron microscope. In: Practical scanning electron microscopy, electron and ion microprobe analysis, pp 211–262

Oatley CW (1972) The scanning electron microscope. Cambridge University Press, Cambridge

Spence JCH (2013) High-resolution electron microscopy, 4th edn. Oxford University Press, Oxford

Stanescu MM, Bolcu D, Ciuca I, Dinita A (2014) Non uniformity of composite materials reinforced with carbon and carbon-kevlar fibers fabric. Materiale Plastice 51(4):355–358

Ursache Ş, Cerbu C, Hadăr A (2024) Characteristics of carbon and kevlar fibres, their composites and structural applications in civil engineering-a review. Polymer 16(1):127
Woodruff DP, Delchar TA (1994) Modern techniques of surface science. Cambridge Solid State Science Series USA. Cambridge University Press, Cambridge

Chapter 4
Application of Scanning Electron Microscopy for the Study of Fragments of Radio-Absorbing Composite Materials and Coatings

This chapter presents a review of the methods of measuring materials' electrophysical parameters in the UHF range. It considers theoretical foundations of electromagnetic radiation interaction with composite materials and shows the application of analytical expressions obtained with reliance on well-known theoretical approaches to calculating the parameters of RF absorbent materials.

An approach is developed for computing the parameters of frequency characteristics of RF absorbent coatings and materials of the components of special-purpose structures. It is based on using the results of theoretical analytical research into the interaction of electromagnetic waves with composite RF absorbent materials and on specific experimental data obtained with electron microscopy and X-ray microanalysis techniques.

The approach proposed allows a highly accurate study of the internal structure, including geometric parameters, of fillers and coating layers, the quantitative chemical compositions of coating materials, which, in turn, makes it possible to model the physical processes occurring during the interaction of radio waves and the material of RF absorbent coating in a wide frequency range.

Examples of calculating the operation range width of RF absorbent coatings and the limit wavelengths of that range obtained with the application of the proposed approach are demonstrated.

4.1 Peculiarities of Investigating Frequency Characteristics of RF Absorbent Composite Materials and Coatings

Studying and designing RF absorbent materials necessitate both experimental work and calculation-based modeling of their radio-technical characteristics by solving problems on the interaction of electromagnetic waves with materials and structures.

V. L. Bogdanov et al., *Identification of Special-Purpose Structures by Their Fragments Based on Scanning Electron Microscopy*, Advanced Structured Materials 233,
https://doi.org/10.1007/978-3-031-88827-4_4

Along with simple calculations of, e.g., single-layer absorbent materials, in the case of complex structures the application of relevant computation apparatus is required.

Current experimental methods of measuring electrophysical parameters of absorbent materials in ultra-high-frequency (UHF) range can be conditionally subdivided into four groups: resonator methods, coaxial probe method, transmission line method, and free space method, which with certain assumptions can be considered as a kind of measurement methods in transmission lines. The choice of the most appropriate measurement method in each particular case depends on numerous factors: the frequency range required, the expected values of dielectric and magnetic permeabilities, the demanded precision of measurements, the properties and type of material, the size of the specimen investigated, etc. Experimental methods of studying RF absorbent materials and coatings will be considered in more detail below.

Professionals give a lot of attention to developing RF absorbent materials based on composites in which dielectric matrices consisting of high-molecular polymer compounds contain various fillers. Such fillers can be metal particles, ferrites, resistive threads, electrically conductive inclusions of complex shapes, carbon-containing fibers and nanotubes, etc. To determine the parameters of absorbent materials, understand the physical processes occurring during electromagnetic wave interaction with these materials, as well as develop effective absorbers, various theoretical approaches are used. In this sense, of great interest are studies of multi-layer and gradient RF absorbent materials since the materials of exactly this class have good absorbing properties in a wide frequency range from 3 to 35 GHz.

Of the theoretical methods employed to study an arbitrary layered structure, one of numerous interpretations of the so-called transfer-matrix method as well as recurrence methods and impedance method are often used.

It should be noted that despite the effectiveness of the recurrency (numerical) methods mentioned, as before, analytical methods remain highly relevant.

The abovementioned experimental and theoretical numerical methods of investigating RF absorbent coatings are especially effective when they are applied by the developers of those coatings. In this case, they know in advance many parameters of the research object, e.g., the chemical composition of the coating material or the layer thickness of a multi-layer coating.

However, in a number of cases there is a need to investigate samples of RF absorbent coatings developed by other organizations, including the coatings used in special-purpose products. During the study of such materials, it is nearly impossible to prepare specimens of the necessary shape and fulfill necessary requirements for conducting investigations with the known experimental methods of measuring materials' electrophysical parameters in the UHF range. Hence, it would be appropriate to consider the prospects of using a somewhat different research procedure.

The procedure proposed to investigate RF absorbent coatings employed in the samples of special-purpose structures relies on electron microscopy and X-ray microanalysis methods. Unlike radio-engineering techniques, where the parameters of RF absorbent coatings (coefficients of radio waves reflection and absorption, values of dielectric and magnetic permeabilities, etc.) are measured without the initial information on the internal structure and chemical composition of coatings, these methods permit high-precision studies of their internal structures, including geometric parameters of coating fillers and layers, determination of quantitative chemical composition of the coating material, which, in turn, allows modeling of physical processes occurring during the interaction of radio waves with the material of RF absorbent coating in a wide frequency range. At the next stage, with reliance on the results of known theoretical analytical studies concerning the interaction of electromagnetic waves with RF absorbent composite materials, as well as the specific results of experimental studies of RF absorbent coatings, which were obtained with electron microscopy and X-ray microanalysis methods, researchers determine the working range width for the RF absorbent coating and, finally, based on the physical mechanisms of radio wave absorption in the given material, find the limit wavelengths of the range. The data obtained enable them to formulate some recommendations on electromagnetic radiation parameters for the reliable identification of special-purpose structures.

Section 4.1.1 of this chapter presents a review of the methods of measuring materials' electrophysical parameters in UHF range.

Section 4.1.2 considers theoretical foundations of electromagnetic radiation interaction with materials, including composite and multi-layer ones, along with analytical expressions for calculating RF absorption parameters that were obtained with reliance on well-known theoretical approaches.

Further subunits, on the basis of the achieved analytical expressions achieved for RF absorbent materials and using specific results of experimental studies of RF absorbent coatings, which were obtained with electron microscopy and X-ray microanalysis methods, provide examples of calculating working range widths of RF absorbent coatings and limit wavelengths of the range, which permits the formulation of some recommendations concerning electromagnetic radiation parameters for reliable identification of special-purpose structures.

4.1.1 Methods of Measuring Electrophysical Parameters of Materials in UHF Range

There is a large variety of methods to measure electrophysical parameters of materials in UHF range (Parkhomenko et al. 2019; Jilani et al. 2012; Hyde and Havrilla 2008; Dudeck and Buckley 1992; Ghodgaonkar et al. 1990; D'jakonova and Kazancev 2019): those using coaxial cables and probes, waveguide transmission lines; various resonator, waveband techniques; etc. are also common. A brief review of main experimental methods of measuring material electrophysical properties and parameters at UHF is given below.

Resonator method of measuring materials' electrophysical parameters is a special case of scalar methods of measuring materials' parameters (D'jakonova et al. 2017; Lyubchenkov 2009). They are highly sensitive, provide separate measurements of complex dielectric and magnetic permeabilities, and are applicable in a wide frequency range: from several MHz to tens of GHz. These methods are based on measuring the properties of oscillatory systems (oscillatory system quality factor and resonance frequency) when samples of the materials studied are inserted into them (Janezic and Baker-Jarvis 1999). The resonator has high quality factor and resonates at certain frequencies. A material specimen placed into the resonator cavity influences the resonance frequency f, as well as the resonator quality factor Q. Since the resonance frequency and quality factor of oscillating systems can be measured with high precision, resonator methods of measuring substance parameters are used at UHF, mainly for measuring dielectric specimens with low dielectric losses or in the cases where high sensitivity is required (Keysight Technologies 2017). For studying dielectric permeability of a material, its specimen is to be placed into the antinode of electric field, for studying material's magnetic permeability parameters—into the antinode of magnetic field.

Consider in more detail how resonator methods are employed on the example of determining complex magnetic permeability of material specimen by means of cavity resonator.

First, at the first stage, the resonance frequency and quality factor of the system are measured in the initial state without specimen presence. Then a specimen is inserted into the oscillating system, and the same system parameters are measured again. Next, real (ε') and imaginary (ε'') parts of dielectric permeability can be determined by such formulas (Zajicek et al. 2006):

$$\varepsilon' = 1 + \frac{V_c(f_c - f_s)}{2V_s f_s}, \quad \varepsilon'' = \frac{V_c}{4V_s}\left(\frac{1}{Q_s} - \frac{1}{Q_c}\right), \tag{4.1}$$

where f_c is resonance frequency of the empty resonator, f_s is resonance frequency of the resonator with the specimen placed into it, Q_c is quality factor of the empty resonator, Q_s is quality factor of the resonator containing the specimen measured, V_c is volume of empty resonator, and V_s is volume of the specimen.

Thus, when a specimen is inserted into the resonator, resonance frequency is shifted to the low-frequency region determined by the real part of the dielectric permeability of the substance studied. Resonator transmission bandwidth is increased by the specimen, which, in turn, is determined by the imaginary part of dielectric permeability of the specimen studied. The resonance frequency and quality factor of the empty resonator can be calculated by its geometric dimensions. It is also very important to determine the volume of the studied specimen accurately.

The same procedure is used to determine the magnetic permeability of the specimen. Since the maximums of electric and magnetic fields in the resonator are shifted in respect of each other, magnetic field antinode corresponds to electric field zero, and vice versa. The mutual influence of material's dielectric and magnetic properties is very weak, e.g., magnetic permeability close to unity can be measured with high precision even if sample dielectric permeability is high. In this respect, resonator techniques have certain advantages over other measurement methods, which will be discussed below.

High sensitivity of resonator techniques is due to the repeated passage of electromagnetic wave inside the resonator. Its average path is approximately equal to the product of resonator length by its quality factor Q. Therefore, the wave passes through the resonator Q times, and field amplitude in the place of specimen location also increases by Q times. As a result, the response from the specimen inserted into the resonator is amplified, which increases measurement sensitivity. On the other hand, this reduces the permissible range of the measured values of imaginable parts of dielectric and magnetic permeabilities. Besides, an important disadvantage of all resonator methods of measurements is their narrowband characteristics, i.e., measurements can only be carried out at a single dedicated (fixed) frequency or at a series of discrete frequencies in the case when multi-mode resonator is used. Also, there are certain limitations related to the specimens measured, viz., it is necessary for them to be rather thin and have relatively low losses. Otherwise, it would be difficult to detect resonance when the specimen measured is placed into a resonator system.

Coaxial probe method is employed solely for measuring dielectric properties of the materials whose magnetic permeability is equal to unity. It belongs to the group of non-destructive control methods, so it is used when there is a need to study dielectric material parameters of a ready product or if it is impossible to make a test specimen.

Usually, an open-ended coaxial probe is used. In fact, it is a transmission line with one end cutoff. The measurement of material's complex dielectric permeability is carried out by immersing the probe into a liquid or by touching a flat surface of a solid material with it (Zajicek et al. 2006). The fields generated on the opened end of the probe penetrate into the material and change as a result of the contact with the material investigated. Then, the reflection coefficient of such line, which is a complex value, can be measured and correlated with the value of dielectric permeability (Ovsov 2016) and that way to measure the latter.

Before the measurement, it is necessary to perform the complete procedure of measuring system calibration. A common method of calibrating (Zajicek et al. 2006; Ovsov 2016) consists in taking standard short circuit and idling measures, and using a liquid with known values of dielectric permeability as the matched load.

Open-ended coaxial probes are intended, primarily, for measuring liquid materials since in this case the probe is completely immersed into a liquid environment. When this method is used to measure solid materials, there is an inevitable air gap between the probe surface and the specimen measured. Three-stage calibration to some extent compensates for errors that can happen in reflected signal measurements. The main source of error in the measurements by coaxial probe method is the instability of UHF cable phase characteristics as well as the presence of air gaps between the specimen measured and probe structural elements.

In the case of measuring the dielectric permeability of solid materials by coaxial probe method, the specimen surface must be quite smooth (the gap between the specimen and the probe must not be larger than 50 μm) and it must be sufficiently thick for the main electromagnetic field energy to propagate in the material rather than the air gap between the probe and the specimen. Current instruments relying on coaxial probe method are provided with specialized software, ensure high measurement speed, have portable design, a wide frequency range (10 MHz–67 GHz), depending on the type of probes used, and are easily calibrated (Degiorgi et al. 2017).

As a result, the main advantages of the coaxial probe method are the possibility of measuring electrophysical parameters of liquid and powdered substances, simplicity and easiness of calibration and measurements, including non-destructive analysis. Another advantage is a wide frequency range of probe operation—from tens of megahertz to tens of gigahertz. The main disadvantage of coaxial probe technique is its limited precision as compared to transmission line, free space, or cavity resonator methods.

Transmission line methods (Rohde and Schwarz 2006) are widely used for determining complex dielectric and magnetic permeabilities of materials in UHF range and consist in establishing the parameters (reflection, absorption, and transmission coefficients) of the measurement line that contains the specimen investigated. A typical measuring installation for implementing transmission line method consists of transmission line and software tools for calculating dielectric and magnetic permeabilities. Usually, a section of rectangular waveguide or a coaxial overhead line is used as the transmission line.

At present, there are many various techniques for determining the values of dielectric and magnetic permeabilities of material specimen by measuring transmission line parameters (Nicolson and Ross 1970). One of the most common is Nicolson–Ross–Weir model (Weir 1974; You et al. 2015), which is applied for waveguide transmission line. According to it, on the basis of experimentally measuring the coefficient of reflection R of d thick specimen and transmission coefficient T (reflection and transmission coefficients are measured in waveguide planes located on the opposite sides of the specimen), the values of complex ε and μ are calculated by means of the following expressions (Rozanov et al. 2002):

$$\mu = \frac{\lambda_0}{\sqrt{1 - (\lambda_0/\lambda_c)^2}} \left(\frac{1+R}{1-R}\right) \left[\frac{i}{2\pi d} \ln\left(\frac{1}{T}\right)\right]$$
$$\varepsilon = \frac{\lambda_0}{\mu} \left\{\frac{1}{\lambda_c^2} - \left[\frac{1}{2\pi d} \ln\left(\frac{1}{T}\right)\right]^2\right\}, \tag{4.2}$$

where λ_0 is wavelength in the free space and λ_c is wavelength in the waveguide.

As can be seen from (4.2), waveguide transmission lines are determined by the type (cross section) of the waveguide used, therefore, their wave range is banded.

Coaxial transmission line technique is applicable to a wider frequency range than waveguide ones, however, it requires toroidal specimens, which are more difficult to make. For maximal filling of the measured coaxial line, the specimen has the shape of a disk with a hole in its center. Complex permeabilities of the specimen are determined by the results of measuring its reflection coefficient in the short-circuited line mode, and also in the line that imitates idling. To eliminate measurement errors resulting from the measuring path imperfections, measuring system calibration is applied. Calibration is performed by determining correction factors in the formula describing the disparity of the reflection coefficient measured in the case of imperfect path and its true value (Politiko et al. 2018).

Summarizing the above, it can be noted that measurements in transmission lines is one of the most universal methods of measuring electrophysical parameters of materials. It does not require specialized measuring equipment, and the specimens measured have small overall dimensions that depend on the type of transmission line employed. Besides, this method permits measurements of anisotropic materials and those with high magnetic losses.

However, it should be remarked that despite measuring path calibration the method of measuring material parameters in transmission lines has a significant potential systemic error caused by the presence of air gaps between transmission line elements and the specimen. Such gaps, in which electric field is concentrated, lead to systematic underestimated values of dielectric permeability. It would be noted that the measurements of magnetic permeability are free of the abovementioned errors since the presence of small gaps does not change magnetic field distribution in the specimen area (Veselago 1968).

Free space method is in many cases used as the primary one for measuring radiophysical parameters of absorbent coatings.

The methods of measuring the parameters of RF absorbent materials that were considered above, despite all their advantages mentioned, have a number of key limitations, which in many cases restrict their application. First, during coaxial and waveguide measurements, there are inevitable gaps between the specimen and measuring elements, which results in uncontrollable inaccuracies and errors of measurements. Second, fundamental difficulties appear in measuring the parameters of a large class of metamaterials (Semenenko et al. 2019) that contain inclusions of significant dimensions in the form of fibers, wire spirals, rings, and dielectric resonators. Third, significant practical difficulties appear in measuring electrophysical characteristics of widely used geometrically heterogeneous materials, e.g., pyramid-shaped RF absorbent materials based on polyurethane foam. In this connection, the method of non-contact measurements in free space is better both in controlling electromagnetic parameters of magneto-dielectric materials, including heterogeneous ones, in the production process where the possibility of non-contact measurements can be determined, and in scientific research (Varadan et al. 1991).

Some aspects of free space method have been borrowed from optics, so it is also referred to as quasioptical. The idea of quasioptical method consists in examining electromagnetic waves transmitted and reflected from the environment, which are gathered in narrow bunches by means of respective guiding devices (emission sources and receivers). As a rule, free space method relies on horn directional antennas for energy focusing on the material specimen and registering the intensity of the wave reflected from the specimen or the wave transmitted through it. This method is non-contact and can be applied to materials that are to be studied at high temperatures and in aggressive environments (Challa et al. 2008).

Despite its simplicity, the technical implementation of quasioptical method runs into quite a number of difficulties in the case when the wavelength of electromagnetic radiation proves to be comparable to the sizes of the specimens studied and the antennas used. This leads to various diffraction phenomena that interfere with measurements. To reduce measurement errors resulting from diffraction on specimen edges and multiple reflections (re-reflections and mutual reflections) between antennas and path heterogeneities, various calibration procedures are employed (Bartley and Begley 2005, 2012; Akhter and Akhtar 2016). Besides, methods of signal digital filtration in time domain (Abeles 1948) are actively used.

One of the main advantages of free space method over other measurement techniques is its applicability in field conditions, when there is no possibility of conducting measurements in laboratory on stationary measuring equipment. In fact, such task often appears, e.g., when RFAC is to be applied on large-sized objects and coating control is usually performed in the process of its application to the object. Besides, radiophysical control is necessary after the application of coatings for assessing their quality during exploitation of those objects. In the majority of cases, there is no necessity and feasibility of determining material parameters of coatings on the objects, while the most informative parameter measured during radiophysical control is frequency dependence of coating reflection coefficient. In this case, to carry out radiophysical control by free space method directly on the product, a portable measuring kit is used.

In connection with analyzing the method of studying the characteristics of RF absorbent materials in free space, it would be advisable to consider such device as anechoic chamber. Anechoic chamber is a special research room inside which RF waves are not reflected from its walls, floor, or ceiling. Such effect is achieved due to RF absorbent material covering the walls, floor, and ceiling inside it. Porous pyramid-shaped absorbers are frequently used as RF absorbent material. The cover is fabricated to absorb RF waves arriving at the material from all possible directions inside the room. Regular RF anechoic chambers have absorption level up to 40 dB in 1 GHz–40 GHz frequency range. To prevent the influence of any external signals on the source of RF waves and the receiver, such chambers are usually designed in such a way that respective external signals cannot get inside the chamber. Transmitting and receiving antennas, as well as other equipment situated in the chamber, are also coated with a RF absorbent material. RF anechoic chambers are used in building diagrams of radiation directivity, studying electromagnetic compatibility and building diagrams of electromagnetic waves scattering. Measurements can be carried out on full-sized objects, including unmanned aerial vehicles, or on reduced models. In these cases, the processes of electromagnetic waves scattering from the objects investigated occur in the same way as in the air space.

Thus, as it was noted above, all main methods of measuring radiophysical parameters of RF absorbent materials in UHF range, used at present, can be conditionally subdivided into four groups: resonator methods, coaxial probe method, transmission line method, and free space method. The choice of the most appropriate measurement method in each specific case depends on numerous factors: the frequency range required, expected values of dielectric and magnetic permeabilities, necessary precision of measurements, material type and properties, the size of the specimen studied, etc.

So, resonator methods of measurements have high precision and are the best for materials with low losses. Yet, their significant limitations are narrow bands (measurements are carried out at a single fixed frequency or at several allocated frequencies in the case of multi-mode resonator) and the infeasibility of simultaneously determining both dielectric and magnetic permeabilities. Coaxial probe method is applicable for much wider bands and is the best for liquids and bulk materials. The principal limitation is its suitability for measuring dielectric permeability of homogenous substances only. Transmission line methods permit measurements in a broad frequency band, they are applicable to anisotropic and high loss materials. However, very stringent requirements are imposed on the specimen fabrication quality, as the specimen investigated should completely, without gaps, fill transmission line cross section since that strongly affects the accuracy of measurements.

Using free space measurement method allows researchers to avoid practically all the abovementioned limitations of other techniques. It is non-contact and belongs to the group of non-destructive methods, can be employed for materials with both low and high values of dielectric and magnetic permeabilities, and permits measurements at high temperatures and in aggressive environments. Free space method can be used both for flat specimens and for measuring parameters of geometrically heterogeneous RF absorbent materials, even of metamaterials, which is an important advantage in research. It should also be noted that one of the main advantages of free space method in comparison with other techniques is its applicability for measurements in field conditions, when it is not possible to carry out measurements in laboratory on stationary facilities.

4.1.2 Theoretical Fundamentals of Electromagnetic Radiation Interaction with Materials, Including Composite and Multi-layer Ones

Transfer-matrix method. Recurrency methods

Considerable attention of materials science specialists is given to the development of RF absorbent materials based on composites in which dielectric matrices consist of high-molecular polymer compounds and contain various fillers. Metal particles, ferrites, resistive threads, electrically conductive inclusions of complex-shaped particles, carbon-containing fibers, nanotubes, etc. are used as fillers. Various theoretical approaches are employed to determine absorbent material parameters, understand physical processes occurring during electromagnetic wave interaction with those materials, and to design effective absorbers. In this sense, investigations of multi-layered and gradient RF absorbent materials are of great interest since the materials of this class have good absorption properties in a wide frequency range of 3–35 GHz.

As to theoretical methods applied to studying an arbitrary layered structure, one of the numerous interpretations of the so-called transfer-matrix methods is employed (Thomson 1950; Haskell 1953; Lowe 1995; Morozov et al. 2001; Landau and Lifshitz 1984), which was first used for optical waves (Thomson 1950). A detailed present-day analysis of matrix approaches is made in review (Stolyarov 1988).

For further calculations, notions of the characteristic matrix of a layer, which has its physical parameters, and the characteristic matrix of a multi-layered structure are introduced. The characteristic matrix of a multi-layered structure is equal to the product of characteristic matrices of its layers: $M = M_1 M_2 \times \cdots \times M_N$, where N is the number of layers.

To measure matrix M, the well-known matrix multiplication rule is used. The calculation begins with determining $M_{N-1} M_N$ product, by which matrix M_{N-2} is then multiplied and so on. Reflection and transmission coefficients are expressed by means of matrix M elements.

In addition to that, researchers widely use other recurrency methods, whose essence is to determine the relation between physical values in neighboring layers. For example, it was shown in Morales and Nuevo (1991), Ponomarenko and Lagunov (2021) that in the case of perpendicular wave propagation, recurrency ratios relate reflection coefficients in the $j + 1$-st and j-th layers. By using these recurrency ratios the required number of times, reflection, and transmission coefficients for the whole layered system can be obtained.

In theoretical studies, it is sometimes convenient to rely on impedance method for calculating reflection coefficients of multi-layer structures (Morozov et al. 2001). In this case, the notion of layers' wave impedance is used.

The main underlying principle of recurrency approaches consists in obtaining a system of algebraic equations for amplitudes of the reflected wave and the wave that has been transmitted through the absorbent layer, with the use of boundary conditions, which in the case of electromagnetic waves meets the conditions of electric and magnetic field's continuity on layer boundaries. The distinctive feature of recurrency method is that when the number of layers in the system exceeds five (including semi-infinite media to the right and left of the actual layered structure), both approaches become numerical in essence. This results from the fact that although analytical expressions for reflection and transmission coefficients can be obtained, they become too cumbersome to be used in practice. When numerical methods are employed, parameters of the layers can be easily and quickly altered by computer modeling, which changes reflection and transmission characteristics of the layered system and makes it possible to predict characteristics of RF absorbent materials at the stage of their development.

It should be stressed that despite the effectiveness of the above recurrency (numerical) methods, the employment of analytical methods retains its relevance. Below, presented in detail are theoretical approaches to investigating RF absorbent materials, which produce analytical expressions permitting the calculation of radiophysical parameters of absorber coatings and materials.

Physical content of the notions of complex dielectric and magnetic permeabilities
Parameters of electric and magnetic components of electromagnetic waves are important in considering (with the account of absorption taken) the processes of electromagnetic radiation reflection at and transmission through the boundaries of different media, e.g., metal–dielectric, dielectric–dielectric, and metal–metal. The analysis of the physical content of dielectric and magnetic permeability notions enables one

to get an insight into physical processes that occur in RF absorbent coatings and materials.

In an absorbing medium, conductivity is not equal to zero ($\sigma \neq 0$), and during wave propagation there would be current $j = \sigma E$ (Balagurov 2015) in it. The medium is also characterized by dielectric ε and magnetic μ permeabilities, which in the general case are complex values. If a space charge in the medium is equal to zero, the strengths of the electric and magnetic fields of harmonic wave are described by the following Maxwell equations (Nikolsky and Nikolskaya 1989; Tamm 1979):

$$\begin{aligned} rot\ H &= j + i\omega D,\ rot\ E = -i\omega B, \\ div\ H &= 0,\ \varepsilon\varepsilon_0\, div\ E = \rho = 0, \\ D &= \varepsilon\varepsilon_0 E,\ B = \mu\mu_0 H,\ j = \sigma E, \end{aligned} \tag{4.3}$$

where H and E are strengths of magnetic and electric fields, and ε and μ are relative dielectric and magnetic permeabilities.

If the strength of electric field changes in accordance with the harmonic law, $E(t) = E_0 e^{i\omega t}$ (ω is the frequency of electromagnetic field), the first Maxwell equation is of the form:

$$rot\ H = i\omega\varepsilon_0\left(\varepsilon - i\frac{\sigma}{\omega\varepsilon_0}\right)E_0, \tag{4.4}$$

where the expression in brackets can be considered as dielectric permeability that acquires complex values when the environment is conductive, i.e., $\dot{\varepsilon} = \varepsilon - i\dfrac{\sigma}{\omega\varepsilon_0}$, where $\dot{\varepsilon}$ is complex dielectric permeability.

Then Maxwell equations take the form:

$$rot\ H = i\omega\varepsilon_0\dot{\varepsilon}E_0,\ rot\ E = i\omega\mu_0\dot{\mu}H_0, \tag{4.5}$$

where $\dot{\mu}$ is complex magnetic permeability appearing by analogy with complex dielectric permeability.

Dielectric and magnetic permeabilities will be complex values even when the medium is not conductive ($\sigma = 0$). Their complex nature results from the inertia of polarization and magnetization processes. The physical content of this phenomenon consists in the processes of medium polarization and magnetization occurring at a finite rate, which causes a certain time delay of the electric field displacement vector $\mathbf{D}(\mathbf{r}, t)$ relative to electric field strength vector $\mathbf{E}(\mathbf{r}, t)$, and delay of magnetic field induction vector $\mathbf{B}(\mathbf{r}, t)$ relative to magnetic field strength vector $\mathbf{H}(\mathbf{r}, t)$.

In an ideal dielectric, at the field strength $E(t) = E_0 e^{i\omega t}$, induction is determined by the expression $D = \varepsilon\varepsilon_0 E_0 e^{i\omega t}$. But in the actual fact, the process of substance polarization does not follow the change in electric field strength immediately, and induction, being a time function, is retarded, i.e., induction oscillations occur with a certain phase lag θ and $D = \varepsilon\varepsilon_0 E_0 e^{i(\omega t - \theta)} = \varepsilon\varepsilon_0 e^{-i\theta} E_0 e^{i\omega t} = \varepsilon\varepsilon_0(\cos\theta - i\sin\theta)E_0^{i\omega t}$.

Thus, the expression $\varepsilon\varepsilon_0(\cos\theta - i\sin\theta)$ should be taken as the true dielectric permeability. The inertia of magnetization processes is described in the same way.

Due to the inertia of polarization and magnetization processes, as well as taking into account electric and magnetic losses, permeabilities $\dot{\varepsilon}$ and $\dot{\mu}$ should be considered as complex values: $\dot{\varepsilon} = \varepsilon' - i\varepsilon''$, $\dot{\mu} = \mu' - i\mu''$. Minuses are placed before imaginary parts, which has energy-related treatment.

Macroscopically, the abovementioned physical mechanisms of imaginary part appearance in complex dielectric and magnetic permeabilities (finite conductivity, polarization, and magnetization delay) are not different from each other since each of them, due to heat release, cause energy losses of the wave propagating in the substance.

Dielectric and magnetic permeability modules are expressed by the following relations:

$$|\varepsilon| = \sqrt{\varepsilon'^2 + \varepsilon''^2} \text{ and } |\mu| = \sqrt{\mu'^2 + \mu''^2}. \tag{4.6}$$

Here, the phase lag of displacement vector $\mathbf{D}(\mathbf{r}, t)$ relative to electric field strength vector $\mathbf{E}(\mathbf{r}, t)$ and magnetic field induction vector $\mathbf{B}(\mathbf{r}, t)$ relative to magnetic field strength vector $\mathbf{H}(\mathbf{r}, t)$ are determined by such expressions:

$$\theta_\varepsilon = arctg\left({}^{\varepsilon''}\!/_{\varepsilon'}\right), \theta_\mu = arctg\left({}^{\mu''}\!/_{\mu'}\right).$$

If the dielectric has conductivity that in the general case can depend on frequency, complex dielectric permeability of the medium takes the form:

$$\dot{\varepsilon} = \varepsilon' - i\left(\varepsilon'' + \frac{\sigma(\omega)}{\omega\varepsilon_0}\right). \tag{4.7}$$

If time lag effects are neglected, i.e., $|\varepsilon''| \ll {}^{\sigma}\!/_{\omega\varepsilon_0}$, (4.7) implies that complex dielectric permeability has the real part coinciding with electrostatic dielectric permeability, while the imaginary part is determined by substance conductivity and will be inversely proportional to frequency ω if conductivity does not depend on frequency ($\sigma(\omega) = const$).

The complex nature of dielectric and magnetic permeabilities of the medium determines the complex nature of wave number k. Since $k = \frac{\sqrt{\varepsilon\mu}\omega}{c}$, in the case of complex dielectric and magnetic permeabilities, $k = k' - k''$.

Now the formula for wave can be written as

$$E(t; z) = E_0 e^{i(\omega t - kz)} = E_0 e^{-k'' z} e^{i(\omega t - k' z)}.$$

Thus, as the wave propagates deeper into the substance, the field strength will weaken exponentially due to energy absorption (on account of dielectric and magnetic losses and the losses caused by the conductivity present). As a result, heat is released in the medium of wave propagation.

The notion of medium complex dielectric permeability allows the propagation of electromagnetic waves in various media to be analyzed from a unified perspective of Maxwell complex equations, which is of great scientific and practical importance for solving a large number of engineering tasks.

The complex nature of dielectric and magnetic permeabilities is related to two rather important effects—the complex nature of wave impedance in the medium of electromagnetic waves propagation and skin effect.

Transmission of electromagnetic waves through interfaces of different media. Wave impedance

Consider an example of two non-conductive medias 1 and 2 with different values of dielectric and magnetic permeabilities $\varepsilon_1, \varepsilon_2$ and μ_1, μ_2. Wave impedances (ratios of electric and magnetic field strengths for two respective media) are of the form:

$$Z_1 = \sqrt{\frac{\mu_1\mu_0}{\varepsilon_1\varepsilon_0}},\ Z_2 = \sqrt{\frac{\mu_2\mu_0}{\varepsilon_2\varepsilon_0}}. \tag{4.8}$$

At the strengths of electric and magnetic fields E_1, H_1 of the reflected wave and the strengths of electric and magnetic fields E_2, H_2 of the transmitted wave, the following expressions for electromagnetic wave reflection and transmission coefficients in terms of amplitude (r, t) are valid:

$$r = \frac{E_1}{E_0} = \frac{Z_2 - Z_1}{Z_2 + Z_1},\ t = \frac{E_2}{E_0} = \frac{2Z_2}{Z_2 + Z_1}, \tag{4.9}$$

where E_0 is electric field strength of the incident wave.

Such expressions are obtained for energy reflection and transmission coefficients:

$$R = \frac{E_1 H_1}{E_0 H_0} = r^2,\ T = \frac{E_2 H_2}{E_0 H_0} = t^2, \tag{4.10}$$

where H_0 is magnetic field strength of the incident wave.

It should be noted at once that in the case of equal wave impedances of the two media ($Z_1 = Z_2$), reflection coefficient r is equal to zero.

Now consider wave impedance for the electromagnetic wave of a conductive medium ($\sigma \neq 0$).

For the electric field strength of a wave propagating along the axis, the following is valid (Nikolsky and Nikolskaya 1989; Tamm 1979):

$\frac{\partial^2 E}{\partial z^2} = i\omega\mu\mu_0(\sigma + i\omega\varepsilon\varepsilon_0)E$, $\gamma^2 E = i\omega\mu\mu_0(\sigma + i\omega\varepsilon\varepsilon_0)E$, where γ is propagation constant (operator) and $\gamma = \sqrt{i\omega\mu\mu_0(\sigma + i\omega\varepsilon\varepsilon_0)}$.

From Eq. (4.5)

$$\frac{\partial H_y}{\partial z} = -(\sigma + i\omega\varepsilon\varepsilon_0)E_x,\ \frac{\partial E_x}{\partial z} = -i\omega\mu\mu_0 H_y,$$

$\gamma E_x = i\omega\mu\mu_0 H_y$ and $Z_W = \frac{E_x}{H_y} = \frac{i\omega\mu\mu_0}{\gamma} = \frac{i\omega\mu\mu_0}{\sqrt{i\omega\mu\mu_0(\sigma + i\omega\varepsilon\varepsilon_0)}}$ is obtained.

Then, in the general case, the following formula for wave impedance can be written:

$$Z_W = \sqrt{\frac{i\omega\mu\mu_0}{\sigma + i\omega\varepsilon\varepsilon_0}}. \tag{4.11}$$

It determines wave impedance of the absorbent medium which is conductive and has dielectric properties that depend on dielectric permeability.

For an insulator, or dielectric, when $\sigma \to 0$, we obtain

$$Z_W = \sqrt{\frac{\mu\mu_0}{\varepsilon\varepsilon_0}}. \tag{4.12}$$

The characteristic feature of complex dielectric permeability of metals $\dot{\varepsilon} = \varepsilon' - i\frac{\sigma}{\omega\varepsilon_0}$ is a significantly larger value of the imaginary part $i\frac{\sigma}{\omega\varepsilon_0}$ than the real ε' part, which is usually neglected for this reason, i.e., it is considered that

$$\dot{\varepsilon}_M \cong -i\frac{\sigma(\omega)}{\omega\varepsilon_0}. \tag{4.13}$$

The values of conductivity for good and bad conductors can differ by two orders. For example, copper with its specific conductivity $\sigma_{Cu} = 6.0 \times 10^7\ \Omega^{-1}\text{m}^{-1}$ is referred to as a good conductor, while bismuth—which is a semimetal with $\sigma_{Bi} = 8.0 \times 10^5\ \Omega^{-1}\text{m}^{-1}$—as a bad one. For the modules of imaginary parts of complex dielectric permeability of this metal pair, by using formula (4.13) we obtain for light waves (frequency $f = 3 \cdot 10^{14}$ Hz):

$$\frac{\sigma_{Cu}}{\omega\varepsilon_0} \approx 2000 \text{ and } \frac{\sigma_{Bi}}{\omega\varepsilon_0} \approx 30.$$

This implies that for all metals in the entire RF range and partially in optical range $\frac{\sigma}{\omega\varepsilon_0} \gg 1$, which justifies the neglection of the real part in comparison with the imaginary part when calculating dielectric permeability with formula (4.13).

Since $\sigma \gg i\omega\varepsilon\varepsilon_0$, wave impedance for metals is

$$Z_M = \sqrt{\frac{i\omega\mu\mu_0}{\sigma}}. \tag{4.14}$$

Physical content of surface effect. Skin effect

When considering RF absorbent materials based on composite with the filler in the form of finely dispersed metal particles, of considerable importance is the ratio of such parameters as particle size, electron free path, electromagnetic radiation wavelength, and the main characteristic of metal absorption ability—skin layer thickness.

Consider skin effect physics. If there is direct current in a conductor, current density is the same in all points of conductor cross section. A different phenomenon is observed for alternating current. In this case, current density is not the same across the conductor: it is the largest on conductor surface and the smallest on its axis. This irregularity grows with increasing conductor thickness and alternating current frequency, and at very high frequencies, current practically exists only in a thin surface layer, which gave the phenomenon its name—skin effect.

The above phenomenon is accounted for by the emergence of vortex electric field of electromagnetic induction.

We will consider it on the example of cylinder-shaped conductor that carries alternating current along its axis. Alternating current inside the conductor generates a magnetic field whose lines lie in the plane perpendicular to current direction. If the current increases, the increasing magnetic field induces a vortex electric field which at the conductor surface is directed like the current, but in the conductor axis—opposite to the current. This field will increase the current on the surface and reduce it in the axis. The effect is present at any frequency but it becomes especially noticeable at high frequencies.

Due to skin effect, electric current at high frequencies flows predominantly through the surface layer of the conductor, which leads to the reduction of its effective cross section. In this connection, the notion of skin layer thickness δ on which current density reduces fivefold is introduced.

Skin effect also appears during electromagnetic wave interaction with a conductive medium when the electromagnetic wave propagates from a non-conductive medium into a conductive one. Then the skin layer size is defined as the distance from the interface of the two media to the point where the strengths of electric and magnetic fields of the wave are fivefold lower.

The alternating magnetic field of the wave generates a vortex electric field in the internal layers. The vortex electric field, in turn, excites vortex currents in the internal layers. The magnetic field of these vortex currents inside the conductor will be directed against the primary excitatory magnetic field and will weaken it.

Now, we consider the fundamentals of skin effect theory. We assume, like in Tamm (1979), that fields and currents change in accordance with time harmonic law, i.e., they are proportional to $\exp(-i\omega t)$. We also assume that the electromagnetic wave propagates along Z-axis and is strictly transverse, in other words, $H_y = H$ and $E_x = E$. All other components of magnetic field strength and components of electric field strength are equal to zero ($H_x = H_z = 0; E_y = E_z = 0$). Then for $rot H$, we obtain

$$\nabla \times H = \begin{vmatrix} i & j & k \\ \frac{\partial}{\partial x} & \frac{\partial}{\partial y} & \frac{\partial}{\partial z} \\ 0 & H_y & 0 \end{vmatrix} = -\frac{\partial H_y}{\partial z} = (\sigma + i\omega\varepsilon\varepsilon_0)E_x. \tag{4.15}$$

For $rot\ E$ we obtain

$$\nabla \times E = \begin{vmatrix} i & j & k \\ \frac{\partial}{\partial x} & \frac{\partial}{\partial y} & \frac{\partial}{\partial z} \\ E_x & 0 & 0 \end{vmatrix} = \frac{\partial E_x}{\partial z} = -i\omega\mu\mu_0 H_y. \tag{4.16}$$

By differentiating (4.16) by Z and substituting the expression obtained into (4.15), we have

$$\frac{\partial^2 E_x}{\partial z^2} = i\omega\mu\mu_0(\sigma + i\omega\varepsilon\varepsilon_0)E_x. \tag{4.17}$$

In high conductivity metals, up to the frequencies of hundreds of MHz, the following inequality holds $\left|{}^{\sigma}/_{\omega\varepsilon\varepsilon_0}\right| \gg 1$, then

$$\frac{\partial^2 E_x}{\partial z^2} + i\omega\mu\mu_0\sigma E_x = 0. \tag{4.18}$$

Electric field strength, spreading into the conductor depth along axis Z, will drop proportionally to $e^{-z/\delta}$, i.e., $E_x(z) \sim e^{-z/\delta}$, where δ is skin layer thickness. At the distance of δ from the interface of two media, field amplitude drops e-fold. At the distance of several δ, the field is negligible, e.g., at the depth of 10δ, it is weakened e^{10}-fold.

The solution of Eq. (4.18) for electric field strength is of the form:

$$\delta = \sqrt{\frac{2}{\omega\mu\mu_0\sigma}} = \frac{1}{\sqrt{\pi\mu\mu_0\sigma f}}. \tag{4.19}$$

For example, for copper at $f = 10$ GHz, $\delta = 0.66 \times 10^{-3}$ mm.

Similar equations can be written for magnetic field. Thus, high-frequency electromagnetic fields in a conductor are concentrated near its surface.

Reflection of flat harmonic electromagnetic wave from ideally conductive plane

RF absorbent coatings are often applied on the surfaces of various special-purpose products that are highly conductive and due to skin effect, electromagnetic wave practically does not penetrate deep inside those surfaces. In this connection, certain aspects of electromagnetic wave reflection from an ideally conductive surface are of special interest.

We will consider reflection of a flat harmonic wave incident from vacuum along axis z perpendicularly to ideally conductive flat surface separating vacuum ($z \leq 0$) and a conductor ($z \geq 0$).

Such problem is a special case of the problem on electromagnetic wave reflection from the interface of two media, when complex dielectric permeability of the second medium has the imaginary part that is many times larger than the real part, which is typical of conductive bodies.

Medium conductivity causes absorption of electromagnetic wave propagating in it. Absorption occurs due to the emergence of conduction currents and Joule heat release in the conductive medium. The absorption increases with an increase in medium conductivity. For an ideally conductive medium, skin layer thickness, as can be seen from expression (4.19), tends to zero when conductivity increases. As a result, electromagnetic field is absent inside an ideally conductive body. Hence, an electromagnetic wave incident on an ideally conductive medium is completely reflected from its surface. The sources of the reflected electromagnetic wave are surface currents circulating on the surface of ideally conductive body.

To determine the reflected electromagnetic wave, it is necessary that its amplitude and phase be related to respective parameters of the incident wave. For that, it is required to use the continuity condition for the tangential component of electric field on the ideal conductor surface. Unlike in the problem on electromagnetic wave reflection from the interface of two arbitrary media, only one continuity condition can be used in this case. This results from the fact that electromagnetic field inside an ideally conductive body is equal to zero. For that reason, the coefficient of electromagnetic wave transmission into the conductor is known—obviously, it is equal to zero. Hence, to solve the problem on electromagnetic wave reflection from an ideally conductive plane, only the coefficient of reflection from that plane is to be determined.

In the left-hand half-space ($z \leq 0$), the incident E_0 and reflected E_1 waves propagate. Since for transmitting waves $E_2 = 0$ and $H_2 = 0$, the equation of continuity of electric field tangential component on $z = 0$ plane takes the form:

$$E_2 = E_0 + E_1 = 0.$$

This implies that during reflection from an ideally conductive surface, there is a reflected electromagnetic wave propagating in the opposite direction in respect of the incident wave, which has the same amplitude of the vectors of electric and magnetic field strengths, while the phase of these vectors' oscillations differs from the phase of incident wave oscillations by 1800.

The same conclusion can be made on the basis of the following considerations. When $\sigma \to \infty$, as can be seen from (4.14) and (4.19), $\delta = 0$ and $Z_M = Z_2 = 0$. Then, using (4.9), $r = -1$.

Electromagnetic wave transmission through substance layer

Absorption of electromagnetic wave energy in RF absorbent materials, as noted above, occurs due to dielectric, magnetic, and conduction losses. At the same time, during magnetic wave incidence on the material there is reflection from the interface of two media—the environment and the material itself. The greater the difference of wave impedances of those media, the larger is the value of reflection coefficient.

So, when developing absorbent coatings, the key task is coordinating the absorbing structure and the environment for the integral effect of reflection to be minimal.

There are several ways of reducing the reflection from the first interface of the RF absorbent coating. The simplest way of producing the coordinating layer is the resonance one, when the adjusting layer $\lambda/4$ thick (where λ is the wavelength of the incident radiation) is placed in front of the absorber layer.

For wideband absorbent materials, a technique of reducing reflection is used, which is based on the fact that the wave impedance of non-conductive adjusting layer is determined by the expression $Z = (\mu\mu_0/\varepsilon\varepsilon_0)^{1/2}$. By choosing the necessary ratio of magnetic and dielectric permeabilities μ/ε, Z can be made equal to free space impedance and, taking into account expression (4.9), reflection coefficient $r = 0$. Lately, porous non-magnetic materials are often used as the matrix of an RF absorbent composite material since due to the presence of unfilled pores, they have the value of dielectric permeability near unity and, as a result, their wave impedance is close to free space impedance. That is, matrix material also plays the role of the adjusting layer.

In connection with the above, it will be appropriate to consider the process of electromagnetic wave transmission through a substance layer.

Consider an incidence of a flat electromagnetic wave on a substance layer d thick with the wave impedance Z_2, which is adjacent to two half-infinite media with wave impedances Z_1 and Z_3 (Landau and Lifshitz 1984). The incident wave is partially reflected from both the first and the second interfaces and penetrates into the third medium. The main properties of the substance layer, in terms of absorbent coatings and materials, can be investigated with taking into account the reflection coefficient in the first medium.

Taking into consideration that at $z > d$ the third medium is infinite we obtain reflection coefficient for the first interface as

$$\dot{r} = \frac{Z_2\,(Z_3 - Z_1) + i\,\left(Z_2^2 - Z_1 Z_3\right)\tan k_2 d}{Z_2\,(Z_3 + Z_1) + i\,\left(Z_2^2 + Z_1 Z_3\right)\tan k_2 d}, \tag{4.20}$$

where $k_2 = 2\pi/\lambda_2$, λ_2 is the wavelength in the second medium.

The complete wave transmission through the layer in the simplest case is possible if the expressions in brackets in (4.20) are equal to zero. That is, $Z_1 = Z_2 = Z_3$. In actual fact, this inequality corresponds to three identical media when interfaces are absent. But it is possible that physically different media have the same wave impedance. This is realized when μ/ε ratio is the same for all three media.

The condition of another possibility of complete wave transmission through a substance layer can be determined by previously dividing the numeral and the nominative of (4.20) by $\tan k_2 d$. Then the real part of the numeral at different Z_1 and Z_3 can be equal to zero due to $\cot k_2 d = 0$, while the imaginary part—at $Z_2^2 = Z_1 Z_3$. This means that the layer thickness has to be a multiple of an uneven number of wavelength quarters ($d = (2n - 1)\lambda_2/4$, $n = 1, 2, 3, \ldots.$). The physical content of such layer property is related to the waves reflected from each interface being in this case the same in terms of amplitude and phase. The antiphase properties of waves reflected from interfaces are provided by the phase shift that is additional for the second wave, at the distance $2d$, $2k_2 d = (2n - 1)\pi$. Therefore, as a result of their interference the reflected wave will be absent in the first medium. It is this property that is used to produce adjusting resonance layers. Since the properties considered are related to a certain wavelength and are resonant, when oscillation frequency deviates from the calculated one the reflection coefficient increases rather quickly.

It should be noted that if layer thickness is set as constant while the wavelength is changed, the discrete set of wavelengths at which reflection coefficient will be equal to zero is $\lambda_{2n} = 4d/(2n-1)$, where $n = 1, 2, 3, \ldots\ldots$ Thus, as can be seen from this simple example, on the wavelength scale, transmission bands alternate with reflection bands. This property is also general for more complex periodic structures, multi-layer ones in particular.

Calculation of parameters of gradient RF absorbent materials

When developing wideband RF absorbent materials, researchers give special attention to gradient and multi-layer composite-based coatings. In these materials, dielectric and magnetic permeabilities gradually (gradient materials) or step-wisely (multi-layer structures) change from values close to those of the environment to the values providing the required absorption level. Such absorbers can be very wide-banded at a small value of reflection coefficient and small thickness but they are the most difficult to implement in practice. The advantage of gradient materials as compared to multi-layer ones is the absence of reflection on interlayer boundaries.

It was proposed in Morozov et al. (2001) that the gradient layer should be placed on a metal mirror (ideally conductive plane) and half-space above it should be empty. A flat electromagnetic wave polarized along axis y is perpendicularly incident on this layer; field components depend on coordinates x. Then from Maxwell equations $rot\ \mathbf{H} = -i\omega\varepsilon_0\varepsilon\mathbf{E}$, $rot\ \mathbf{E} = i\omega\mu_0\mu\mathbf{H}$ we have

$$\frac{\partial E_y}{\partial x} = i\omega\mu_o\mu H_z, \ \frac{\partial H_z}{\partial x} = i\omega\varepsilon_0\varepsilon E_y. \tag{4.21}$$

In the domain above the layer

$$E_y = e^{ik_o x} + r \cdot e^{-ik_0 x}, \ H_z = \frac{1}{Z_0}\left(e^{ik_0 x} - r \cdot e^{-ik_o x}\right), \tag{4.22}$$

where $Z_0 = \sqrt{\mu_0/\varepsilon_0}$, r is reflection coefficient in electric field amplitude. Characteristic impedance is a function of coordinate x and is determined as

$$Z = E_y/H_z. \tag{4.23}$$

By substituting (4.22) into (4.23) when $x = 0$, we obtain impedance on the gradient layer surface (input impedance):

$$Z_1(0) = \frac{1+r}{1-r} Z_0. \tag{4.24}$$

From (4.23), for impedance in the gradient domain, we obtain

$$\frac{dZ}{dx} = \frac{H_z \frac{\partial E_y}{\partial x} - E_y \frac{\partial H_z}{\partial x}}{H_z^2}. \quad (4.25)$$

By substituting (4.21) into (4.25), we obtain

$$\frac{dZ}{dx} = -i\omega\varepsilon_0\varepsilon(x) Z^2 + i\omega\mu_0\mu(x). \quad (4.26)$$

On the ideally conductive plane ($x = d$), field component $E_y = 0$, therefore $Z(d) = 0$.

In view of the continuity of field tangential components on the boundary $x = 0$, the impedance on this boundary is also continuous:

$$Z(0) = Z_1(0) = \frac{1+r}{1-r} Z_0. \quad (4.27)$$

Then for reflection coefficient:

$$r = \frac{Z(0)/Z_0 - 1}{Z(0)/Z_0 + 1}. \quad (4.28)$$

Thus, by obtaining the solution $Z(x)$ of differential equation (4.26) with the initial condition $Z(d) = 0$ and by calculating $Z(0)$, it is possible, by using formula (4.28), to find the reflection coefficient of electric field amplitude from the gradient layer.

It should be noted that Eq. (4.26) does not have an analytical solution for arbitrary functions $\varepsilon(x)$ and $\mu(x)$, but it can be obtained by numerical methods. Numerical calculations of the coefficient of reflection from a gradient layer at normal wave incidence can be carried out by impedance method by considering the gradient layer, which consists of a large number of rather thin layers with coordinate-independent permeabilities inside the layers. For the wave incident at any angle, including perpendicular incidence, matrix method can be used. The application of this method is considered in Morozov et al. (2001) on the example of d-thick non-magnetic gradient layer located on a metal substrate, with an empty half-space situated above it. It was assumed that depth x was counted from the layer boundary with free space ($0 < x < d$). It was also assumed that the real and imaginary parts of the dielectric permeability depend on x in accordance with the law:

$$\varepsilon'(x) = 1 + a\left(\frac{x}{d}\right)^2, \varepsilon''(x) = b\left(\frac{x}{d}\right), \quad (4.29)$$

where a and b are constants. As can be seen from (4.29), on the surface of the layer $x = 0$, complex dielectric permeability is equal to free space permeability. With the increase of depth x, the module of function $|\varepsilon(x)|$ increases. Such "gradual transition" permits obtaining small reflection from the free space boundary with the gradient absorbent layer at the wavelengths that are not too large in comparison with the thickness of d layer.

In Morozov et al. (2001), the gradient layer was subdivided into N layers of $\Delta = d/N$ thicknesses. The value of coordinate x inside the layer number n was denoted as x_n (the layers were numbered from that adjacent to the surface $x = 0$). Then for x_n, we have

$$x_n = \frac{\Delta}{2} + (n - 1)\,\Delta, n = 1, 2, \ldots, N,$$

and for dielectric permeability in the layers:

$$\varepsilon_n = 1 + a\left(\frac{x_n}{d}\right)^2 + ib\left(\frac{x_n}{d}\right).$$

The choice of number N is made by a numerical experiment, i.e., by increasing N to the value at which reflection coefficient stops changing within the set precision of its determination.

The application of matrix method at $a = 5$, $b = 3$ allowed finding the dependence of amplitude reflection coefficient on relative wavelength λ/d at various incidence angles (Morozov et al. 2001). The dependences mentioned showed that with an increase in the wavelength, the tendency to reflection coefficient increase was observed. The calculations were carried out when the gradient layer was subdivided into $N = 30$ layers, which proved sufficient for computing the reflection coefficient with the absolute precision of 0,01.

Effective permeability of composites

Usually, by composite material (composite) one means the medium that consists of inclusion particles (filler) situated in a binder material (matrix). If the inclusions are located orderly, the medium is sometimes called an artificial dielectric. Powdered metals, ferrites, or conductor segments can be used as fillers. Non-absorbent dielectrics with low dielectric permeability, e.g., epoxy, rubber glue, etc., are commonly used as matrix. The filler is placed in a liquid matrix, and after solidification a solid composite is produced. If the sizes of inclusions and the distances between them are rather small in comparison with the electromagnetic radiation wavelength in the medium, composite can be considered as a homogeneous material with certain effective dielectric and magnetic permeabilities. Composite filler permeabilities are denoted as ε_f and μ_f, those of the matrix—as $\varepsilon_m$and μ_m, and effective permeabilities of the composite—as ε_{ef} and μ_{ef} (Bespalova and Kondrashov 2014).

For example, the calculation of effective dielectric permeability of the composite based on cylindrical dipoles is carried out by Odelevsky formula (Morozov et al. 2001). For the case when the main axes of the inclusions are directed along the composite electric field (axis x), the formula for effective dielectric permeability is of the form:

$$\varepsilon_{ef} = \varepsilon_m \left[1 + \frac{p}{(1 - p)\,n_x + \frac{\varepsilon_m}{\varepsilon_f - \varepsilon_m}}\right], \tag{4.30}$$

where n_x is coefficient of inclusions depolarization along the axis and p is volume concentration of inclusions. Taking into account that $\varepsilon_f = i\sigma/\omega$ (σ is electric conductivity of inclusions) and the polarization of an elongated ellipsoid of revolution α_x is determined by expression:

$$\alpha_x = v\frac{\varepsilon_m}{\varepsilon_m\left(\varepsilon_f - \varepsilon_m\right) + n_x},$$

the formula for ε_{ef} will be of the following form:

$$\varepsilon_{ef} = \varepsilon_m\left[1 + \frac{p}{v\frac{\varepsilon_m}{\alpha_x} - pn_x}\right], \tag{4.31}$$

where $n_x = \left(\frac{b}{a}\right)^2\left\{\ln\left(\frac{2a}{b}\right) - 1\right\}$ at $b/a \gg 1$ (Bespalova and Kondrashov 2014); a, b are half-axes of the ellipsoid of revolution; the long axis a is directed along axis x, which is parallel to the external field; v is volume of the ellipsoid of revolution.

For the case of chaotically oriented elongated ellipsoids of revolution, whose polarization in the direction transverse in respect of the long axis can be neglected, in (4.31), substitution should be made, where K is orientation factor; $K = 1/2$ at chaotic orientation in the plane parallel to the field in the composite (Morozov et al. 2001).

When calculating effective dielectric permeability of a composite based on cylindrical dipoles, which is true, e.g., for composites based on carbon fibers, volume $v = \pi r^2 (2h)$, where $a = h$ and $b = \sqrt{\frac{3}{2}}r$ (Morozov et al. 2001).

Composites based on conductive fibers

For imparting RF absorbent properties to a composite, a dispersed electrically conductive filler is to be introduced into its composition. The main electrophysical characteristics of the material in which its conductive phase is in a dispersed state are determined by filler microstructure, the sizes of its particles, and the volume fraction of its polymer matrix. Particles of various shapes and sizes are used as a conductive filler. Particle shapes are conditionally subdivided into spherical (soot, metal powders, metallized micro-glass spheres), laminar (exfoliated graphite, metal powders after a special treatment), and needle-like (carbon fibers, metallized fibers). The choice of one or another electrically conductive filler is determined by effective complex dielectric permeability of the composite, depending on filler concentration. Taking into consideration filler dispersion, the concentration should be optimized not only from the viewpoint of achieving the required level of reflectivity reduction and also by regarding physical, mechanical, and technological limitations during production and use of RF absorbent materials.

Taking into account that in RF absorbent materials and coatings with electrically conductive fillers, energy absorption only occurs due to electric losses, to achieve the pre-set values of electromagnetic wave reflection coefficient, the optimal filler concentration should be used. When critical filler concentration is achieved, composite material conductivity increases step-wisely, which leads to an increase in the imaginary part of dielectric permeability. Despite the fact that energy losses in a substance are determined by this very component, an excessive increase of the imaginary part leads to a higher coefficient of incident electromagnetic wave reflection from material external boundary, which is due to a mismatch of the wave impedances of free space and absorbent material media, and, as a result, reduces the effectiveness of absorber functioning.

Introducing fiber fillers of various lengths and diameters into the polymer matrix permits absorption range to be widened and reflection coefficient to be reduced, as a rule, without deteriorating other performances. A common type of electroconductive fillers is carbon fiber (Bespalova et al. 2015; Vlasenko et al. 2025), which consists almost entirely of pure carbon (up to 99.5% of its mass).

Carbon fibers have a number of valuable electrophysical and physico-chemical properties: low density and electrical conductivity, along with a low thermal expansion coefficient. Besides, carbon fibers are compatible with a majority of polymers and exceed other fibrous fillers in rigidity and moisture resistance. With that, as investigations show, the range of possible values of the dielectric permeability of RF absorbent materials with carbon fibers as electrically conductive fillers differs broadly and can be regulated during material molding by selecting fiber concentration and its electrical resistivity, which can be varied widely.

For the case when conductive fibers of length l and diameter d are used as the filler, Odelevsky formula (4.30) for effective dielectric permeability is of the following form (Bespalova et al. 2015; Vlasenko et al. 2025):

$$\varepsilon_{ef} = \varepsilon_m + \frac{\varepsilon_m k_v}{F\left(1 + \frac{\varepsilon_m}{F(\varepsilon_f - \varepsilon_m)}\right)}, \tag{4.32}$$

where ε_m is dielectric permeability of the composite matrix, ε_f is dielectric permeability of the filler, and k_v is filler volume concentration. Depolarization coefficient F at $l/d \gg 1$ is expressed with a high approximation accuracy by the formula:

$$F = \ln\left(2\frac{l}{d}\right) \Big/ \left(\frac{l}{d}\right)^2. \tag{4.33}$$

Relative complex dielectric permeability of a conductor or a resistive material, which the filler is, can be presented as

$$\varepsilon_f = \varepsilon_f' + i\frac{\sigma}{\omega\varepsilon_0}, \tag{4.34}$$

where σ is specific conductivity of the filler (resistivity $\rho = 1/\sigma$), ω is circular frequency, ε_0 is absolute dielectric permeability in vacuum and, practically, in free space. For practical use, the imaginary part can be written in the following way (Vlasenko et al. 2025): $60\lambda/\rho$ (here l is measured in centimeters, ρ—in Ωcm).

For composites, in which carbon fibers are used as filler $\sigma/\omega\varepsilon_0 \gg \varepsilon'_=$ and $\sigma/\omega\varepsilon_0 \gg \varepsilon_<$. Finally, for ε_{ef} we will have

$$\varepsilon_{ef} = \varepsilon_m + \frac{D}{1 - \left(i\frac{\lambda_R}{\lambda_0}\right)}, \tag{4.35}$$

where $\lambda_R = \frac{\varepsilon_m \rho}{60F}$, $D = \frac{\varepsilon_m k_v}{F}$, ρ is filler resistivity, $\lambda_0 = 2\pi A/\omega$ is electromagnetic radiation wavelength.

The degree of electromagnetic wave absorption in a layer of thickness d is determined by the multiplier $e^{-\frac{2\pi d}{\lambda} Im(\sqrt{\varepsilon\mu})}$. This multiplier with increasing wavelength at $\lambda \le \lambda_R$ changes only slightly (taking into account ε_{ef}), but decreases rapidly at a further increase (Bespalova et al. 2015). Therefore, wavelength λ_R can be chosen as the maximum wavelength $\lambda_{\max}$ of the operating range, i.e., $\lambda_{\max} = \lambda_R$.

In an actual material, filler fibers are mostly distributed chaotically: there are fibers perpendicular to the electric field strength vector (they practically do not affect the process of material polarization), the majority of fibers are not parallel to the electric field vector, a certain part of the fibers can be intertwisted. Due to that, formula (4.35) takes the form:

$$\varepsilon_{ef} = \varepsilon_m + \frac{D\alpha_1\alpha_2}{1 - \left(i\frac{\lambda_R}{\lambda_0}\right)^{1-\alpha_3}}. \tag{4.36}$$

Coefficients α_2 and α_3 are determined experimentally for a particular material and a particular technology. Coefficient α_1 approximately takes into account the orientation of electric field strength vector in respect of fiber directions and is equal to

0.333—at chaotic fiber orientations;

0.5—at fiber orientations in the plane parallel to electric field strength vector;

1.0—at fiber orientations in the direction of electric field strength vector.

The degree of electromagnetic wave absorption in the layer of thickness d is determined by multiplier—$e^{-\frac{2\pi d}{\lambda} Im(\sqrt{\varepsilon\mu})}$. This multiplier changes insignificantly with an increase of wavelength at $\lambda \le \lambda_R$ (taking into account ε_{ef}), but decreases rapidly at its further increase. Therefore, for the maximum wavelength $\lambda_{\max}$ of the RF absorbent coating operation range, wavelength λ_R can be chosen, i.e., $\lambda_{\max} = \lambda_R$ (Bespalova et al. 2015).

Electromagnetic radiation absorption by finely dispersed metal particles and composites based on them

Lately, the results of theoretical research into the physical properties of finely dispersed particles (including nanoparticles) have become increasingly widespread in various engineering areas. This is due to the theoretically predicted properties of numerous finely dispersed objects (metal particles in particular) being unique and enabling researchers to design and develop effective RF absorbent materials.

In composite RF absorbent materials, finely dispersed metal particles, specifically, those of iron, are widely used as fillers. Iron, besides electromagnetic wave absorption on account of its conductivity, demonstrates significant absorption properties due to magnetic losses (in this case, the value of the imaginary part of complex magnetic permeability μ'' is rather large).

In the case when spherical particles of conductive or magnetic material are used as composite filler and electromagnetic radiation frequencies are such that skin layer size is commensurate with particle sizes, skinning process significantly influences the frequency dependence of reflection coefficient (Anzulevich 2013; Meshram et al. 2004). The process of electromagnetic radiation absorption by a composite consisting of a dielectric matrix with inclusions of ferromagnetic material balls, at the frequencies lower and higher than that of natural resonance frequency is determined by skin effect. Here, absorption maximums occur when skin layer thickness is commensurate with the size of inclusions.

The problem addressed in Kraev (2017) studies the absorption of external electromagnetic field energy by spherical non-magnetic metal objects of the same radius R at an arbitrary ratio of particle radii and the main characteristic of metal's absorption capability—skin layer thickness δ. An ensemble of metal particles the average distance between which permits the study of electromagnetic radiation scattering and absorption by each individual particle is considered.

It should be noted that particle geometry produces a significant effect on absorption. It is shown below that it is the accounting for particle shape that in the general case makes it possible to calculate the dependence of absorption cross section on the ratio between δ and R. Besides, absorption characteristics essentially depend on the ratio between electron free path l_{ee} and particle linear size R. Work (Kraev 2017) analyzes the case when $l_{ee} \ll R \ll \lambda$, where l is electromagnetic radiation wavelength.

By using quasi-classical kinetic equation for electron function, the following expression was obtained for particle absorption cross section:

$$\sigma = \frac{64\pi^3 e^2 c}{3\omega\hbar^3} \cdot \frac{m^2 u_F^3 \tau_0^2 \omega_L}{\left(1 + \omega^2\omega_L^2\tau_0^2/\left(2\omega + \omega_L\right)^2\right)^2 \left(\omega_L + 2\omega\right)} \cdot F\left(\omega\right), \tag{4.37}$$

where e is electron charge, ε_F is Fermi energy, ω is electromagnetic radiation frequency, and ω_L is Larmor frequency of electron precession in alternating electromagnetic field.

Function $F(\omega)$ is too cumbersome for practical use and is calculated by means of numerical integration but in boundary cases this function can be determined without applying numerical integration. For high frequencies ($\omega\tau_{ee} \gg 1$), where τ_{ee} is relaxation time in the case of electron–electron collisions, the analysis of expression (4.36) with reducing $F(\omega)$ function shows that absorption cross section has its maximum at the frequency

$$\omega_{\max} = c^2 / \left(2\pi \sigma R^2\right). \tag{4.38}$$

By comparing the formula for skin layer size $\delta = c/\sqrt{2\pi\sigma\omega}$ and (4.38), the condition for maximum absorption is obtained: $R = \delta$.

Unlike Kraev (2017), works (Galst'yan and Ravaev 1992; Gladkov 2011) consider electromagnetic wave absorption by spherical magnetic particles.

In Gladkov (2011), to analyze electromagnetic radiation absorption by spherical inclusions present in the composite matrix as the filler, an approach different from Kraev (2017) is used. It is shown that in the particles whose sizes are much smaller than the wavelength of the radiation used, radiation absorption is due to magnetic losses, i.e., to the effective non-zero imaginary part of magnetic permeability that appears in such materials.

It is assumed in this work that despite dense packing, there is no direct contact between particles, and free space is filled with composite matrix. A numerical solution of the differential equation for the wave electric component was carried out in the form:

$$\nabla \times \mu^{-1} (\nabla \times E) - k_0^2 \left[\varepsilon - i\frac{\sigma}{\omega\varepsilon_0}\right] E = 0, \tag{4.39}$$

where ε, μ, and σ are local dielectric permeability, magnetic permeability, and conductivity, respectively. It is shown that maximum of the dependence of magnetic permeability imaginary part on electromagnetic radiation frequency is reached when the following condition is fulfilled:

$$R/\delta = 2. \tag{4.40}$$

It is also shown in Galst'yan and Ravaev (1992) that when condition (4.40) is fulfilled, the imaginary part of magnetic permeability has a maximum and rises abruptly with an increase in composite filler concentration.

In Elsukov (2009), the process of pulsed UHF heating of composites with inclusions of spherical metal particles is considered. It is shown that maximum electromagnetic wave absorption occurs at the maximum value of the imaginary part of magnetic permeability and with condition (4.40) fulfilled (in the same way as this is in Galst'yan and Ravaev (1992), Gladkov (2011).

For magnetic materials, the skin layer value is determined by the expression:

$$d = \frac{c}{2p\sqrt{m_{st} f s}}, \tag{4.41}$$

where μ_{st} is static magnetic permeability, $\omega = 2\pi f$, σ is electric conductivity of particle material. By using (4.40) and (4.41), we obtain the following formula for resonance frequencies of electromagnetic radiation absorption which is due to skin effect (skinning):

$$f = \frac{c^2}{4\pi^2 \mu_{st} \sigma R^2}. \tag{4.42}$$

If the composite filler has particles of different radii, it follows from (4.42) that the maximum frequency of RF absorbent operating range $f_{\max}$ corresponds to the minimal particle size, while minimal frequency $f_{\min}$ corresponds to the maximum size of the particles.

Frequency dependence of magnetic permeability of magnetic materials. Snoek's and Asche constants

It is known that the type of frequency dependence of magnetic permeability of magnetic materials used in developing RF absorbent materials depends on domain boundary movements, skin effect, and natural ferromagnetic resonance (FMR). Domain boundary movements are usually too inertial to contribute to frequency dispersion in UHF range, so natural magnetic resonance is the main spectrum-forming mechanism in UHF range. Natural FMR is determined by the dynamics of magnetic moment movement in zero external magnetic field under the influence of internal field anisotropy.

Without taking into account the magnetic losses related to domain boundary movements and losses due to skin effect, the frequency dependence of magnetic permeability is of the following Lorentzian form:

$$\mu(f) = 1 + \frac{(\mu_{st} - 1)}{1 - if/f_{rel} - (f/f_{res})^2}, \tag{4.43}$$

where μ is dynamic magnetic permeability, μ_{st} is static magnetic permeability, f_{rel} is relaxation frequency, f_{res} is the resonance frequency of natural FMR, and f is the frequency of electromagnetic wave acting on magnetic material. The frequency of natural FMR is related to internal field of magnetic anisotropy H_A by the following expression:

$$f_{res} = \gamma' H_A, \tag{4.44}$$

where $\gamma' = \gamma/2\pi$, γ is gyromagnetic ratio, $\gamma' = 2.8$ MHz/E (Adenot et al. 2002).

According to Lorentz law of frequency dispersion (4.43), magnetic field permeability increases along with frequency to FMR frequency. Above f_{res}, it drops significantly, which is unacceptable for the majority of engineering applications. In actual materials, magnetic field permeability usually decreases with frequency in the entire frequency range due to magnetic losses. So, the suitability of a material to UHF applications can be assessed approximately on the basis of μ_{st} and f_{res} values. Static permeability can be accepted as the assessment of dynamic permeability below FMR frequency; therefore, it characterizes material quality within working frequency range. FMR frequency determines the limit of operating frequency range above which the material does not demonstrate magnetic properties.

Static permeability and resonance frequency are related to each other by Snoek's law. In the case when magnetic material consists of single-domain grains that, as the first approximation, are spherical, Snoek's law is of the form (2004):

$$(\mu_{st} - 1)\, f_{res} = \frac{2}{3}(4\pi \gamma' M_O), \tag{4.45}$$

where M_O is material saturation magnetization, the right-hand part of this expression is referred to as Snoek's constant—K_S.

It is important that Snoek's constant K_S only depends on saturation magnetization of magnetic material, while magnetic anisotropy field and other structure-dependent values are not involved in this constant. Unlike value M_O, which is determined by material composition only, H_A is a function of magnetic and crystalline structures, which, in turn, are dependent on the technological details of material production and processing. Changes of anisotropy field due to introducing additives, annealing, material treatment after fabrication, along with other technological procedures are considered in Chai et al. (2010), Hendrych et al. (2009), Yang et al. (2008), and Baclet et al. (1996). These methods permit material anisotropy to be varied widely, so values of μ_{st} and f_{res} usually depend on technological process parameters. Vice versa, value Ks is structurally independent, it is only determined by material composition. The majority of ferrites (except those having hexagonal structure) are characterized with Snoek's constant in the range from 2 to 5 GHz (Bespalova and Kondrashov 2014). Ferromagnetic metals and alloys can have higher values of K_S due to higher saturation magnetization. For example, for iron, M_O=1700 Gs and K_S= 40 GHz.

Snoek's law is valid for numerous magnetic materials. Independence of Snoek's constant from structure parameters allows its wide use in analyzing data on UHF properties of materials. Snoek's law limits the feasibility of producing materials both with high values of dynamic magnetic permeability and high magnetic losses at UHF. It should be noted that Snoek's law in its (4.45) form is not universal for all ferromagnetic particle types. In particular, it is not applicable to thin ferromagnetic films.

Thin ferromagnetic films can have high magnetic permeability at higher frequencies as compared to bulk materials and a close static magnetic field permeability, which makes them promising for developing RF absorbent materials.

Static permeability and resonance frequency for thin ferromagnetic films are related by the following expression:

$$(\mu_{st} - 1)\, f_{res}^2 = (4\pi\ \gamma' M_O\ )^2\ , \tag{4.46}$$

whose right-hand part, like that in (4.45), is only determined by value M_O. By analogy with Snoek's constant, Asche constant K_A is introduced (Rozanov and Osipov 2006; Chambers and Tennant 1994) as the right-hand part of expression (4.46), with averaging factor involved into it, if necessary.

Similar to (4.45), expression (4.46) limits allowable values of static magnetic permeability in ferromagnetic materials. Formulas (4.45) and (4.46) differ by the fact that in (4.46) material saturation magnetization being squared in (4.46), due to which (4.46) allows much higher magnetic permeability values. For example, if FMR frequency is equal to 1 GHz and M_O = 1700 Gs, (4.45) leads to μ_{st} = 40, and (4.46)—to μ_{st} = 3600.

Asche constant is the evaluation of the part of magnetic moments that precess in alternating magnetic field and, respectively, contribute to peak magnetic losses related to FMR. Asche constant can also be useful for evaluating limit characteristics of UHF devices that rely on magnetic materials, including RF absorbent ones.

For a multi-layer composite material whose filler consists of magnetic particles shaped as thin films and their own magnetic moment is located within the film area, expression (4.46) can be re-written as Rozanov (1999):

$$(\mu_{st} - 1)\, f_{res}^2 = (1/3)\, p\, (4\pi\, \gamma' M_O\,)^2 . \qquad (4.47)$$

With regard to the requirement that the material for UHF applications must be non-conductive, i.e., be a composite with a dielectric matrix, (4.47) takes into account inclusions concentration denoted as p. For composites based on thin ferromagnetic films (Naito and Suetake 1971), $p \approx 0.1$ can be considered as the realistic evaluation of limit concentration.

Natural FMR frequency f_{res} is practically the same for a bulk material, thin film, and multi-layer composite material in which ferromagnetic films are used as fillers. Resonance frequency also practically does not depend on concentration p, and for iron (Fe), it is equal to 3 GHz.

According to (4.3), the real part of magnetic permeability is approximately equal to μ_{st} up to the frequencies of f_{res} order, while at higher frequencies μ drops rapidly. Thus, f_{res} can be considered as the frequency of magnetic permeability cutoff (Lyubchenkov 2009).

In recent years, the development of effective microwave absorbers has become highly relevant, due to an increase in the number of various electromagnetic radiation sources and extension of the frequency range used. As compared to ferrites with spinel and garnet structures, ferrite with hexagonal (magnetoplumbite) structure has more expressed magnetic anisotropy, high saturation magnetization, and high frequency of natural ferromagnetic resonance (e.g., the frequency of natural ferromagnetic resonance in unsubstituted barium ferrite is 45–50 GHz). Hence, ferrites with magnetoplumbite structure are promising as microwave absorbers in high-frequency range. Barium ferrite is a representative of this class of compounds, which is already known for its microwave absorber applications (Torukhanov 2018). It is known that dielectric and magnetic properties of barium ferrite can be modeled by substituting Fe^{3+} and Ba^{2+} for other ions (Rozanov et al. 2005; Kittel 1948). For example, substitution of Fe^{3+} ions with those of Al^{3+} makes it possible to increase magnetic anisotropy from 17 to 32.5 Ke, which, respectively, leads to an increase in the resonance frequency of natural ferromagnetic resonance (FMR) from 47.6 to 90 GHz (Kittel 1948). It is shown in Rozanov et al. (2005) that substitution of iron ions for In^{3+} ions results in a sharp drop in the frequency of natural FMR, while the substitution of iron and barium ions for cobalt and titanium ions leads to a decrease of anisotropy field and FMR frequency to 10 GHz. The effect of shapes and sizes of hexagonal ferrite particles on magnetic properties and electromagnetic absorption characteristics is also known. It was shown in Shcherbakov (2018) that the values of electromagnetic losses for nanosized ferrite particles can be larger than those in micrometer size particles. The main limitation of using ferrite microparticles is a decrease in the value of saturation magnetization, which is an important characteristic in the development of microwave absorbers.

Snoek's law is not valid for hexagonal ferrites, as well as for thin magnetic films, and static magnetic permeability for hexagonal ferrites is determined via Asche constant by the following expression (Acher et al. 1996):

$$K_A = (\mu_{st} - 1) f_{res}^2 = \left(\gamma' 4\pi M_O\right)^2 \left(1 + \frac{H_\theta + H_\varphi}{4\pi M_O}\right), \quad (4.48)$$

where M_O is material saturation magnetization, H_θ is anisotropy field directed perpendicularly to the hexagonal plane, and H_φ is anisotropy field that determines magnetic moment direction in the hexagonal plane.

It is necessary for composite materials to take into account filler concentration (Rozanov 1999), and formula (4.48) for composites is of the form:

$$(\mu_{st} - 1) f_{res}^2 = p\left(\gamma' 4\pi M_O\right)^2 \left(1 + \frac{H_\theta + H_\varphi}{4\pi M_O}\right) = pK_A, \quad (4.49)$$

where the concentration of inclusions p is taken into account.

Operating range width for RF absorbent materials and coatings

The key characteristic of RF absorbent materials is their operating range width since coatings that have a narrow minimum for reflection coefficient at one frequency will have limited applicability irrespective of this minimum depth. When the wavelength varies within this range, the coefficient of electromagnetic wave reflection from the absorbent coating does not exceed the value chosen.

As noted above, multi-layer and gradient absorbent coatings have significant operating ranges. Multi-layer structures often consist of several alternating systems—dielectric-conductive film (the so-called Jaumann screen (Chambers and Tennant 1996; Ruck et al. 1970)). One of the first varieties of such coatings was developed and used back in the years of World War II. The RF absorbent coating made according to Jaumann scheme was 75 mm thick and had operating wavelength range from 3 to 30 cm. The structure of its material consisted of seven layers of graphite-filled conductive paper separated by intermediate polyvinyl chloride foam layers.

It should be noted that with a large number of layers, multi-layer coatings can be considered as gradient ones in which values ε' and ε'' depend on the distance to a metal substrate, monotonously rising from values 1 and 0, respectively, on the front edge of the coating to high values on its back edge. When calculating gradient coatings, interference effects are usually neglected. The coatings produced are wideband ones but are thicker in comparison with other types of absorbent coatings.

A standard task of absorbent coating development is to achieve as low as possible level of reflection coefficient in the widest possible wavelength range $\lambda_{min}..., \lambda_{max}$ with the minimal coating thickness. Operating range width is determined as $\Delta\lambda = \lambda_{max} - \lambda_{min}$ (or frequency range width $\Delta f = f_{max} - f_{min}$) and within it power reflection coefficient does not exceed the set value (usually 10 dB).

These requirements are mutually contradicting, as real coatings can only have low reflection coefficients in limited frequency ranges. So, the width of operating wavelength range $\Delta\lambda$ is related to coating thickness d and the level of the predetermined amplitude reflection coefficient R_0. This dependence is explicitly known for certain simple cases only, e.g., for a uniform interference coating. If $R_0 \ll 1$ and the frequency dependence of material parameters can be neglected, value $\Delta\lambda$ is given by the following ratio:

$$\frac{\Delta\lambda}{\lambda_0} = \frac{2R_0}{(\pi d/\lambda_0)\cdot|\varepsilon-\mu|}, \tag{4.50}$$

where λ_0 is operating wavelength midrange. If $|\varepsilon| \gg |\mu|$, coating thickness and operating range width are small values in comparison with the wavelength ($d \ll \lambda_0$, $\Delta\lambda \ll \lambda_0$), then, taking into account that in interference coatings

Input expression with the square root (4.50) is reduced to such expression

$$\Delta\lambda/d \approx (32/\pi)\, R_0\mu'. \tag{4.51}$$

Formula (4.51) confirms the fact that wideband characteristics of a single-layer coating are improved with increasing magnetic permeability of the layer. For this reason, obtaining materials with high magnetic permeability values is considered to be the key factor in developing effective thin absorbent coatings.

Of great practical interest are coatings with wider operating ranges. Wide range, as compared to that of single-layer resonance coatings, is achieved by using gradient and multi-layer coatings.

Integral correlation for the reflection coefficient of a multi-layer magneto-dielectric absorbent material on a metal substrate is of the form (Lagarkov et al. 2005):

$$\left|\int_0^\infty \ln|R(\lambda)|\,d\lambda\right| \le 2\pi^2 \sum_j \mu_{st.j} d_j, \tag{4.52}$$

where d_j, $\mu_{st.j}$ are the thickness and static magnetic permeability of the j-th structure layer.

It should be noted that there are fundamental limitations on operating range width for RF absorbent coatings. If a single-layer coating has amplitude reflection that is not inferior to R_0 within operating range $\lambda_{min} \ldots, \lambda_{max}$, expression (4.52) transforms into

$$\frac{\Delta\lambda}{d} \le \frac{2\pi^2\mu_{st}}{|\ln R_0|}, \tag{4.53}$$

where $\Delta\lambda = \lambda_{max} - \lambda_{min}$. Inequality (4.53) can be used to evaluate the limit operating range width of the coating with the set thickness and reflection coefficient level. This inequality can also be used for broadband single-layer and multi-layer coatings in which d_j and $\mu_{st.j}$ are the same.

With any law of dielectric permeability frequency dispersion for single-layer wideband dielectric coating ($\mu_{st} = 1$) and multi-layer dielectric coating, irrespective of the number of layers, the thickness smaller than $\Delta\lambda/17.2$ cannot be obtained at the level of $R_0 = 0.316$ (at such R_0 level, power reflection coefficient does not exceed 10 dB).

It can be seen from (4.47) that composite RF absorbent materials based on thin ferromagnetic films can have the highest values of UHF magnetic permeability of all magnetic materials and, as a result, the widest operating range $\Delta\lambda$ at the same absorbent layer thickness d. For example, for the frequency of natural ferromagnetic resonance, which is equal to 3 GHz, and saturation magnetization $4\pi M_0 = 2.15T$, with taking into account the limit composite filling with spherical iron particles p=0.4, $\mu_{st} = 6.3$ is obtained from (4.45). For hexagonal ferrites, with the account of smaller saturation magnetization than that of ferromagnetics and strong anisotropy field H_θ, the right-hand part of (4.48) usually is 150—200 GHz, which by using (4.9) gives the value of μ_{st} = 3.5–4.5 at p = 0.5. For composites based on thin ferromagnetic films, the realistic limit concentration can be considered to be $p = 0.1$, at which it follows from (4.47) that μ_{st} = 14.3 (Rozanov 1999).

It can be seen that operating range value $\Delta\lambda$ at the same absorbent layer width d in composites based on thin ferromagnetic films is approximately three times larger than $\Delta\lambda$ in composites based on hexagonal ferrites and approximately twice as large as that in composites based on spherical particles.

Inequality (4.53) imposes restrictions on operating range width of RF absorbent materials and coatings at the set RF absorbent layer width and it can be used for evaluating the operating range limit value in the coating with a set thickness and reflection coefficient level. But the knowledge of operating range value does not allow the determination of operating range location on the wavelength or frequency scale. So, for complete information on the range, it is necessary to find λ_{max} or f_{min}. As shown above, the upper limit of operating range of magnetic RF absorbent coatings is determined by the frequency of natural ferromagnetic resonance (FMR), i.e., $\lambda_{max} = \lambda_{res}$. If λ_{max} is known, by means of expression (4.53), λ_{min} can be calculated as well.

The above inequality is also valid for dielectric RF absorbent materials (μ_{st}=1). For example, for composites, with regard to carbon fibers, $\Delta\lambda$ width can be determined. Then, by making use of the fact that for the maximum wavelength λ_{max} of the operating range of RF absorbent coating based on carbon fibers, wavelength λ_R can be chosen and maximum wavelength can be found $\lambda_{max} = \lambda_R = \frac{\varepsilon_m \rho}{60F}$ (Volkovskaya et al. 2017). Also, λ_{min} is calculated in the same way as that for magnetic materials and, therefore, complete information on operating range can be obtained.

4.2 Results of Studying Fragments of RF Absorbent Composite Materials and Coatings

Section 4.2 is concerned with the peculiarities of calculating the parameters of different types of RF absorbing coatings made of composite materials. The results of studying the structures and chemical compositions of these coatings are given in Chap. 3 (Sect. 3.3.1). However, additional experimental studies were conducted with a view to improving the accuracy of calculating the parameters of some RF absorbent coatings and getting a deeper insight into the properties of the specimens examined.

4.2.1 Results of Studying Fragments of 9M727 Iskander Cruise Missile Body Coating

Some of the most effective magnetic fillers for composite materials used in RF absorbent coatings are carbonyl iron and conductive fibers, carbon fibers in particular. The main advantage of such composites is the possibility to significantly vary their electromagnetic parameters, depending on the shape, sizes, filler concentration, and other factors.

The analysis of studying the structure and chemical composition showed that two types of RF absorbent coatings are used in this special-purpose product. The first coating has the working layer approximately 600 μm thick and is a composite material with dense ball-shaped iron inclusions of 1–6.5 μm diameter in the bulk (Fig. 4.1).

The other coating has the working layer about 1700 μm thick and is a composite material in whose bulk high-density carbon fiber inclusions about 7 μm thick are observed (Fig. 4.2).

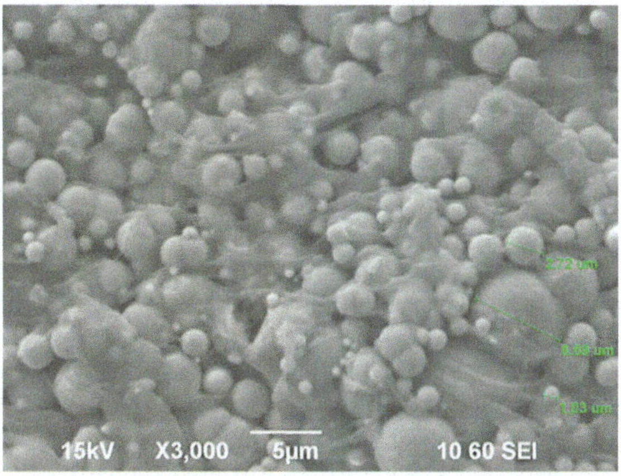

Fig. 4.1 Secondary electron image of the section of coating material layer with iron ball-shaped inclusions (scanning electron microscopy)

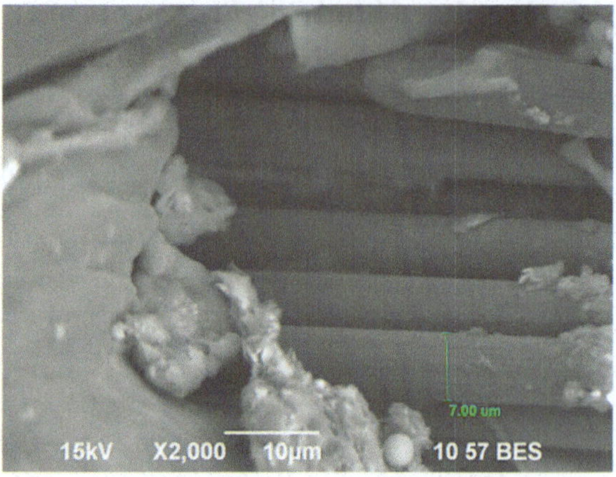

Fig. 4.2 Backscattered electron image of the section of coating material layer with carbon fiber inclusions (scanning electron microscopy)

Now, we analyze the first type of RF absorbent coating (with ball-shaped iron-particle filler).

In the case when composite filler consists of spherical particles of a conductive or magnetic material and the frequencies of electromagnetic radiation are such that the skin layer thickness is comparable with particle sizes, the skin effect significantly affects the frequency dependence of reflection coefficient (Elsukov 2009; Kraev 2017). The process of electromagnetic radiation absorption by the composite consisting of a dielectric matrix and inclusions of ferromagnetic material balls at the frequencies below and above the natural magnetic resonance frequency (MRF) is determined by skin effect. Absorption maximums are observed in the case when skin layer thickness is comparable with the sizes of inclusions.

In Gladkov (2011), the absorption of external electromagnetic field energy by metallic (non-magnetic) spherical objects of radius R at an arbitrary ratio of particle radius and skin layer thickness (the main characteristic of metal's absorbing capacity). Section 4.1 presents expression (4.38) for the dependence of the maximum cross section of particle absorption on frequency $\omega_{\max}$. By comparing the formula for skin layer thickness $\delta = c/\sqrt{2\pi\sigma\omega}$ and (4.38), the condition for maximum absorption $R = \delta$ is obtained.

According to Volkovskaya et al. (2017), Gladkov (2011), for the case of absorption of electromagnetic wave with spherical magnetic particles, the maximum of the dependence of the imaginary part of magnetic permeability on electromagnetic radiation frequency is satisfied on condition presented by expression (4.40). When condition (4.40) is satisfied, the imaginary part of magnetic permeability rises dramatically with an increase in composite filler concentration (Volkovskaya et al. 2017). In Galst'yan and Ravaev (1992), the process of pulsed UHF heating of composites containing spherical metallic particles was considered. It was shown that maximum electromagnetic wave absorption occurs at the maximum value of the imaginary part of magnetic permeability and upon the fulfillment of condition (4.40) (the same as in Volkovskaya et al. (2017), Anzulevich (2013)).

For magnetic materials, the skin layer thickness is determined by expression (4.41), where μ_{st} is static magnetic permeability, $\omega = 2\pi f$, and σ is the electrical conductivity of particle material (for iron $\sigma = 10^7$ S/m).

Using (4.40) and (4.41), we obtain expression (4.42) for resonance frequencies of electromagnetic radiation absorption related to skin effect.

The key issue for RF absorbent coatings is working range width, since coatings that have a narrow minimum for the reflection coefficient at a single frequency have limited applicability irrespective of the minimum depth. A standard task in developing absorbent coatings is to achieve as low a value of reflection coefficient as possible in the widest wavelength range $\lambda_{\min} \ldots, \lambda_{\max}$. Working range width is determined in terms of wavelengths as $\Delta\lambda = \lambda_{\min} - \lambda_{\max}$ or in terms of frequencies as $\Delta f = f_{\max} - f_{\min}$. Within this range, reflection coefficient for power does not exceed the set value (usually, it is 10 dB).

Expression (4.42) implies that in our case the maximum working range frequency fmax of the RF absorbent coating corresponds to the minimum size of particles (R = 0.5 μm), while its minimum size fmin corresponds to the maximum particle size (R = 3.25 μm).

Static magnetic permeability is determined by Snoek's law (2019) which for magnetic material is of the following form:

$$(\mu_{st} - 1) f_0 = K_S, \tag{4.54}$$

where f_0 is the frequency of natural magnetic resonance (3 GHz for iron) and K_S is Snoek's constant (for iron it is 40 GHz). From (4.54) we obtain $\mu_{st} = 14.3$.

Using (4.42), we can calculate limit frequencies and wavelengths of the working range for minimum and maximum sizes of iron inclusions. For particle radius 0.5 μm, $f_{\max} = 28$ GHz ($\lambda_{\min} = c/f_{\max} = 1.1$ cm); for particle radius 3.25 μm, $f_{\min} = 0.7$ GHz ($\lambda_{\max} = c/f_{\min} = 43$ cm).

The analysis of the data obtained shows that working range measured in wavelengths is quite considerable and is approximately 42 cm. However, when calculating the range width, it is necessary to take into account the basic limitation for this parameter, which is related to the thickness of RF absorbent coating (Rozanov 1999).

For composite magnetic RF absorbent materials, there is a limitation of parameters within the working range $\lambda_{\min} \ldots, \lambda_{\max}$, which is determined by expression (4.53). In inequality (4.53), $\Delta\lambda = \lambda_{\max} - \lambda_{\min}$ is the working range, d is RF absorbent layer thickness, μ_{st} is effective static magnetic permeability of the composite, which is calculated using expression (4.54) for Snoek's law but taking into account the concentration of composite filler p (Rozanov and Osipov 2006). That is, when using (4.54) for computing μ_{st}, it is necessary to multiply K_S by the concentration value p, which for composite magnetic RF absorbent coatings is equal to 0.4. Inequality (4.53) imposes limitations on the width of working range of RF absorbent materials and coatings at the set RF absorbent layer width; it can be used to evaluate the limit width of the working range for the coating of the set thickness and reflection coefficient level.

Table 4.1 Results of calculating the parameters of RF absorbent material

λ_{max}, cm	λ_{min}, cm	$\Delta\lambda$, cm	f_{max}, GHz	f_{min}, GHz	Δf, GHz
7	1	28	24.5	4	20.5

The coefficient of reflection from the absorbent layer for amplitude $R_0 = 0.316$ on condition that absorption coefficient for power does not exceed 10 dB. For the coating under study $d = 600\ \mu m = 0.06$ cm. We obtain from (4.54), with the account of concentration $p = 0.4$, $\mu_{st} = 6.3$, and from (4.53) $\Delta\lambda \approx 6$ cm is obtained. Therefore, the working range, limit frequencies, and limit wavelengths should be corrected significantly.

Taking into account the assumption that the average working range frequency $((f_{max} + f_{min})/2 = 14.5$ GHz) is not changed and operating range narrowing is the same on the left and on the right from the average frequency, we find corrected predicted parameters of the RF absorbent coating that is under study (Table 4.1).

4.2.2 *Results of Studying Fragments of Wing Edge Coating of 9M727 Iskander Cruise Missile*

The studies of the chemical composition and structure of a fragment of the leading wing edge of the 9M727 missile of Iskander-K system have shown that this coating is a double-layer composite material with a brown layer from 2.2 to 1.8 mm thick and a dark layer 600 μm thick. In terms of chemical composition and properties, the distinction between these layers is rather arbitrary. Brown and dark layers have the identical chemical composition of matrices. The principal chemical elements of the matrices are as follows: 13% Si (silicon), 26% O (oxygen), 40% C (carbon), and 16% Cl (chlorine). The concentrations of chemical elements in the matrices of the abovementioned layers differ by a few percent. In both layers, there are zones with particles that have significant content of Fe (iron)—up to 30%. However, a much larger number of zones with higher iron content are observed in the dark layer. Based on the fact that iron is a ferromagnetic, a fragment of the coating on the leading edge of the wing of 9M727 cruise missile of *Iskander-K* system can be classified as a magnetic composite RF absorbent material with the absorbing layer width $d = 2600\ \mu m$.

Similar to the specimen analyzed in the previous subsection, in this case, the basic limitation on the absorbent coating working range width exists as well, which for magnetic RF absorbent materials (including composite ones) is expressed by relation (4.53). Effective static magnetic permeability μ_{st} of the composite is also determined by Snoek's law, with the account being taken of the density of composite filler (iron particles).

Taking into consideration (4.53), at the set width of the RF absorbent material layer, it is possible to evaluate limit parameters of the coating working range width with the set thickness and reflection coefficient. However, the data merely on the working range size do not permit establishing the working range location on the wavelength or frequency scale. Therefore, getting complete information on this range necessitates determination of $\lambda_{\max}$ or $f_{\min}$.

The upper limit of the working range of magnetic RF absorbent coatings is determined by the resonance frequency of the natural ferromagnetic resonance (FMR), above which the material does not have magnetic properties. The resonance frequency of natural FMR for Fe is 3 GHz. Therefore, the abovementioned resonance frequency should be considered as the upper frequency limit of the working range of RF absorbent coating of the leading edge of the wing of 9M727 cruise missile of *Iskander-K* system, i.e., $f_{res} = f_{\max} = 3\text{GHz}$ and $\lambda_{\min} = c/f_{\max} = 10$ cm. To evaluate $\lambda_{\max}$, we obtain from (4.53)

$$\lambda_{\max} = \frac{2\pi^2\mu_{st}}{|\ln R_0|}d + \lambda_{\min}. \tag{4.55}$$

Snoek's law (Parkhomenko et al. 2019; Lyubchenkov 2009) for magnetic composite material is

$$(\mu_{st} - 1)f_{res} = pK_S, \tag{4.56}$$

where p is composite filler concentration and K_S is Snoek's constant, which depends solely on the saturation magnetization of the magnetic material, which is only determined by the material composition, and for iron $K_S = 40$ GHz. For Fe, $f_{res} = 3$ GHz, $p = 0.4$ for composites with magnetic particles filler, thus, from (4.49) we obtain $\mu_{st} = 6.3$.

The reflection coefficient for amplitude from the absorbent layer $R_0 = 0.316$ on condition that power absorption coefficient does not exceed −10 dB. For the coating under study $d = 2600\ \mu\text{m} = 0.26$ cm, $\lambda_{\min} = 10$ cm, $\mu_{st} = 6.3$. From (4.55) we obtain $\lambda_{\max} \approx 38$ cm. Taking into account that $f_{\min} = c/\lambda_{\max} = 0.8$ GHz, the predicted frequency characteristics of the RF absorbent layer of the wing edge material in 9M723 missile of *ISKANDER-M* system, which is a magnetic composite RF absorbent material, can be summarized in Table 4.2.

Table 4.2 Results of calculating the parameters of RF absorbent material

$\lambda_{\max}$, cm	$\lambda_{\min}$, cm	$\Delta\lambda$, cm	$f_{\max}$, GHz	$f_{\min}$, GHz	Δf, GHz
38	10	28	3	0.8	2.2

4.2.3 Results of Studying Fragment of Two-Layer Coating of Nose Fairing of 9M727 Iskander Cruise Missile

Analysis of the results of studying the structure and chemical composition of a fragment of the nose fairing of the 9M727 cruise missile of the Iskander-K complex with a two-layer coating showed that this sample is a composite material with a thickness of 6000 μm, which consists of two layers of different structure and chemical composition. The thickness of the black layer is about 1650 μm. The black layer is a composite material that contains inclusions in the form of particles of various sizes (5–100 μm). These inclusions contain iron (up to 60% Fe) and barium (up to 18% Ba). Besides, the particles contain cobalt (Co) and titanium (Ti). The presence of these elements in the particles permits to conclude that the inclusions are hexagonal barium ferrite $BaFe_{12}O_{19}$, in which some of Fe^{3+} and Ba^{2+} ions are substituted with Co and Ti ions.

In recent years, the development of effective microwave absorbers has become an urgent problem because of the increasing number of sources of electromagnetic radiation and the broadening of the frequency range used. Compared to ferrites with spinel and garnet structures, ferrites with hexagonal structure (magnetoplumbite) have more pronounced magnetic anisotropy, significant saturation magnetization, and a high frequency of natural ferromagnetic resonance. Barium ferrite without substituted ions has the natural ferromagnetic resonance frequency of 45–50 GHz. Due to these properties, ferrites with the magnetoplumbite structure are promising microwave absorbers for use in the high-frequency range.

Barium ferrite is known as an effective microwave absorber (Meshram et al. 2004), its dielectric and magnetic properties can be adjusted by substituting Fe^{3+} and Ba^{2+} ions with other ions (Torukhanov 2018; Shcherbakov 2018). The substitution of Fe^{3+} with Al^{3+} allows increasing the magnetic anisotropy field from 17 to 32.5 kOe that leads to an increase in the frequency of natural ferromagnetic resonance from 47.6 to 90 GHz (Shcherbakov 2018). At the same time, the substitution of Fe^{3+} ions with In^{3+} ions causes a sharp decrease in the resonant frequency of the natural ferromagnetic resonance (FMR), and the substitution of Fe^{3+} and Ba^{2+} ions with cobalt Co^{2+} and titanium Ti^{4+} ions reduces the magnetic anisotropy field and the FMR frequency to 10 GHz (Torukhanov 2018). The shape and size of hexagonal ferrite particles are known to effect the magnetic properties and electromagnetic absorption. It was shown in Rozanov et al. (2005) that nanoscale ferrite particles can provide higher values of electromagnetic losses compared to micrometer-sized particles. However, the main disadvantage of nanoparticles is a decrease in the saturation magnetization, which is an important parameter in the development of microwave absorbers.

It is worthwhile to note that the black coating layer is not located under the brown layer along the entire lower part of the studied sample, but wavelike penetrates the brown layer volume. This permits to consider the entire studied composite as a radiation-absorbent coating with a total thickness of 6000 μm.

As noted in Sect. 4.1, the resonant frequency of the natural FMR is related to the internal magnetic anisotropy field Hk by relation (4.44), where $\gamma' = \gamma/2\pi$, and γ is the gyromagnetic ratio, $\gamma' = 2.8$ MGz/Oe (Kittel 1948).

According to the law of frequency dispersion of magnetic materials, the magnetic permeability μ increases proportionally to the frequency up to FMR. The frequency drops significantly above the resonant frequency (f_{res}), which is unacceptable for most technical applications. In real materials, the magnetic permeability usually decreases with frequency over the entire frequency range due to magnetic losses. Therefore, the suitability of a material for ultra-high-frequency (UHF) applications can be roughly estimated basing on μ_{st} and f_{res}. The static permeability μst can be considered an estimate of the dynamic permeability at frequencies below the FMR frequency, so it characterizes the quality of the material within the working frequency range. In other words, the FMR frequency determines the limit of the working frequency range, above which the material does not have magnetic properties. As shown above, the resonant frequency of the natural FMR for barium ferrite in which iron ions are substituted with cobalt and titanium ions is 10 GHz. Therefore, the upper frequency limit of the working range of the radiation-absorbent coating of the nose fairing fragment of the 9M727 *Iskander-K* cruise missile should be the indicated resonant frequency, i.e., $f_{res} = f_{\max} = 10$ GHz.

Similar to thin ferromagnetic films, Snoek's law does not hold for hexagonal ferrites, and the static magnetic permeability is determined by expression (4.48) with Asher's coefficient (Adenot et al. 2002). In (4.48), M_O is the saturation magnetization of the material, H_Θ is the anisotropy field directed normally to the hexagonal plane, and H_φ is the anisotropy field that determines the direction of the magnetic moment in the hexagonal plane. Taking into account the concentration of the filler (Rozanov and Osipov 2006), formula (4.48) for composites takes the form (4.49), where p is the concentration of inclusions.

Inequality (4.53) limits the working range width of radiation-absorbent materials at a given width of the radiation-absorbent material layer, and can be used to estimate the maximum working range width of a coating with given thickness and reflection coefficient.

As noted above, in our case, $f_{res} = f_{\max} = 10$ GHz, so $\lambda_{\min} = c/f_{\max} = 3$ cm. To estimate $\lambda_{\max}$ in (4.53), we use relation (4.55).

Taking into account that K_A for hexagonal ferrites does not exceed 200 GHz^2 (Rozanov and Osipov 2006), and the maximum value for $p \approx 0.5$ in the composites with hexagonal barium ferrite as a filler (Rozanov and Osipov 2006), $f_{res} = 10$ GHz and $d = 6000$ µm, the amplitude coefficient of reflection from the absorbing layer $R_0 = 0.316$, provided that the power absorption coefficient does not exceed -10 dB, we find $\lambda_{\max} \approx 23$ cm from (4.49) and (4.55). Since $f_{\min} = c/\lambda_{\max} = 1.3$ GHz, the frequency characteristics of the coating material (a magnetic composite broadband radiation-absorbent material) of the nose fairing fragment of the 9M727 cruise missile of the *Iskander-K* complex are summarized in Table 4.3 (the maximum possible working ranges $\Delta\lambda$ and Δf are indicated).

Taking into account that K_A for hexagonal ferrites does not exceed 200 GHz^2 (Rozanov and Osipov 2006), and the maximum value for $p \approx 0.5$ in the composites

Table 4.3 Results of calculations of parameters of radiation-absorbent material

λ_{max}, cm	λ_{min}, cm	$\Delta\lambda$, cm	f_{max}, GHz	f_{min}, GHz	Δf, GHz
23	3	20	10	1.3	8.7

with hexagonal barium ferrite as a filler (Rozanov and Osipov 2006), $f_{res} = 10$ GHz and $d = 6000$ μm, the amplitude coefficient of reflection from the absorbing layer $R_0 = 0.316$, provided that the power absorption coefficient does not exceed −10 dB, we find $\lambda_{max} \approx 23$ cm from (4.49) and (4.55). Since $f_{min} = c/\lambda_{max} = 1.3$ GHz, the frequency characteristics of the coating material (a magnetic composite broadband radiation-absorbent material) of the nose fairing fragment of the 9M727 cruise missile of the *Iskander-K* complex are summarized in Table 4.3 (the maximum possible working ranges $\Delta\lambda$ and Δf are indicated).

4.2.4 Results of Studying Fragments of Nose Fairing Coating of 9M727 Iskander Cruise Missile (Brown Sample 1 and Black Sample 2)

4.2.4.1 Sample 1

The analysis of the results of studying the structure and chemical composition of sample 1 showed that it is a composite coating with a thickness of 3120 μm with a polymer matrix filled with particles of various sizes (0.5–10 μm); the particle contains high amounts of iron (over 50% Fe) and barium (over 20% Ba). Sample 1 consists of three layers of different structures and chemical compositions. The top layer is about 120 μm thick, the middle layer is 2830 μm thick, and the bottom layer is 170 μm thick. The highest volume density of the filler particles is observed in the middle layer of the coating. Besides Fe and Ba, the particles contain cobalt (Co) and titanium (Ti). The presence of these elements in the particles permits to conclude that the inclusions are hexagonal barium ferrite $BaFe_{12}O_{19}$, in which Fe^{3+} and Ba^{2+} ions are partially substituted with Co and Ti ions.

Similar to the fragment of the nose fairing of the 9M727 cruise missile of the *Iskander-K* complex with the two-layer coating, sample 1 is a magnetic radiation-absorbent composite coating with a thickness of $d = 2830$; the filler is barium ferrite in which some Fe ions are substituted with Co and Ti ions. The resonant frequency of natural FMR for barium ferrite containing cobalt and titanium ions is 10 GHz. Therefore, the upper frequency limit of the working range of the radiation-absorbent coating of sample 1 should be equal to the indicated resonant frequency, i.e., $f_{res} = f_{max} = 10$ GHz, then $\lambda_{min} = c/f_{max} = 3$ cm. To estimate λ_{max} in (4.53) for magnetic broadband radiation-absorbent materials, we use relation (4.55).

Table 4.4 Results of calculations of parameters of radiation-absorbent material

λ_{max}, cm	λ_{min}, cm	$\Delta\lambda$, cm	f_{max}, GHz	f_{min}, GHz	Δf, GHz
12.5	3	9.5	10	2.5	7.5

Taking into account that K_A for hexagonal ferrites does not exceed 200 GHz^2, and the maximum value for $p \approx 0.5$ in the composites with hexagonal barium ferrite as a filler, $f_{res} = 10$ GHz and $d = 2830$ μm, the amplitude coefficient of reflection from the absorbing layer $R_0 = 0.316$, provided that the power absorption coefficient does not exceed −10 dB, we find $\lambda_{max} \approx 12.5$ cm from (4.49) and (4.55). Since $f_{min} = c/\lambda_{max} = 2.4$ GHz, the frequency characteristics of the coating material (a magnetic composite broadband radiation-absorbent material) of the nose fairing fragment (sample 1) of the 9M727 cruise missile of the *Iskander-K* complex are summarized in Table 4.4 (the maximum possible working ranges $\Delta\lambda$ and Δf are indicated).

4.2.4.2 Sample 2

The analysis of the results of studying the structure and chemical composition of sample 2 showed that this sample is a composite coating with a thickness of 3230 μm, which consists of two layers with different structures and chemical compositions. The thickness of the opaque black coating layer is about 175 μm and the thickness of the shiny black coating layer is about 3055 μm. The latter layer contains in its volume particles of various sizes (5–50 μm) with a rather high density. X-ray microanalysis showed that these particles contain iron (Fe) and barium (Ba) in high concentrations. Besides, the particles contain cobalt (Co) and titanium (Ti). The presence of these elements in the particles permits to conclude that the inclusions are hexagonal barium ferrite $BaFe_{12}O_{19}$, in which Fe^{3+} and Ba^{2+} ions are partially substituted with Co and Ti ions.

Similar to sample 1, sample 2 is also a magnetic composite coating with a radiation-absorbent 3055-μm-thick layer that contains a filler—barium ferrite with Fe ions partially substituted with Co and Ti ions.

The data used for calculating the parameters of sample 1 can be used to calculate the frequency characteristics of the coating of sample 2. The only difference is in the width of the radiation-absorbent layer (it is thicker in sample 2).

Using (4.49) and (4.55), we find $\lambda_{max} \approx 13.3$ cm. Since $f_{min} = c/\lambda_{max} = 2.3$ GHz, the frequency characteristics of the coating material (a magnetic composite broadband radiation-absorbent material) of the nose fairing fragment (sample 2) of the 9M727 cruise missile of the *Iskander-K* complex are summarized in Table 4.5 (the maximum possible working ranges $\Delta\lambda$ and Δf are indicated).

Table 4.5 Results of calculations of parameters of radiation-absorbent material

$\lambda_{\max}$, cm	$\lambda_{\min}$, cm	$\Delta\lambda$, cm	$f_{\max}$, GHz	$f_{\min}$, GHz	Δf, GHz
13.3	3	10.3	10	2.3	7.7

4.2.5 *Results of Studying Fragments of Thermal Barrier Coating of 9M723 Missile Body of Iskander-M Complex*

Studies of the chemical composition and structure of the thermal barrier coating of the body of *Iskander-M* complex 9M723 missile showed that the coating is a composite material that contains several layers with different structures and chemical compositions. The top layer of the coating with a thickness of 124 μm is a composite with a polymer matrix that contains 60% C and 29% O, and doped with Mg, Si, Cl; the matrix is filled with particles which contain 40% Fe and 15% Cr.

Thus, the coating of the body of 9M723 missile of the *Iskander-M* complex is a thermal barrier coating, as it contains the porous layer and the foamed glass layer. The layer reinforced with glass fiber layers increases the strength of the coating, and the ferromagnetic iron (Fe) particles in the top layer allow attributing the studied coating not only to the thermal barrier materials, but also to the magnetic radiation-absorbent coatings with a 124-μm-thick radiation-absorbent layer.

For magnetic composite radiation-absorbent materials, a fundamental limitation is imposed on the width of the working range of the absorbing coating (4.53), where $\Delta\lambda = \lambda_{\max} - \lambda_{\min}$ is the working range, *d* is the thickness of the radiation-absorbent coating, and μst is the effective static magnetic permeability of the composite, taking into account the concentration of the composite filler *p*.

The static permeability is related to the resonant frequency (fres) of the natural ferromagnetic resonance (FMR) by Snoek's law (1948), which for a composite material with magnetic inclusions has the form (4.56) (Rozanov and Osipov 2006), where K_S is Snoek's coefficient which depends only on the saturation magnetization of the magnetic material and is determined only by the composition of the material. For iron, $K_S = 40$ GHz, $f_{res} = 3$ GHz, and $p = 0.4$ for composites with a filler in the form of magnetic particles. From (4.56), we find $\mu_{st} = 6.3$.

The frequency (fres) is considered the maximum frequency of the working range for magnetic radiation-absorbent materials and coatings. For Fe $f_{res} = 3$ GHz, so $f_{res} = f_{\max} = 3$ GHz, and $\lambda_{\min} = c/f_{\max} = 10$ cm.

Formula (4.55) allows estimating $\lambda_{\min}$.

The amplitude coefficient of reflection from the absorbing layer is $R_0 = 0.316$, provided that the power absorption coefficient does not exceed -10 dB. For the coating under study, we have $d = 124$ $\mu m = 0.0124$ cm, $\lambda_{\min} = 10$ cm, and $\mu_{st} = 6.3$, so (4.55) gives $\lambda_{\max} \approx 10.2$ cm. Since $f_{\min} = c/\lambda_{\max} = 2.9$ GHz, we find $\Delta f = 0.1$ GHz and $\Delta\lambda = 0.2$ cm.

Such small values of the working range and the thickness of the radiation-absorbent layer lead to electromagnetic wave energy losses, but the magnitude of

Table 4.6 Results of calculations of parameters of radiation-absorbent material

λ_{max}, cm	λ_{min}, cm	$\Delta\lambda$, cm	f_{max}, GHz	f_{min}, GHz	Δf, GHz
10.2	10	0.2	3	2.9	0.1

these losses is insignificant. Besides, the working range is narrower even compared to that of resonant narrowband radiation-absorbent coatings. Therefore, it can be assumed that the presence of iron particles in the layer may be associated with both the formation of a radiation-absorbent layer and the technological processes of deposition of the coating, which are not aimed at obtaining absorbing properties. The results of calculations of the parameters of the thermal barrier coating of the 9M727 *Iskander* cruise missile body are listed in Table 4.6.

4.2.6 Results of Studying Aerodynamic Rudder Fragment of 3M-14 Cruise Missile of Kalibr Complex

The analysis of the results of studying the structure and chemical composition of a fragment of the aerodynamic rudder of the 3M-14 cruise missile of the Kalibr complex showed that this sample is a composite consisting of two layers of different structures and chemical compositions. The sample contains a thin top layer with a thickness of 10 μm and a base layer with a thickness of 6 mm. The main layer of the composite material is a polymer (matrix) reinforced with layers of fiberglass fabric to increase strength and heat resistance. In total, more than 25 such layers were observed with a fiberglass thickness of about 11 μm. The layers of the artificial fabric contain particles (0.5–100 μm) with a high content of iron (up to 45% Fe). The number of layers with iron-containing particles corresponds to the number of layers of artificial fabric (about 25).

Figure 4.3 shows an image of a section in the zone of the main layer of the fragment of the aerodynamic rudder in X-ray characteristic radiation (spatial distribution of chemical elements along the section of the sample under study). The white dots correspond to the presence of the chemical element indicated under the image.

Thus, the presence of layers containing particles with a high Fe content (iron is a ferromagnet) permits to classify the material of the aerodynamic rudder fragment of the 3M-14 cruise missile of the *Kalibr* complex as a magnetic composite multi-layer radiation-absorbent material.

To form broadband radiation-absorbent coatings, materials with a high dynamic magnetic permeability are required. Before analyzing the radiation-absorbent properties of the fragment of the aerodynamic rudder of the 3M-14 cruise missile of the *Kalibr* complex, let us briefly analyze the frequency dispersion of magnetic permeability in the microwave range, and determine the limitations that link the parameters of the frequency dispersion of magnetic permeability of materials. It should be noted

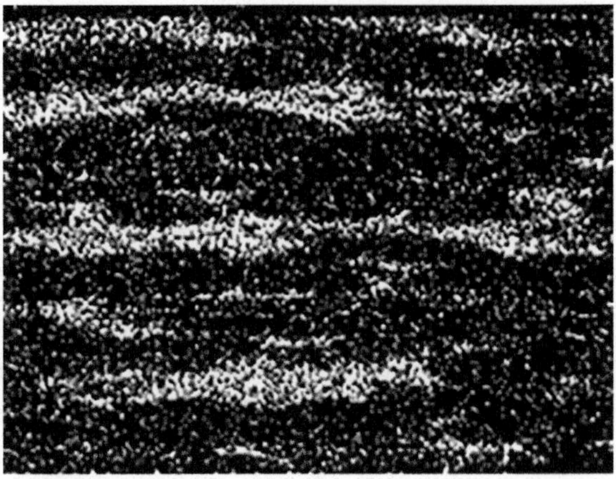

Fig. 4.3 Image of section in zone of main layer of aerodynamic rudder fragment in X-ray characteristic radiation. White dots indicate areas with high Fe content

that thin ferromagnetic films (FM) have high values of microwave magnetic permeability.

It is known that the type of frequency dependence of the magnetic permeability of magnetic materials used for radiation-absorbent materials is determined by the movement of domain walls (boundaries), skin effect, and natural ferromagnetic resonance (FMR). The motion of domain walls is usually too inertial to contribute to the frequency dispersion in the microwave range, so in this range, the main mechanism that forms the magnetic spectra is the natural ferromagnetic resonance. The natural FMR is determined by the dynamics of magnetic moment motion in a zero external magnetic field under the influence of the internal anisotropy field.

Without taking into account the magnetic losses associated with the movement of domain boundaries and losses due to the skin effect, the frequency dependence of the magnetic permeability has the Lorentzian form (4.43), where μ is the dynamic magnetic permeability, μ_{st} is the static magnetic permeability, f_{res} is the resonance frequency of the natural FMR, and f is the frequency of the electromagnetic wave that affects the magnetic material. The resonance frequency of the natural FMR is related to the internal magnetic anisotropy field H_A by the relationship (4.44), where $\gamma' = \gamma/2\pi$ and γ is the gyromagnetic ratio.

According to the Lorentzian law of frequency dispersion (4.43), the magnetic permeability increases with frequency up to the FMR frequency. Above fres, it drops significantly, which is unacceptable for most technical applications. In real materials, the magnetic permeability typically decreases with frequency over the entire frequency range due to magnetic losses. Therefore, the suitability of a material for microwave applications can be roughly estimated basing on the values of μ_{st} and f_{res}. The static permeability can be taken as an estimate of the dynamic permeability below the FMR frequency, so it characterizes the quality of the material within the working frequency range. The FMR frequency determines the limit of the working frequency range, above which the material does not have magnetic properties.

The static permeability and resonant frequency are related to each other by Snoek's law. When the magnetic material consists of single-domain grains of a spherical shape in the first approximation have, Snoek's law is written as (4.45) (Snoek 1948), where M_0 is the saturation magnetization of the material. The right side of this expression is called Snoek's coefficient (K_s). This coefficient depends only on the saturation magnetization of the magnetic material; it does not depend on the magnetic anisotropy field and other structure-dependent parameters. Unlike M_0, which is determined only by the composition of the material, H_A is a function of the magnetic and crystal structure, which, in turn, are affected by the material production and processing technology. Changes in the anisotropy field due to introducing additives, annealing, and processing the material after production and other technological techniques were considered in Bekker et al. (2004), Chai et al. (2010), Hendrych et al. (2009), and Yang et al. (2008). These techniques allow changing the anisotropy of the material within a wide range, so the values of μ_{st} and f_{res} usually depend on the processing parameters. On the contrary, the value of K_s is structurally independent and is determined only by the material composition. For most ferrites (except for those with hexagonal structure), Snoek's coefficient is in the range from 2 to 5 GHz (Naito and Suetake 1971). Ferromagnetic metals and alloys can have higher K_s values due to higher saturation magnetization. For example, iron has $M_O = 1700$ G and $K_S = 40$ GHz.

Snoek's law is valid for many magnetic materials. The independence of Snoek's coefficient from structural parameters allows it to be widely used in the analysis of data on microwave properties of materials. Snoek's law limits the possibilities of producing both materials with high dynamic magnetic permeability and high magnetic losses at microwave frequencies. It should be noted that Snoek's law in the form (4.45) is not valid for all types of ferromagnetic particles. In particular, it is not applicable for thin ferromagnetic (FM) films.

Thin FM films can have a high magnetic permeability at higher frequencies compared to bulk materials with a comparable static permeability, so they are promising for developing radiation-absorbent materials.

The static permeability and resonance frequency of thin FM films are related by expression (4.46), the right side of which, as in (4.45), is determined only by the M_O value. By analogy with Snoek's coefficient, Asher's coefficient K_A (Baclet et al. 1996) is introduced as the right side of expression (4.46) with the averaging factor added to it if necessary.t

Similar to (4.45), expression (4.46) limits the permissible values of the dynamic magnetic permeability in FM materials. These expressions differ: in (4.46), the saturation magnetization of the material is squared, so (4.46) allows for much higher values of magnetic permeability. For example, if the FMR frequency is 1 GHz and $M_O = 1700$ G, (4.45) gives $\mu_{st} = 40$, whereas (4.46)—$\mu_{st} = 3600$.

Asher's coefficient is an estimate of the fraction of magnetic moments that precess in an alternating magnetic field and, accordingly, contribute to the magnetic loss peak associated with ferromagnetic resonance (FMR). Asher's coefficient can also be used to estimate the limiting characteristics of microwave devices that use magnetic materials, including radiation-absorbent materials.

For a multi-layer composite material with a filler comprised of magnetic particles in the form of thin films whose intrinsic magnetic moment lies in the film plane, (4.45) can be represented as (4.47).

Taking into account the fact that the material for microwave applications should be non-conductive, i.e., a composite with a dielectric matrix, (4.46) takes into account the concentration of filler particles p. For composites based on thin ferromagnetic films (Lagarkov et al. 2005), $p \approx 0.1$ can be considered a realistic estimate of the limiting concentration.

The resonance frequency of the natural FMR f_{res} is practically the same for a bulk material, a thin film, and a multi-layer composite material in which ferromagnetic films are used as a filler. The resonance frequency is also practically independent of the concentration p and is equal to 3 GHz for iron (Fe).

According to the real part of the magnetic permeability is approximately equal to μ_{st} up to the frequencies of the order of f_{res}; at higher frequencies μ drops sharply. Hence, f_{res} can be considered the cutoff frequency of the magnetic permeability (Rozanov and Osipov 2006). Since the studied fragment of the aerodynamic rudder of the 3M-14 cruise missile of the *Kalibr* complex is a multi-layer composite radiation-absorbent material with films (iron layers) as a filler, the resonance frequency f_{res} should be considered the maximum frequency of the working range of this material, which is equal to 3 GHz.

In our case $f_{res} = f_{\max}$, and $\lambda_{\min} = c/f_{\max}$ (where c is the speed of light).

According to Rozanov (1999), there is a fundamental limitation in the width of the working range of absorbing coating, which is expressed by the integral relation (4.52) for the reflection coefficient of a multi-layer magneto-dielectric absorbing material.

For broadband multi-layer radiation-absorbent materials (in which $\mu_{(st.j)}$ is the same for all layers), expression (4.47) takes the form (4.53) and can be used to estimate the maximum width of the working range of a coating with given thickness and reflection coefficient.

Expressions (4.47) and (4.53) allow calculating μ_{st}, $\lambda_{\max}$, and $\Delta\lambda$, as well as the working frequency range Δf.

For iron (Fe), $K_s = 2/3(4\pi\gamma' M_O) = 40$ GHz, $K_A = 3600$ GHz $f_{res} = 3$ GHz, and for a composite filled with FMR films, $p = 0.1$.

Substituting these data into (4.47), we obtain $\mu_{st} = 14.3$.

To estimate $\lambda_{\max}$, we use expression (4.55).

The amplitude coefficient of reflection from the absorbent layer is $R_0 = 0.316$, provided that the power absorption coefficient does not exceed -10 dB. For the fragment under study, we have $d = 0.6$ cm, $\lambda_{\min} = c/f_{\max} = 10$ cm. From (4.55), we find $\lambda_{\max} \approx 155$ cm, and $f_{\min} = c/\lambda_{\max} = 0.2$ GHz. The frequency characteristics of the material of the aerodynamic rudder fragment of the 3M-14 cruise missile of the *Kalibr* complex, which is a magnetic composite multi-layer and broadband radiation-absorbent material, are summarized in Table 4.7 (the maximum possible working ranges $\Delta\lambda$ and Δf are indicated).

Table 4.7 Results of calculations of parameters of radiation-absorbent material

λ_{max}, cm	λ_{min}, cm	$\Delta\lambda$, cm	f_{max}, GHz	f_{min}, GHz	Δf, GHz
155	10	145	3	0.2	2.8

4.3 Formulating Recommendations Concerning Electromagnetic Radiation Parameter

In Sects. 4.1 and 4.2, a procedure for calculating parameters of some frequency characteristics of RF absorbent coatings and materials was proposed.

This procedure of studying RF absorbent coatings applied in special-purpose structures relies on employing electron microscopy and X-ray microanalysis. These techniques permit highly accurate examination of the inner structure of RF coatings, including geometric parameters of their fillers and layers, as well as determining quantitative chemical compositions of coating materials. This, in turn, provides the opportunity of modeling the physical processes that occur during radio wave interactions with the material of RF absorbent coating in a wide frequency range.

At the next stage, using the results of theoretical analytical studies regarding the interaction of electromagnetic waves with composite RF absorbent materials and specific experimental data obtained with electron microscopy and X-ray microanalysis, the working range width of RF absorbent coating is determined. Taking into account the physical mechanisms of electromagnetic wave's absorption in the particular material, researchers also establish limit wavelengths of the abovementioned wave range.

The results obtained make it possible to formulate some recommendations regarding electromagnetic radiation parameters for effective identification of special-purpose structures. In fact, the above procedure allows establishing the wave range of electromagnetic radiation within which the probability of identifying special-purpose structures is rather low.

It was noted in Sect. 4.1 that there are fundamental limitations of the working range width for RF absorbent coatings. These limitations are established by expression (4.53), which serves as the basis for estimating and calculating the maximum coating working range width, computing limit frequencies and wavelengths of this range at the specified coating thickness, the value of static magnetic permeability and the set level of backscatter coefficient that is not exceeded within the working range. That is, the wave range width is determined as $\Delta\lambda = \lambda_{max} - \lambda_{min}$ (or the range width frequencies $\Delta f = f_{max} - f_{min}$) and within which reflection coefficient for power does not exceed the set value (usually—10 dB, which corresponds to the reflection coefficient of 10%).

It should be emphasized that expression (4.53) is quite universal and can be used for dielectric and magnetic RF absorbent coatings, in particular, for broadband single- and multi-layer ones.

Section 4.2 presents the application of the abovementioned procedure for computing the operating range width as well as limit wavelengths and frequencies of the range for magnetic RF absorbent materials used in specific special-purpose structures. Computation results are shown in Table 4.1. The working range of a RF absorbent coating or material is determined exactly as the region with low probability of identifying special-purpose structures.

Geometric parameters of RF absorbing coatings and materials, such as the thicknesses of RF absorbent layers and dimensions of the particles of composite coating fillers, were determined with electron microscopy. The quantitative chemical compositions of coating materials and fillers were studied with X-ray microanalysis. The values of effective static magnetic permeability were determined according to Snoek's and Ashe's laws. As noted above, to compute the range of unlikely identification of special-purpose structures and limit wavelength and frequency values of the range, the statement on the fundamental limitation of the width of RF absorbent coatings and materials (expression 4.53) was used. All stages of the abovementioned computations are presented in Sect. 4.2.

The data presented in Table 4.1 give evidence that the use of composites with fillers that have different magnetic properties permit the development of RF coatings with variable ranges of low-probability of identifying special-purpose structures.

Using the composite with the filler that consists of BaFe12O19 hexagonal ferrite particles (positions 3, 4, 5 in Table 4.1), in which a part of Fe^{3+} Ba^{2+} ions is substituted with Co and Ti ions and which has the natural ferromagnetic resonance (FMR) frequency of 10 GHz (it is this frequency that is the upper limit of the unlikely identification), it is possible to shift the range on the frequency scale toward higher frequencies and on the wavelength scale—toward lower frequencies.

To increase the maximum frequency of the range is also possible in the case when spherical particles of a conductive or magnetic material are used as the composite filler, while the electromagnetic radiation frequencies are such that the size of skin layer is comparable with particle dimensions. Then the skinning process significantly influences the frequency dependence of the reflection coefficient. The process of absorbing electromagnetic radiation by a composite consisting of a dielectric matrix and ball-shaped inclusions of ferromagnetic material at the frequencies that differ from those of the natural magnetic resonance are determined by the skin effect. Here, absorption maximums occur when the skin layer thicknesses are comparable with the sizes of inclusions. In such a case, the maximum frequency of unlikely identification of f_{max} corresponds to the minimum sizes of ferromagnetic ball-shaped inclusions. Such conditions are implemented in the process of producing the coating indicated in f_{max} position 1 of Table 4.1 (in our case, this is the maximum range frequency) (Table 4.8).

It should be emphasized that the width of the unlikely identification range is determined by the RF absorbent layer, the effective static magnetic permeability, and the chosen level of reflection coefficient, which is not exceeded within the range. So, it can be expected that the widest range of unlikely identification would be observed in considering the RF absorbent material indicated in position 7 of Table 4.1. This material has the highest values of ${\mu_{st}}^{ef} = 14.3$ and $d = 6000\,\mu\text{m}$. The high ${\mu_{st}}^{ef}$

Table 4.8 Results of calculating the operation wave range width and limit wavelengths and frequencies of the range of magnetic RF absorbent materials

Name of special-purpose structure	RF absorbent layer thickness, (μm)	Value of effective static magnetic permeability (relative units)	Range of unlikely identification of special-purpose structures (cm, GHz)	Limit wavelengths and frequencies of the operation range (cm, GHz)
1. Coating of 9M727 cruise missile of *Iskander-K* system	$d = 600\,\mu\text{m}$	$\mu_{st}^{ef} = 6.3$	$\Delta\lambda = 6$ cm, $\Delta f = 20.5$ GHz	$\lambda_{max} = 7$ cm, $\lambda_{min} = 1$ cm; $f_{max} =$ 24.5 GHz, $f_{min} = 4$ GHz
2. Wing-edge coating of 9M727 cruise missile of *Iskander-K* system	$d = 2600\,\mu\text{m}$	$\mu_{st}^{ef} = 6.3$	$\Delta\lambda = 28$ cm, $\Delta f = 2.2$ GHz	$\lambda_{max} = 38$ cm, $\lambda_{min} = 10$ cm; $f_{max} = 3$ GHz, $f_{min} = 0.8$ GHz
3. Double-layer nose cone coating of 9M727 cruise missile of *Iskander-K* system	$d = 6000\,\mu\text{m}$	$\mu_{st}^{ef} = 2$	$\Delta\lambda = 20$ cm, $\Delta f = 8.7$ GHz	$\lambda_{max} = 23$ cm, $\lambda_{min} = 3$ cm; $f_{max} = 10$ GHz, $f_{min} = 1.3$ GHz
4. Fragment of nose cone coating of 9M727 cruise missile of *Iskander-K* system (specimen 1)	$d = 2830\,\mu\text{m}$	$\mu_{st}^{ef} = 2$	$\Delta\lambda = 10$ cm, $\Delta f = 7.5$ GHz	$\lambda_{max} = 12.5$ cm, $\lambda_{min} = 3$ cm; $f_{max} = 10$ GHz, $f_{min} = 2.5$ GHz
5. Fragment of nose cone coating of 9M727 cruise missile of *Iskander*-K system (specimen 2)	$d = 3055\,\mu\text{m}$	$\mu_{st}^{ef} = 2$	$\Delta\lambda = 10.3$ cm, $\Delta f = 7.7$ GHz	$\lambda_{max} = 13.3$ cm, $\lambda_{min} = 3$ cm; $f_{max} = 10$ GHz, $f_{min} = 2.3$ GHz
6. Heat-insulating body coating of 9M727 cruise missile of *Iskander-M* system	$d = 124\,\mu\text{m}$	$\mu_{st}^{ef} = 6.3$	$\Delta\lambda = 0.2$ cm, $\Delta f = 0.1$ GHz	$\lambda_{max} = 10.2$ cm, $\lambda_{min} = 10$ cm; $f_{max} = 3$ GHz, $f_{min} = 2.9$ GHz
7. Fragment of aerodynamic steering system of 3M-14 *Kalibr* cruise missile	$d = 6000\,\mu\text{m}$	$\mu_{st}^{ef} = 14.3$	$\Delta\lambda = 145$ cm, $\Delta f = 2.8$ GHz	$\lambda_{max} = 155$ cm, $\lambda_{min} = 10$ cm; $f_{max} = 3$ GHz, $f_{min} = 0.2$ GHz

value is due to the fact that in the case of multi-layer RF absorbent material with the ferromagnetic thin-film filler, to calculate $\mu_{st}{}^{ef}$, Ashe's law is used, which, in comparison with Snoek's law, allows much higher magnetic permeability values.

As follows from the analysis of positions 1–5 in Table 4.1, in manufacturing individual parts of the same structure, different RF absorbent coatings that have different ranges of unlikely identification were used.

We would consider separately position 6 of Table 4.1. Such small operating range values and RF absorbent layer thickness that are shown in position 6 of Table 4.1 lead to the losses of electromagnetic wave energy, but the amount of those losses is insignificant. Besides, the operation range width is even less than that in the resonance narrowband RF absorbent coatings. Therefore, it can be assumed that the presence of iron particles that cause absorption in the layer could be related to both the formation of RF absorbent layer and to the technological processes employed in producing the coating as a whole, which are not aimed at obtaining absorption properties.

In the practical application of the results of computing the unlikely identification range and limit range frequencies, certain circumstances that could lead to computation errors should be taken into account. For example, it should be taken into consideration that theoretical treatments of the process of electromagnetic waves absorption were often performed for normal waves falling on the absorbent layer, which often cannot be implemented in practice. Besides, there exists some uncertainty concerning the values of filler concentrations in the used composite RF absorbent coatings and materials, as well as the values of resonance frequencies of natural ferromagnetic resonance.

At the same time, despite certain limitations of the above procedure proposed for assessing the width and limit frequencies of the range, the authors consider it possible and advisable to use the results obtained that are shown in Table 4.1 for determining electromagnetic radiation parameters in the practical work on effective identification of special-purpose structures.

References

Abeles F (1948) Sur la Propagation des Ondes Electromagnetiques dans les Milieux Stratifies. Ann Phys (Paris) 3:504–520

Acher O, Boscher C, LeGuellec P, Baclet P, Perrin G (1996) Investigation of the gyromagnetic permeability of amorphous Cofenimosib manufactured by different techniques. IEEE Trans Magn 32(5):4833–4835. https://doi.org/10.1109/20.539167

Adenot AL, Acher O, Taffary T, Longuet L (2002) Sum rules on the dynamic permeability of hexagonal ferrites. J Appl Phys 91(10):7601–7603. https://doi.org/10.1063/1.1447505

Akhter Z, Akhtar MJ (2016) Free-space time domain position insensitive technique for simultaneous measurement of complex permittivity and thickness of lossy dielectric samples. IEEE Trans Instrum Meas 65(10):2394–2405. https://doi.org/10.1109/TIM.2016.2581398

Anzulevich AP et al (2013) Electromagnetic modeling of a composite of conductive spherical inclusions. News Samara Sci Cent Russ Acad Sci 15(4):769–772 [in Russian]

Baclet P, Perrin G, Acher O, Peuzin JC (1996) Variations of the anisotropy energy versus sputtering gas pressure in ferromagnetic thin films: a magneto-elastic effect. IEEE Trans Magn 32(5):4737–4739. https://doi.org/10.1109/20.539135

Balagurov BY (2015) Electrophysical properties of composites: macroscopic theory. LENAND, Moscow [in Russian]

Bartley PG, Begley SB (2005) Improved free-space s-parameter calibration. In: Proceedings of the 2005 IEEE instrumentation and measurement technology conference, Ottawa, Canada

Bartley PG, Begley SB (2012) A new free-space calibration technique for materials measurements. In: Proceedings of the instrumentation and measurement technology conference, Graz, Austria, pp 47–51

Bekker V, Seemann K, Leiste H (2004) A new strip line broad-band measurement evaluation for determining the complex permeability of thin ferromagnetic films. J Magn Magn Mater 270(3):327–332. https://doi.org/10.1016/j.jmmm.2003.08.033

Bespalova EE, Kondrashov EK (2014) Peculiarities of correction of the fire-safe material for anechoic chambers when changing the parameters of radar absorbent materials. Aviat Mater Technol 2:47–52 [in Russian]

Bespalova EE, Belyaev AA, Shirokov VV (2015) Radio-absorbing materials for high-power microwave radiation [in Russian]. Proc VIAM 3:45–51

Bychenok DS, Plyushch AO, Gorokhov GV, Bychenok VS, Kuzhir PP, Maksimenko SA (2016) Corrugated composite microwave absorbers with carbon fibers. J Tech Phys 86(12):124–128 [in Russian]

Chai G, Yang Y, Zhu J, Lin M, Sui W, Guo D, Li X, Xue D (2010) Adjust the resonance frequency of $(Co_{90}Nb_{10}/Ta)_n$ multilayers from 1.4 to 6.5 Ghz by controlling the thickness of Ta interlayers. Appl Phys Lett 96(1):012505. https://doi.org/10.1063/1.3290252

Challa RK, Kajfez D, Gladden JR, Elsherbeni AZ (2008) Permittivity measurements with a non-standard waveguide by using TRL calibration and fractional linear data fitting. Prog Electromagn Res B 2:1–13

Chambers B, Tennant A (1994) Design of wideband Jaumann radar absorbers with optimum oblique incidence performance. Electron Lett 30(18):1530–1532. https://doi.org/10.1049/el:19941023

Chambers B, Tennant A (1996) Optimised design of Jaumann radar absorbing materials using a genetic algorithm. IEE Proc. Radar Sonar Navig 143(1):23–30. https://doi.org/10.1049/ip-rsn:19960316

Degiorgi M, Costa F, Monorchio A, Manara G (2017) Iterative non-ambiguous estimation of dielectric permittivity from broadband transmission/reflection measurements. In: 2017 progress in electromagnetic research symposium—Spring (PIERS), St. Petersburg, pp 817–821

D'jakonova OA, Kazancev YN, (2019) A resonator for measuring the dielectric constant of thin films. Zhurnal Radioelektroniki (Journal of Radio Electronics) 7:1–8 [in Russian]

D'jakonova OA, Kazancev YN, Kalenov DS, (2017) Measuring system for determining the electromagnetic characteristics of materials by the resonator method using scalar circuit analyzers. Zhurnal Radioelektroniki (Journal of Radio Electronics) 7:1–12 [in Russian]

Dudeck KE, Buckley LJ (1992) Dielectric material measurement of thin samples at millimeter wavelengths. IEEE Trans Instrum Meas 41(5):723–725. https://doi.org/10.1109/19.177352

Elsukov EP et al (2009) Influence of the shape, chemical and phase composition of Fe-based particles on the microwave characteristics of composites with a dielectric matrix. J Tech Phys 79(4):125–130 [in Russian]

Galst'yan EA, Ravaev AA (1992) Thermal effect of pulsed microwave radiation on structurally inhomogeneous materials. J Tech Phys 62(1):42–53 [in Russian]

Ghodgaonkar DK, Varadan VV, Varadan VK (1990) Free-space measurement of complex permittivity and complex permeability of magnetic materials at microwave frequencies. IEEE Trans Instrum Meas 39(2):387–394. https://doi.org/10.1109/19.52520

Gladkov SO (2011) To the theory of energy absorption of electromagnetic radiation by fine metal particles. Appl Mech Tech Phys 52(2):3–14 [in Russian]

Haskell NA (1953) The dispersion of surface waves on multilayered media. Bull Seismol Soc Am 43(1):17–34. https://doi.org/10.1785/BSSA0430010017

Hendrych A, Zivotsky O, Postava K, Pistora J, Kraus L, Kubınek R (2009) Magnetic behaviour and domain structure in as-quenched, annealed, and stress-annealed Cofecrsib Ribbons. J Magn Magn Mater 321:3771–3777. https://doi.org/10.1016/j.jmmm.2009.07.032

Hyde M IV, Havrilla MJ (2008) A nondestructive technique for determining complex permittivity and permeability of magnetic sheet materials using two flanged rectangular waveguides. Progress Electromagnet Res 79:367–386. https://doi.org/10.2528/PIER07102405

Janezic MD, Baker-Jarvis J (1999) Full-wave analysis of a split-cylinder resonator for nondestructive permittivity measurements. IEEE Trans Microw Theory Tech 47(10):2014–2020. https://doi.org/10.1109/22.795077

Jilani MT, Rehman MZ, Khan AM, Khan MT, Ali SM (2012) A brief review of measuring techniques for characterization of dielectric materials. Int J Inf Technol Electr Eng 1(1):1–5

Keysight Technologies (2017) Basics of measuring the dielectric properties of materials. Appl Note, 5989-2589EN

Kittel C (1948) On the theory of ferromagnetic resonance absorption. Phys Rev 73(2):155–161. https://doi.org/10.1103/PhysRev.73.155

Kraev ID et al (2017) Study of the effect of dispersity of functional particles of carbonate iron on the radio-absorbing characteristics of the composite based on them. Aviation Mater Technol 1(46):51–60

Lagarkov AN, Osipov AV, Rozanov KN, Starostenko SN (2005) Microwave composites filled with thin ferromagnetic films. In: Proceedings of symposium R of ICMA, Singapore, pp 74–77

Landau LD, Lifshitz EM (1984) Electrodynamics of continuous media. Pergamon Press, Oxford

Lowe MJS (1995) Matrix techniques for modeling ultrasonic waves in multilayered media. IEEE Trans Ultrason Ferroelectr Freq Control 42(4):525–542. https://doi.org/10.1109/58.393096

Lyubchenkov AV (2009) Analysis of methods for measuring electrodynamic parameters of materials absorbing microwave electromagnetic radiation. Bull Voronezh State Tech Univ 5(9):17–19 [in Russian]

Meshram MR, Agrawal NK, Sinha B, Misra PS (2004) Characterization of M-type barium hexagonal ferrite-based wide band microwave absorber. J Magn Magn Mater 271(2–3):207–214. https://doi.org/10.1016/j.jmmm.2003.09.045

Morales JJ, Nuevo MJ (1991) A general method for calculating reflection and transmission coefficients in multiple slabs. Am J Phys 59(12):1140–1143. https://doi.org/10.1119/1.16625

Morozov GV, Maev RG, Dreik GVF (2001) Multiple reflection method for electromagnetic waves in layered dielectric structures. Quant Electron 31(9):767–773 [in Russian]

Naito Y, Suetake K (1971) Application of ferrite to electromagnetic wave absorber and its characteristics. IEEE Trans Microw Theory Tech 19(1):65–72. https://doi.org/10.1109/TMTT.1971.1127446

Nicolson AM, Ross GF (1970) Measurement of the intrinsic properties of materials by time-domain techniques. IEEE Trans Instrum Meas 19(4):377–382. https://doi.org/10.1109/TIM.1970.4313932

Nikolsky VV, Nikolskaya TI (1989) Electrodynamics and radio wave propagation. Nauka, Moscow [in Russian]

Ovsov D (2016) Study of the characteristics of dielectric antenna coatings using a vector circuit analyzer. Electron Sci Technol Bus 4(00154):102–105 [in Russian]

Parkhomenko MP, Kalenov DS, Eremin IS et al (2019) Usage and experimental tests of the modified waveguide method for measurement of the complex permittivity and permeability of materials. Phys Wave Phen 27(4):299–306. https://doi.org/10.3103/S1541308X19040101

Politiko AA, Semenenko VN, Chistyaev VA, Kibets SG, Garanov VA (2018) Microwave spectra of dielectric and magnetic permeability of composites based on carbonyl iron in the ultra-wide frequency band. In: VI All-Russian Microwave Conference, pp 45–49 [in Russian]

Ponomarenko VI, Lagunov IM (2021) Absorbers of electromagnetic waves. POLYPRINT, Simferopol [in Russian]

Rohde & Schwarz (2006) Measurement of material dielectric properties. Appl Note, RAC0607-0019

Rozanov KN, Li ZW, Chen LF, Koledintseva MY (2995) Microwave permeability of Co_2Z composites. J Appl Phys 97(1):013905. https://doi.org/10.1063/1.1827911

Rozanov KN, Osipov AV (2006) Limitations on microwave magnetic permeability of materials. News of F. Skorina Gomel State Univ 6(39):28–31 [in Russian]

Rozanov KN, Simonov NA, Osipov AV (2002) Measurement of magnetic permeability of ferromagnetic films at ultra-high frequencies. Radio Eng Electron 47(2):229–338 [in Russian]

Rozanov KN (1999) Fundamental limitations for the width of the working range of radio-absorbing sheds. Radio Eng Electron 44(5):526–530 [in Russian]

Ruck G, Barrick DE, Stuart WD, Krichbaum CK (1970) Radar cross section handbook. V. 2. Ch 8. Plenum Press, New York

Semenenko VN, Chistyaev VA, Politiko AA et al (2019) Test stand for measuring the free-space electromagnetic parameters of materials over an ultrawide range of microwave frequencies. Meas Tech 62(2):161–166. https://doi.org/10.1007/s11018-019-01601-5

Shcherbakov SV et al (2018) The effect of aluminum substitution on the field of effective magnetic anisotropy and the degree of magnetic texture pf anisotropic polycrystalline hexagonal ferrites of barium and strontium for substrates pf microstrip microwave electronics devices. News of higher educational institutions. Electron Mater 21(3):170–176 [in Russian]

Snoek JL (1948) Dispersion and absorption in magnetic ferrites at frequencies above one Mc/s. Physica 14(4):207–217. https://doi.org/10.1016/0031-8914(48)90038-X

Stolyarov SN (1988) Influence of interference coatings on optical stability of semiconductor laser mirrors. Quant Electron 15(8):1637–1643 [in Russian]

Tamm IE (1979) Fundamentals of the theory of electricity. Mir, Moscow

Thomson WT (1950) Transmission of elastic waves through a stratified solid medium. J Appl Phys 21(2):89–93. https://doi.org/10.1063/1.1699629

Torukhanov AV et al (2018) Messbauer studies and microwave properties of barium hexaferrites with substitution with Al3+ and In3+ ions. Solid State Phys 60(9):1723–1732 [in Russian]

Varadan VV, Hollinger RD, Ghodgaonkar DK (1991) Free-space, broadband measurements of high-temperature, complex dielectric properties at microwave frequencies. IEEE Trans Instrum Meas 40(5):842–846. https://doi.org/10.1109/19.106308

Veselago VG (1968) Electrodynamics of substances with simultaneously negative electrical and magnetic permeabilities. Sov Phys Uspekhi 10(4):509–514

Vlasenko EA, Bokova ES, Dedov AV, Devin KL Polyvinyl chloride-based radar absorbent material. Plast Masses 5–6:41–43 [in Russian]

Volkovskaya II, Semenov VE, Rybakov KI (2017) Effective high-frequency magnetic permeability of compacted metal powder materials. News of universities. Radiophysics 60(10):892–902

Volkovskaya II, Semenov VE, Rybakov KI (2017) Effective high-frequency magnetic permeability of compacted metal powder materials. News of universities. Radiophysics 60(10):892–902

Vovchenko LL, Matsuy LY, Oleinik VV, Launets VL, Zagorodny VV, Normand FL (2011) Resonant nature of the interaction of multilayer carbon nanotubes with millimeter-wave radiation. Nanosyst Nanomater Nanotechnol 9(4):759–769 [In Russian]

Weir WB (1974) Automatic measurement of complex dielectric constant and permeability at microwave frequencies. Proc IEEE 62(1):33–36. https://doi.org/10.1109/PROC.1974.9382

Yang A, Imrane H, Lou J, Kirkland J, Carmine Vittoria C, Sun N, Harris VG (2008) Effects of boron addition to the atomic structure and soft magnetic properties of FeCoB films. J Appl Phys 1 103(7):07E736. https://doi.org/10.1063/1.2838226

You KY, Lee YS, Zahid L et al (2015) Dielectric measurements for low-loss materials using transmission phase-shift method. Jurnal Teknologi (Science & Engineering). 77(10):69–77. https://doi.org/10.11113/jt.v77.6286

Zajicek R, Vrba J, Novotny K (2006) Evaluation of a reflection method on an open-ended coaxial line and its use in dielectric measurements. Acta Polytechnica 46(5):50–54. https://doi.org/10.14311/882

Chapter 5
Using Scanning Electron Microscopy to Identify Special-Purpose Structures by Their Fragments

This chapter is concerned with a comparative analysis of the existing approaches to identifying special-purpose structures, and estimates the advantages and limitations of each of them. The tasks to be solved in the identification of special-purpose structures (means of destruction) are established. It is accepted in the current practice to distinguish three classes of tasks addressed in the identification of special-purpose structures: identifying, classifying, and diagnostic ones. Characteristics of each class of tasks are described.

Special emphasis is placed on using standards and technical normative documentation in the process of identifying special-purpose structures.

A general scheme of the identification study, which includes four main stages, namely, preparatory, analytical, comparative, and summarizing (synthesizing) ones, is proposed. Each stage is analyzed in detail.

An important role in identifying means of destruction belongs to the analysis of the chemical composition and structure of their fragments. The results of comprehensive analysis allow determining the material composition and quantitative characteristics of the compounds it consists of. These results are used to identify metal or alloy grades and determine the compositions of composite materials along with the technologies of their production.

An interactive knowledge base for monitoring the materials of special-purpose structures in the framework of the transdisciplinary cluster of information support to the development of special-purpose structures has been compiled.

The principles of creating a transdisciplinary cluster for information support to the development of means of destruction (TCIS DMD) and monitoring of materials and technologies for special-purpose structures have been formulated, and corresponding artificial neural networks have been constructed.

Supplement presents a general review of high-tech means of destruction that are employed by the Russian Federation in the war against Ukraine, along with their technical characteristics.

After the beginning of the Russian Federation armed aggression against Ukraine, the issue of identifying special-purpose structures (means of destruction) (further—

V. L. Bogdanov et al., *Identification of Special-Purpose Structures by Their Fragments Based on Scanning Electron Microscopy*, Advanced Structured Materials 233,
https://doi.org/10.1007/978-3-031-88827-4_5

SPS (MD)) by analyzing the materials used in their manufacturing has become of utmost importance. This analysis is performed as a part of the complex of identification studies. The results of those studies make it possible:

- to determine the group (subsonic, supersonic, hypersonic) and class (ballistic, cruise missiles, unmanned aerial vehicles (UAV), etc.) of SPS (MD);
- establish the SPS (MD) country of origin;
- assess the level of tactical and technical characteristics, performances and limit capabilities of specific SPS (MD);
- envisage possible trends in their upgrading;
- elaborate proposals concerning the development of methods and recommendations in producing technical means of counteracting the abovementioned SPS (MD).

However, the methodology of studying SPS (MD) and their components as a specific type of research is still at the stage of development and requires systematization and more thorough specialized consideration. This is also necessary for separating these studies from other types of research in which material science expert examinations can also be conducted. In turn, the results of SPS (MD) identification studies can be used in forensic and criminalistic investigations to discover the facts of violating the rules and regulations of warfare, as well as those of international transfer of special-purpose products and technologies.

Thus, the issue of elaborating the methodology of SPS (MD) identification studies and searching for effective means of analyzing their results is coming to the fore.

The general goal of this chapter is to present:

- the analysis of existing approaches to SPS (MD) identification, their advantages and limitations;
- the methodology of compiling an interactive database for monitoring SPS (MD) element base and materials;
- the methodology of creating artificial neural networks for SPS (MD) identification on the basis of electron microscopy analysis of their fragments.

5.1 Existing Approaches to Identification of Special-Purpose Structures, Their Advantages and Limitations

For the identification of SPS (MD), whose structural elements, as a rule, consist of combinations of various metals, alloys, and composites, an adequate definition of its subject is of utmost importance. This is necessary for separating that from forensic examinations, which also study SPS (MD) fragments. Besides, the formulation of the subject and objectives of identification allows specifying the range of tasks which it deals with, and determining the competence and knowledge to be possessed by the researcher who deals with identification of special-purpose structures.

Besides, for determining the content of SPS (MD) identification subject, it is necessary to systematize the tasks addressed in respective studies.

5.1.1 Tasks Addressed in Identifying Special-Purpose Structures and Means of Destruction

At present, it is customary to distinguish three classes of tasks dealt with when identifying SPS (MD): identification, classification, and diagnostic ones (Fig. 5.1).

Classification tasks. The first task is to determine the nature of special-purpose structures, i.e., categorize them as metals, alloys, or composites.

The next one is to attribute them to standardized classes—varieties, subvarieties, metal and alloy grades, composites, product types, etc.

A common feature of classification tasks is attributing the examined fragment (fragments) to a set whose all elements have common properties (a complex of properties) (Shcherbakovsky 1996; Araújo et al. 2024).

In some cases, attributing a metal, alloy, or composite to a certain S&T class is of special importance, as it forms a significant element of the identification subject. For example, in examining SPS (MD) fragments, establishing the material belonging to military or dual-use goods that are not produced in the country manufacturing the SPS (MD) is a cause to investigate the facts of violating international sanctions and establish the legality of their international transfers.

Identification tasks. The tasks are subdivided into two kinds: ascertaining specific individual identity and establishing the belonging to a group.

For SPS (MD) that consists of metals, alloys, or composites, two kinds of individual identification are distinctive: the identification by separated fragments and that by the source of origin.

In identifying SPS (MD), the parts used to recognize the whole (a particular variety, type of MD) are usually product pieces, fragments. Remaining within the framework of the specialized knowledge of the material science expert, the study of materials that make up fragments of SPS (MD), which during operation were exposed to significant and, most frequently, sign-reversing temperatures, pressures, and overloads, the researcher has to involve qualitatively new methods of analysis. Besides, in dealing with identification tasks, of significant importance are determination of material surface topologies, their structures, and chemical compositions at the micro-level.

The task of establishing the whole by its separated parts is solved with reliance on the comprehensive examination of properties of material and morphological properties of its separated parts. The identity is ascertained in the case of detecting a complex of unique properties that emerged during the production, operation, and disintegration of the whole. To assess the identification significance of each individual property and their certain aggregates is only possible if the patterns of their emergence and change under the effect of various factors are known. The properties

CLASSIFICATION TASKS

Determining metallic or non-metallic type of objects;
Establishing objects belonging to the standard class that corresponds to the concept of genus.

IDENTIFICATION TASKS

Establishing specific-individual identity:

- establishing the whole by its individual parts (fragments in particular);
- establishing specific origin sources of the objects made of metals, alloys and composites.

Establishing object belonging to a group:

- establishing group belonging in terms of manufacturing features;
- establishing group belonging in terms of the features of use, operation, storage;
- establishing group belonging in terms of damage features.

DIAGNOSTIC TASKS

Establishing object state:

- determining individual properties of materials;
- determining production method, technology;
- ascertaining the intended purpose of the material, product;
- establishing the initial state of the object (detection of production, decimal numbers).

Establishing certain features in the examined objects

- determining the localization of fragments;
- ascertaining the fact of contact interaction;
- establishing the features of objects' mechanical interaction (forming, changes in morphology, fracture mechanism and type);
- determining the peculiarities of thermal effects (local and bulk effects);
- determining the features of chemical (physico-chemical) action of the aggressive environment, atmosphere.

Establishing cause-and-effect relationships

- determining possible algorithms of forming the properties of means-of-destruction components;
- determining the interrelation of components' properties with functional features of the specific kind of means of destruction.

Fig. 5.1 Tasks addressed in identifying special-purpose structures (means of destruction)

caused by the operation conditions of the whole and those of its separation into parts in the identified event are of special significance for identifying the whole product.

It is possible to recognize fragments as parts of certain special-purpose structures (means of destruction), relying on the property of metals, alloys, and composites to retain information on the external influences that led to objects' disintegration.

A more detailed examination of the microscale topology of material surfaces, their structures and chemical compositions would enable researchers to identify the source of SPS (MD) origin.

In material science or forensic expert studies of SPS (MD) fragments, the facility (plant, factory) where a product or its unit was manufactured, and where its metal, alloy, or composite was produced (metallurgical plant, etc.), the deposit where raw material was mined should be considered as the source of SPS (MD) production.

The task of identifying the country which produced an SPS (MD), its components and materials can be solved systemically by referring to the information fund where data on manufacturing plants are collected, with the account being taken of the features of their technological routes, sources of raw materials origin (mining sites), etc. (Araújo et al. 2024).

Establishing the group belonging of SPS (MD) is a task in itself and is addressed separately, since in the majority of cases means of destruction cannot be regarded as an individual material forms (elements, units, components) with a distinct shape, size, and place in the product as a whole. The components presumed to be identical are numerous—an aggregate of unit or body fragments, parts of fastenings, remains of wiring, thermal insulation, etc.

Consider in more detail the objects to be identified in the framework of a comprehensive study for SPS (MD) attribution to a certain group.

An intermediate identification object to be attributed to a group is a fragment of SPS (MD) (a unit or component product) made of metal, alloy, or composite. Such element performs an independent function as a component of SPS (MD)—e.g., the warhead, control unit, RF transparent fairing, engine, etc. The direct research object is a complex of material properties; however, in this case, objects can differ in composition, shape, size, and functional purpose.

At the same time, SPS (MD) fragments, being parts of products and objects made of metals, alloys, and composites, due to their properties are tangible carriers of information on the effects and processes they were exposed to during manufacturing and direct use. During identification studies of the group belonging to SPS (MD) fragments, researchers establish the material (metal, alloy, composite) that was used, the equipment and main technological processes employed in manufacturing, service (operation) restrictions imposed on the means of destruction (speed, temperature, resistance to compression and bending, etc.), with the account being taken of the materials used in SPS (MD) structure (structure components).

The analysis of the chemical composition of SPS (MD) fragment materials, their physico-mechanical properties, and operation restrictions enables researchers to determine their group belonging to a specific means of destruction and its component part. The direct objects of studies are properties of the fragment materials used by the designer in the component elements of the means of destruction, based on their purpose, application peculiarities, and exposal to external factors during operation. Those could be the features of material properties (refractory properties, heat resistance, RF absorption), welding, similar layers and coatings on fragment surfaces, and their structure.

Taking into account the differentiated approach, the tasks for attributing SPS (MD) to different groups by various features are distinguished: peculiarities of manufacturing, application, operation, co-location, fracture, and separation (Shcherbakovsky 1996; Araújo et al. 2024).

Diagnostic tasks. Diagnostic studies are aimed at establishing the purpose, features of application and certain properties (characteristics) of SPS (MD) and their components by analyzing their fragments. One of the main tasks of diagnostic studies is to assess the object state, which includes examining SPS (MD) composition, structure, shape, and main characteristics (Table 5.1).

The task of determining object characteristics is closely related to those of determining the method, manufacturing technology, and materials of SPS (MD). Comprehensive materials science analysis permits determination of their specific production technologies (casting, rolling, parameters of mechanical, heat, chemical-thermal treatment, welding, the application of heat-resistant and RF absorbent coatings, etc.). As a rule, such studies deal with determining the conformity (non-conformity) of the properties of metals, alloys, and composites to regulatory requirements. Depending on the study purpose, the majority of the abovementioned tasks can be both a final and an intermediate one (Burgot and Jean-Louis Burgot 2023).

From the analysis of the tasks addressed, the subject of identifying special-purpose structures (means of destruction) can be determined. This is establishing the factual data concerning the match of identified objects according to the results of examining their fragments, determining the features of their application, alongside with assessing their characteristics and evaluating SPS (MD) effectiveness.

5.1.2 Existing Methods of Identifying Special-Purpose Structures

A varied complex of SPS (MD) properties necessitates a task of investigating an aggregate of physical, physico-chemical, engineering research methods at the analytical stage of identification study. The diversity of methods leads to certain difficulties in choosing the means of executing the tasks set, which requires the analysis of the methods of studying SPS (MD) fragments.

The classification of research methods of special-purpose structure materials, which are used to identify those, is given below:

1. Organoleptic research methods.
   - 1.1 Visual examination (qualified observation).
     - 1.1.1 Examination without preliminary surface preparation.
     - 1.1.2 Examination with preliminary surface preparation.

Table 5.1 Establishing group belonging of SPS (MD) metals and alloys

Group definition	Material properties		Signaling properties
	Chemical composition	Microstructure	
Group specified by features of product manufacturing, metal alloy production	1. The same quantitative content of principal components, alloying elements and impurities 2. The same qualitative and quantitative composition of coating (coating layers) 3. Deviations of qualitative and quantitative composition from standards	1. The same material structure in the bulk, surface layer, coating 2. The same thickness of surface layers, coating 3. Deviations of structure from standards, presence of defects	1. Parameters of surface, shape, design, which point to morphological features of production mechanisms 2. Marking, its peculiarities 3. Deviations of morphological parameters from standards
Group specified by peculiarities of MD existence, operation, storage	1. Change element composition in the bulk, coating absence 2. Presence of surface contamination	Structure change (work hardening, heat-affected zones, recrystallization)	1. Change in initial shape, dimensions (deformation) 2. Presence of wear, traces of friction, dents, etc. 3. Melting, temper colors
Group specified by peculiarities of MD fracture, disintegration	1. MD material layering, surface layer delamination 2. Metal, alloy saturation, depletion near fragmentation surface	1. Structure change (heat-affected zones, grain deformation, etc.) 2. Corrosion	1. Macro- and micro-relief of fracture surfaces 2. Macro- and micro-deformation of metal 3. Signs of the influence mechanism and MD morphological features on detached parts 4. Traces of short circuiting

2. Physical research methods.

    2.1 Light microscopy techniques.

        2.1.1 Research methods without preliminary object surface preparation.

            2.1.1.1 Observation in reflected light.

            2.1.1.2 Observation under vertical and oblique illumination.

        2.1.2 Research methods with preliminary specimen surface preparation.

- 2.1.2.1 Techniques of light-field (vertical and oblique) lighting.
- 2.1.2.2 Techniques of dark-field lighting.
- 2.1.2.3 Studies in polarized light.
- 2.1.2.4 Phase contrast technique.
- 2.1.2.5 Interference contrast technique.

- 2.2 Electron microscopy methods.
  - 2.2.1 Transmission electron microscopy.
    - 2.2.1.1 Indirect technique (surface replicas).
    - 2.2.1.2 Semi-direct technique (extraction replicas).
  - 2.3 Scanning electron microscopy.
- 2.3 Spectral techniques.
  - 2.3.1 Emission spectral analysis.
    - 2.3.1.1 Atomic emission spectroscopy (AES) with standard excitation sources (arc, spark).
    - 2.3.1.2 Laser microspectral analysis.
  - 2.3.2 Atomic absorption analysis.
    - 2.3.2.1 Flame atomization technique.
    - 2.3.2.2 Electrothermal atomization technique.
  - 2.3.3 X-ray spectral analysis.
    - 2.3.3.1 Analysis by primary emission spectra.
    - 2.3.3.2 X-ray spectral fluorescence analysis.
    - 2.3.3.3 X-ray spectral microanalysis.
  - 2.3.4 Mass spectrometry.
- 2.4 X-ray research methods.
  - 2.4.1 X-ray diffraction analysis.
    - 2.4.1.1 Structure (texture) study techniques.
    - 2.4.1.2 Residual stress analysis.
    - 2.4.1.3 Phase composition analysis.

3. Physico-technical research methods.
   - 3.1 Mechanical properties determination.
     - 3.1.1 Hardness determination (Brinell, Rockwell, Vickers techniques).
     - 3.1.2 Microhardness determination.
   - 3.2 Methods of determining physical properties.
     - 3.2.1 Density determination.
     - 3.2.2 Techniques of electrical property measurements.
     - 3.2.3 Techniques of magnetic property measurements.

4. Chemical and physico-chemical research methods.
   - 4.1 Chemical research methods.
     - 4.1.1 Techniques of selective etching and dyeing with reagents.

- 4.1.2 Gravimetric technique.
- 4.1.3 Titrimetric technique.
- 4.1.4 Gas volumetric technique.
- 4.2 Physico-chemical research methods.
  - 4.2.1 Photometric techniques.
  - 4.2.2 Potentiometric techniques.
  - 4.2.3 Coulometric technique.
  - 4.2.4 Polarographic technique.
  - 4.2.5 Thermometric technique.

5. Special methods of metal and alloy studies.
   - 5.1 Metallographic analysis.
     - 5.1.1 Macrostructure analysis.
     - 5.1.2 Microstructure analysis.
     - 5.1.3 Substructure analysis.
   - 5.2 Fractographic analysis.
     - 5.2.1 Fractographic macroanalysis.
     - 5.2.2 Fractographic microanalysis.

Some of the above methods and techniques find wide application in practice (visual analysis, optical microscopy, emission spectral analysis), others are used less frequently due to the limited amount of research equipment (electron microscopy, atomic absorption analysis, X-ray techniques, etc.). The classification does not include the methods of determining mechanical and physical properties, and techniques that are practically unavailable to researchers (e.g., neutron activation analysis).

Visual inspection belongs to the initial actions in the identification studies. At the preparatory stage, the visual examination of SPS (MD) components is carried out to reveal specific substance and morphological features of the objects made of metals, alloys, and composites, without a detailed analysis of their individual properties.

During visual inspection without preliminary preparation (grinding, etching) of the fragments, their fracture surfaces and deformations are examined to determine the mechanism and nature of failure, load directions, etc. After special surface preparation, visual inspection is carried out by means of microscopic analysis of the material structure to study its internal structure, defects, and detect chemical liquation at different magnifications. The feature of the technique is simple object preparation. A microscope equipped with a module for X-ray spectral microanalysis makes it possible to perform a local chemical analysis of the fragment material.

Spectral analysis techniques are aimed at determining the chemical composition of SPS (MD) fragments. Electron spectroscopy technique is the most common one (Sturm et al. 2019).

Highly sensitive and accurate methods of studying SPS (MD) fragments materials are atomic absorption analysis and X-ray spectral analysis. The latter is employed both for microparticle analysis and for determining the chemical composition of inclusions, layers, and coatings. Good promise is shown by mass spectrometry, which is known for its high sensitivity. Chemical and physico-chemical methods, unlike instrumental ones, provide direct determination of chemical composition. To narrow down the group of objects in terms of peculiarities of features manufacturing and operation, which influenced metal structure and caused internal stresses, X-ray diffraction analysis techniques are used. In identifying SPS (MD), the main purpose of employing physico-technical methods is to determine individual properties of materials: metals, alloys, and composites (Matveev et al. 1986; Odinochkina 1973).

Special techniques of studying metals and alloys are metallographic and fractographic analyses. The aim of metallographic analysis is to visualize, examine, and describe metal internal structures. Metallographic analysis is performed to establish the peculiarities of manufacturing metal products (the kind of thermal, chemical-thermal treatment, and mechanical processing, the presence of protective coating, etc.), determining the nature of thermal or mechanical effects. The task of fractographic analysis is to investigate the features of external structure, relief of broken surfaces (fractures). The mechanism, fracture sequence, the type of applied loads, and other characteristics related to the process of fragment separation can be revealed by the features of fractured structures.

As of now, four groups of methods of SPS (MD) identification are distinguished: morphological analysis, the analysis of composition, structure, and individual properties.

Determining SPS (MD) morphological and substance properties and comparing them with current standards permits establishing the product kind, purpose, and peculiarities of the technologies employed in its manufacturing. If the revealed properties and parameters of the SPS (MD) materials component units and elements are made of do not conform to regulatory requirements, diagnostic procedures to determine the area of using the object material are employed.

5.1.3 Peculiarities of Applying the Standards and Regulatory Technical Documents in Identifying Special-Purpose Structures (Means of Destruction)

The use of regulatory technical documents is a necessary element of SPS (MD) identification research and is caused by a high level of standardization at all stages of their manufacturing, starting from metal, alloy smelting, and ending with the acceptance of finished products.

The principal regulatory document is the standard defining the complex of mandatory norms, rules, and requirements. Essential for identification research is the fact that standards strictly regulate materials' substance characteristics (chemical compositions, mechanical and physical properties, structures), their identification properties (geometrical parameters, shapes, surface roughness, etc.), and the methods of their quality control (testing, analysis, measurements).

Standards are subdivided into the following categories:

- national standards (GOST, DSTU);
- industry standards (OST);
- standards of the republics of former USSR (RST);
- enterprise standards (STP, TU).

National standards are mandatory for all enterprises, organizations, and institutions of the state and local government, of all forms of ownership. Industry standards are mandatory for all enterprises and organizations of a certain industry. Enterprise standards are mandatory for all enterprises (associations) that approved them. Regulatory documents compiled with reliance on the requirements, conditions, and capabilities of economic and political unions, individual countries and ministries, enterprises (TU) contain the data which make it possible to reveal features and peculiarities of products manufacturing at one or the other manufacturing plant.

State standards, standards of the republics of former USSR, and industry standards are stored at libraries, standardization centers and can be used by professionals in their practical work.

Depending on the standardization object, standards can be conditionally subdivided into two groups: object (products and materials) standards and method standards.

The first group contains regulatory documents for all kinds of materials, components, units, and equipment. They determine quality characteristics of objects and their quantitative values. Standards consist of requirements for both finished products and their material.

The other group of standards regulates control techniques by which the set indicators of quality can be determined. The standards are grouped by the nature of material properties controlled by them, e.g., the standards of establishing mechanical properties (measurements of hardness, strength, etc.), the methods of determining the chemical compositions of materials (spectral, chemical, atomic absorption analyses), metallographic techniques of examining the structure of metals, etc.

The norms of standards are characterized with certain narrowness, which should be taken into account during identifying special-purpose structures (means of destruction). The most important limitations are related to normalized qualitative and quantitative indicators of standards. Products, materials can differ both by non-normalized (not mentioned in standards) and by normalized indicators. The restriction in terms of qualitative indicators consists in limiting some substance and signaletic properties of objects made of metals, alloys, and composites.

In view of the considerable diversity of production conditions at various enterprises, which include different raw materials, the condition of equipment, workers' qualification, etc., it is impossible to ensure SPS (MD) production with absolutely identical properties. Accordingly, standards allow for certain tolerances—alterations toward both increase and reduction. Restrictions in quantitative indicators mean that a certain scope, range of quantitative change is set. Many norms of quality indicators are referred to in standards with the note "no more than" if a larger quantity indicator leads to unacceptable deterioration of object properties, or "no less than" if the reduction of properties results from a smaller indicator. This concerns such properties of metal objects as chemical composition, hardness, strength, protective coating thickness, as well as geometric dimensions of the products.

The analysis of regulatory documentation permits a correct assessment of generic, group, and individualizing properties of metals, alloys, composites, and products made of those.

Therefore, the information contained in regulatory technical documents can be used at different stages of identification research. In particular, at the preparatory and analytical stages, the abovementioned informative data are employed for the following purposes.

1. Sampling, choosing the conditions of analysis and technical means (equipment, reagents, etc.) in conformity with standards. This refers to various techniques of identifying substance characteristics: physico-chemical (chemical composition, micro- and macrostructure, etc.), mechanical (hardness, strength, etc.), and signaletic (dimensions, surface roughness, etc.) properties of metals, alloys, composites, as well as and products made of those.
2. Description of material characteristics, parameters, products in the measurement units, and terms recommended by standards. For example, designations of alloys, various internal and external manufacturing defects (which emerge in casting, heat treatment, welding, and other technological processes), types of weld seams, structure banding and fineness, inclusion shapes and types, etc.
3. Establishing the area of material applications (aircraft manufacturing, shipbuilding, instrument making, rocket-and-space engineering, etc.).

The involvement of normative data at the stage of comparing and assessing the results permits the solution of specific tasks of SPS (MD) identification:

- classification ones: referring the studied fragments of objects to a certain set of materials or products, the establishing of the product brand, type by revealing its properties and characteristics regulated by normative documents;
- diagnostics: revealing the process of material production, product manufacturing by its detected properties, following their comparison with the standard requirements;
- establishing the application sphere of SPS (MD) material by the technological features of its production normalized with standards; estimating the time of object manufactured by comparing its properties with the requirements of current or cancelled standards concerning a similar product type;
- identification: establishing the group belonging to the compared objects by the features of their manufacturing.

5.1.4 General Principles of the Methodology of Identification Research of Special-Purpose Structures (Means of Destruction)

A typical methodology pattern of SPS (MS) identification research includes four main stages: preparation, analysis, comparison, and conclusion (synthesis) (Shcherbakovsky and Lapta 2009).

At the preparation stage, the researcher examines the initial materials and fragments, formulates the task, and determines the ways of addressing it. In solving identification tasks, of great importance are SPS (MD) effectiveness (the depth and diameter of the explosion crater, fragmentation field density, armor penetration, etc.), the means of destruction whose identity is to be determined.

To separate SPS (MD) fragments from each other, a description and chart of the identification object fragments is drawn up, with special consideration given to factory markings and inscriptions. The appearance of the fragments and their location are recorded by photography (general, unit, detailed images).

Taking into account the tasks set, the hypotheses advanced and peculiarities of the combat use of the means of destruction, a research plan is drawn up, research schemes and techniques are chosen, with a view to obtaining maximum information about the objects to be identified.

The analytical stage of the expert research consists in a separate study of each identification object. At this stage, with reliance on a system of methods, a complex of substance and signaletic properties of the studied objects is established by examining the features that identify them. The order of detecting, examining, and assessing their properties is determined by the task facing a professional researcher and by specific schemes of SPS (MD) identification studies.

Now, we will list the main principles of the general methodology of SPS (MD) identification research. At first, non-destructive methods are used, and then the techniques that change the object or lead to its disintegration. In the case of a comparative study of micro- and macro-objects, the analysis starts with the latter ones. This permits the elaboration of a particular technique, analysis modes, determination of material properties' homogeneity (variability) in terms of volume or surface. It is also important to determine object macrostructure, chemical composition, and elements' distributions in the bulk of the examined materials.

The revealed properties and compositions of the materials of SPS (MD) fragments are studied in detail at different levels within each type. Such detailing of properties is an effective method of individualizing the objects identified. At the stage of analytical study, through analyzing their features, SPS (MD) properties are revealed in the following sequence:

- generic properties determined by manufacturing features;
- group properties, which are accounted for by the conditions of their production, existence, and destruction;

- individualizing properties determined by the peculiarities of their operation (combat use, resistance to external factors).

An integral part of identification research is a material science study that aims at determining the composition, structure, and other properties of the materials of SPS (MD) fragments.

Thus, at the analytical stage, the following properties of the materials of SPS (MD) fragments are determined:

- morphological (signaletic): general shape, dimensions, design, the presence of surfaces resulting from manufacturing technology, their roughness, relief; the shapes and dimensions of the separate parts of the object, the presence of tracks, traces, dents, deformations, molten zones; macro- and micro-reliefs of fractures surfaces;
- mass, density;
- magnetic properties;
- chemical compositions: those of the core, surface layer (including protective coatings), inclusions of technological origin. In the studies of objects manufactured by casting, melting, to establish the uniformity of elements' distribution, is important. This is due to elements' liquation that accompany solidification. Chemical composition is determined at qualitative, semi-quantitative, or quantitative levels;
- structure and structure topology: structural components, phases, inclusions, cracks, pores, and corrosion. The qualitative description of internal structure is provided: the general nature of the structure, distribution of components, phases, inclusions, their shapes, the presence of texture, banding, surface layer (layers) that are different from the core, corrosion features. Quantitative parameters, the average size of grain, phases, inclusion sizes, the ratio of phase constituents, thickness of surface layer (layers), that of corrosion layer are determined. If necessary, the researcher establishes
- atomic (crystalline) structure of phases, inclusions;
- hardness of object material;
- microhardness of individual phases, inclusions.

The final stages in the general methodology of identification research of metals, alloys, or composites are comparison and assessment of the significance of the revealed properties of objects.

At the stage of comparative studying, the analytical comparison method is usually employed. In its process the researcher establishes the similarity, identity, difference of objects, and phenomena. A comparative study begins with comparing general properties of fragments: color, shape, dimensions, surface relief, material composition, etc. The comparison by signaletic (morphological) properties is performed by traceological methods of comparing, combining, and superposing. These techniques are employed in comparative studies of objects that bear traces of tools, mechanisms, exposure to external phenomena (explosions, thermal, or chemical effects), as well as in studying parts of the whole and searching for interfaces.

The comparison by substance properties consists in establishing similarity and difference relations between objects in such principal forms:

- correspondence (disparity) when objects possess coinciding (different) properties that are provided for by the scientific and technical classification system. Correspondence (disparity) is a basis for the conclusion on the general generic belonging of objects;
- the identity (dissimilarity) in properties that are described qualitatively, or equality (inequality) in those expressed quantitatively.

The identity and equality in terms of detected properties make it possible to identify parts of the whole, establish the source of origin and group belonging of objects, while dissimilarity rules out the positive result of the identification task.

The significance of the detected features is assessed from the viewpoint of them being sufficient for solving the task set. The assessment of the totality of the revealed coincidences and differences is carried out at the final (synthesizing) stage of research.

If after analyzing the results of the entire study the complex of properties is considered to be unique, that permits the conclusion about belonging of the materials compared to a certain whole, a specific source of origin.

However, the formulation of the conclusion on the identity of specific individual material of a SPS (MD) fragment should not be based solely on the results of identification research. The material of SPS (MD) fragment only acts as an intermediate object, while the final object of identification is the product. In the case of identifying a divided whole, the coincidence of properties should be observable in the genera of all orders and the group that corresponds to the peculiarities of product manufacturing and its material. When solving the problem of the same source of origin, one should not forget that the objects being compared, which are identical in group properties that characterize their manufacturing features, can have different properties determined by the conditions of their operation or failure.

At whatever stage the identification study is finished, the kind, subkind, metal or alloy grade, composite (matrix, filler) features, product type should be specified (as the result of establishing its genus belonging). For instance, "The fragments compared belong to a common group, i.e., are parts of class III rods made of 25 GS2 (GOST 5781-82) steel in a single technological process and operated under the same conditions".

5.1.5 Classifying and Diagnostic Studies of Materials of Special-Purpose Structures (Means of Destruction)

The structure of classifying and diagnostic studies of SPS (MD) materials is similar to those of identifying studies and consists of the same four stages: preparation, analysis, comparison, and synthesis.

However, the content of research is different. The purpose of classifying study is to establish the specific class of materials and products. The purpose of diagnostic studies consists in ascertaining the presence of individual properties of special-purpose structures or such properties by which the nature, cause, exposure time, and peculiarities of their use can be judged about.

At the preparation stage, the properties to be detected for solving the task set are defined, along with the ways (methods) of establishing them. To solve classification tasks, as a rule, it is sufficient to determine chemical composition for the studied material, and if a product is studied, its shape and dimensions are to be established in addition. The analytical stage of a diagnostic study can consist of establishing external morphological, structural properties, and chemical composition. For establishing the cause-and-effect relationships, the nature of contact interaction, it is necessary to detect a complex of properties.

At the comparison stage in a classifying study, the properties of the examined object are compared with those of the materials belonging to certain classes. With reliance on the data obtained, a multitude of special-purpose structures (means of destruction) is established. The smallest classification array is the grade of metal, alloy. If necessary, the methods and features of their production are ascertained.

Diagnostic research also compares the properties of the studied object with those of metals, alloys, and composites of certain classes. Besides, in such studies, the properties of the presented materials (structure type, the presence and structure of the surface layer, defects, peculiarities of particles morphology, the nature of fracture, etc.) are compared with the properties of materials exposed to physical (mechanical, thermal) and chemical impacts. If researchers know the emergence patterns for various properties of metals, alloys, and composites, they establish the nature of the impact, the mechanism of properties generation in the studied object.

Ascertaining the grade of metal, alloy, the peculiarities of composite structure deserve special mentioning as the most important point of classification and diagnostic studies. Establishing the grade of the object material is important in identification studies since it permits the range of the objects examined to be narrowed, and the areas of material application to be specified.

Grade is a standard-normalized combination and the number of components (alloying elements and impurities) the alloy (metal) consists of, coupled with mechanical and other properties possessed by the alloy or products made of this alloy in compliance with a certain technology, starting from alloy smelting and ending with processing of finished products. In a great majority of cases, the grade is ascertained on the basis of the metal chemical composition.

However, the standardized chemical composition of alloys are set in rather wide intervals. Not infrequently an overlap of the ranges of element contents in different alloy grades is permissible. It is such overlap of permissible ranges of element concentration values in different alloys that often does not allow the unambiguous attribution of the examined objects to a single grade (Tungol et al. 1990).

In actual alloys and composites, the structure and chemical composition are characterized with macro- and micro-inhomogeneities. Macro-inhomogeneity exists between macroscopic volume areas, parts of the product. Structural macro- and micro-inhomogeneities manifest in uneven distribution of phases, inclusions of various nature, and the existence of heterogeneous layers.

Structural and chemical inhomogeneities are closely related. Without revealing structural micro-inhomogeneity of a material, it is not possible to interpret correctly the results of local chemical analysis. So, in studying microparticles in metals and alloys, the employment of complex methods—local analysis in combination with metallographic analysis (the examination of composite structure topology)—is a necessary methodology (Borisevich et al. 1998).

5.2 Comparative Analysis of Fragment Chemical Compositions and Structures as a Method of Identifying Special-Purpose Structures (Means of Destruction)

Comparative analysis of SPS (MD) fragment's chemical composition is an important stage of their identification, which is carried out to establish the type and grade of metals and alloys, the composition and topology of composite materials, the peculiarities of their production technologies, and to forecast their performances and limitations.

Yet, if identification is understood as the quantitative determination of all chemical elements present in the fragment, the task may prove to be impossible. SPS (MD) fragments, especially those made of composites, are multi-component mixtures of the variants of component combinations that can amount to dozens or even hundreds. However, of significant importance is the fact that object properties, as a rule, do not depend on absolutely all substances it contains, but are only determined by their principal components. So, in the majority of cases, it is not necessary to investigate all additives in detail, it is sufficient to establish their type or the nature of their origin.

To determine the composition of a SPS (MD) fragment, complex chemical analysis is performed. The chemical composition of a substance or material is analyzed in several stages. First, goals are set, then qualitative analysis is carried out, which shows the composition. Quantitative research is conducted when it is necessary to establish the exact content of the components in the material.

The results of complex chemical analysis are the determination of material composition along with quantitative characteristics of the compounds it contains. Later, these results are used to ascertain the grades of steels and alloys, the composition of composite materials and technologies of their production. The results obtained can serve as the basis for identifying the country of origin of the means of destruction, developing proposals toward implementing the technologies of producing domestic analog materials, and forecasting SPS (MD) performance characteristics (Knorozov et al. 1987).

The stages of the chemical analysis of SPS (MD) fragments are

- elaborating a research plan;
- performing qualitative chemical analysis;
- performing quantitative chemical analysis; and

– assessing the results of the fragment material analysis.

At the first stage of elaborating the research plan:

(1) research goal is formulated;
(2) all available data on the studied fragment material, its nature, and origin are collected;
(3) the main substances that could be constituents of the fragment material are determined;
(4) analysis techniques appropriate for the sample of material in question are chosen; and
(5) the procedures of sample preparation based on its capability to dissolve in organic and inorganic solvents, and the ability to react with other chemical substances are developed.

At the second stage (qualitative chemical analysis), it is determined which components are present in the studied fragment of special-purpose structure:

(1) Phase analysis is performed—the analysis of the specimen for the presence of individual phases that differ in their chemical and physical properties.
(2) Elemental analysis is carried out. Element composition—the presence of certain chemical elements in the specimen—is determined. Here, specific chemical substances containing these elements are not identified.
(3) Molecular analysis is performed. Molecular composition—the presence of individual chemical compositions in the specimen—is established. This method is mainly employed for analyzing organic substances, since the same set of chemical elements can be comprised in different organic compounds.

At the third stage, during quantitative chemical analysis, the quantitative content (concentration) of the principal components present in the studied specimen of SPS (MD) fragment material is determined, in which

(1) individual components of the specimen are extracted. Specimen components are separated, concentrated, cleansed, and by chemical transformations are transferred into a form necessary for performing the analysis;
(2) a thorough selection of analysis conditions is carried out;
(3) analytical measuring instruments are calibrated for the analysis of each particular specimen component.

At the fourth stage, the results of material analysis are assessed.

With reliance on the results of studying qualitative and quantitative specimen compositions, the following conclusions are made, which concern

1. Composition, structure, and properties of fragment material components.
2. Specimen belonging to the material that was earlier used by the aggressor country in the means of destruction.
3. Presence of an additive in the material specimen.
4. Material compliance with the European and national standards.

5. The studied material toxicity.
6. Specimen identity to the standard or its uniqueness, i.e., lack of its analogs.

To identify SPS (MD) on the basis of their chemical compositions and the structures of their fragments at the document level, a comparative analysis of regulatory documentation is performed. For widely used objects, there are documented requirements of standards or specifications which the object is to meet and which define the set of parameters that are to be studied for object identification. Hence, for all abovementioned objects the notion "complete chemical analysis"can be applied to the list of chemical and physico-chemical studies that are to be carried out for their identification in compliance with regulatory documentation for the types of metals, alloys, and composites.

At the same time, there are several basic conditions of the successful analysis of the SPS (MD) chemical composition and subsequent identification:

(1) the final goal of research should be clearly stated;
(2) the researcher should have the basic idea of the studied object composition, its nature and origin;
(3) limits of analytical technique capabilities should be taken into account;
(4) one should realize that for complex objects it is not possible to guarantee one hundred percent success, but it is always possible to obtain valuable information that would help in the further solution of the problem.

5.3 Creating an Interactive Knowledge Base for Monitoring Materials of Special-Purpose Structures Within the Framework of the Transdisciplinary Cluster of Information Support for the Development of Means of Destruction

5.3.1 Transdisciplinary Cluster of Information Support for Monitoring Materials of Special-Purpose Structures and Means of Destruction

Physical and mechanical as well as operational properties of materials of special-purpose structures, which primarily include means of destruction (MD), are one of the main factors that determine their tactical and technical characteristics and, ultimately, their effectiveness. In order to analyze and substantiate the directions for the development of new materials and functional elements of weapons, to assess the achieved level of their tactical and technical characteristics, and the level of development and mastery of relevant technologies, it is necessary to collect, generalize, and analyze large volumes of scientific and scientific and technical information on various topics, in particular, on issues of monitoring materials of special-purpose structures. Over time, the amount of information that needs to be processed during research increases significantly, which crucially complicates its processing and generalization. According to the forecast estimates of leading analytical companies, the total world volume of data created and replicated by mankind amounted to more than 3.6 Zettabytes (36 trillion GB) by 2020 and is growing globally annually by 30% (Sardana 2006). However, there are reasonable doubts as to whether such volumes of information can be effectively processed, and, most importantly, adequately perceived and properly understood by decision-makers. The urgent need to respond to this global challenge stimulated new approaches to handling large amounts of information and contributed to the emergence of the Big Data Concept (Sardana 2006; Lyman and Varian 2003).

One of the rational approaches to the processing of large arrays of accumulated information is the transition from data storage and processing to the accumulation and processing of knowledge, which is possible only when modern information systems (ISs) are used for working with knowledge. The creation and development of such systems is a complex task that should be solved only with the use of modern software and special support.

Logic programming languages such as Lisp, Prolog and others were more often used in information systems of the abovementioned type in the past. However, the development of the Web—a worldwide environment that opens access to a huge number of various information resources—has caused an urgent need to create modern, more effective ISs oriented to functioning in various information environments (Rivera et al. 2018), as well as to develop and improve methods and algorithms for distributed knowledge management and processing. The development of modern intellectual analytical tools of various types for solving loosely structured tasks that arise during research on the development of means of destruction in general and the monitoring of materials for special-purpose structures also remains an urgent task.

The abovementioned problems can be solved through the creation or improvement of the existing scientific and methodological apparatus, and the development of specialized software. Besides, the current problems of the development of artificial intelligence (Trends 2020; Ansari et al. 2016) also cannot be solved without these tools. In this direction, an effective tool for the implementation of the developed cognitive information technology for the needs of the effective accumulation and processing of the knowledge system used for researching the problems of the MD development is the creation of a transdisciplinary cluster of information support for the MD development, which contains a knowledge base aimed on monitoring and identification of materials used in MD components and structures.

The main goal of the work is to develop the principles of creating a transdisciplinary cluster of information support for the development of means of destruction (TCIS DMD) and monitoring of special-purpose materials and technologies. To achieve this goal, a number of tasks should be solved, namely, considering transdisciplinary tools as a basis for the effective use of scientific and technical solutions; tasks of TCIS DMD development; the order of displaying narrative descriptions and the algorithm of transdisciplinary cluster functioning; and their architecture.

The important results of research on the stages of development of intellectual analytical tools are

- The concept of building an information system on the basis of Model Driver Architecture, which is designed for the implementation of joint production systems of knowledge (2013). This concept involves the use of an ontological representation of production processes (production ontology) which are used by a platform-independent model in their general sequence.
- Approaches to the design of management information systems (Dubinin and Vyatkin 2012; Kopaygorodsky 2014). In Dubinin and Vyatkin (2012), an approach based on the transformation of ontologies was investigated, i.e., the ontological description of applications based on certain functional blocks. In Kopaygorodsky (2014), the development of semantic information systems, whose architecture feature is the presence of a "variant part" in the form of ontologies of the subject field, methods, and tasks, was considered.
- Technical decision (Zagorulko 2015) regarding the structure of the information system for rational decision-making by experts. The information system is built on specialized shells configured for a certain class of tasks in certain fields of knowledge, where an integrated model with the possibility of ontological representation is applied as the basis of knowledge representation.
- Substantiation of the need to develop and implement a single comprehensive automated system for monitoring the development of technologies and special-purpose materials used in modern means of destruction (Gorbulin 2020), as one of the main and mandatory conditions for the formation of a unified information space.

The consideration of the abovementioned sources allows concluding that the creation and development of information technologies for the needs of the effective accumulation and processing of the knowledge system aimed to the development of special-purpose technologies and materials used in modern means of destruction, which is based on the ontological representation of information, is the most complicated and the least researched issue.

The solution to the problems that arise during the processing of distributed knowledge is proposed through the development and improvement of appropriate methods and algorithms, as well as intelligence analysis tools.

It should also be noted that an important factor in the development of intelligence analysis tools is the presence of interrelated processes of processing, storage, distribution, and use of large amounts of information and knowledge in the form of unstructured documents (Stryzhak et al. 2019; Dovhyi and Stryzhak 2020; Moylan et al. 2002). However, in order to ensure the objectivity of analytical conclusions and decisions made based on the analysis of these documents, it should be taken into account that the mentioned passive knowledge systems relate to different fields, have a significant number of interdisciplinary relationships, and are based on different information technologies and standards.

The system of knowledge, which is large in scope and subjects of nomenclatures, includes such categorical concepts as dictionaries, thesauruses, taxonomies, narratives, narrative discourses, and linguistic corpora (Stryzhak et al. 2019; Dovhyi and

Stryzhak 2020). These categorical concepts belong to the class of loosely structured ones, and according to the nature and type of the presentation to the Big Data class (Strizhak 2014). They are also characterized by multi-dimensionality and multiple latent connections.

One of the ways to solve this problem is the creation and implementation of a transdisciplinary cluster of narrative descriptions of samples of means of destruction (TCIS DMD), and applied technologies and materials. Such a cluster should be understood as a cognitive network-centric management system for spatially distributed processes of accumulation, systematization, storage, and access to various information and knowledge-oriented resources, where narratives of scientific and technical products and examples of MD as a distributed system of knowledge are accumulated. The cluster should provide cross-contextual multiple connectivity of information resources and knowledge systems which will be used for researching problems of MD development. The ontology of the terms used during the development of TCIS DMD is shown in Fig. 5.2 (Chepkov et al. 2022).

TCIS DMD is proposed to be created as a network-centric portal that uses cognitive network procedures of linguistic semantic analysis of network-distributed text documents contained in repositories, knowledge bases and databases, archives, etc., and transdisciplinary procedures for their evaluation. The evaluation will be carried out according to the properties of the content, which provides a dynamic taxonomy and aggregation of network text documents related to the development of means of destruction, appropriate technologies, and materials.

Analysis of the development and application of samples of means of destruction allows determining the tasks of TCIS DMD, and monitoring of technologies and materials (Fig. 5.3).

The solution of the abovementioned tasks is carried out according to a certain algorithm whose structure diagram is shown in Fig. 5.4.

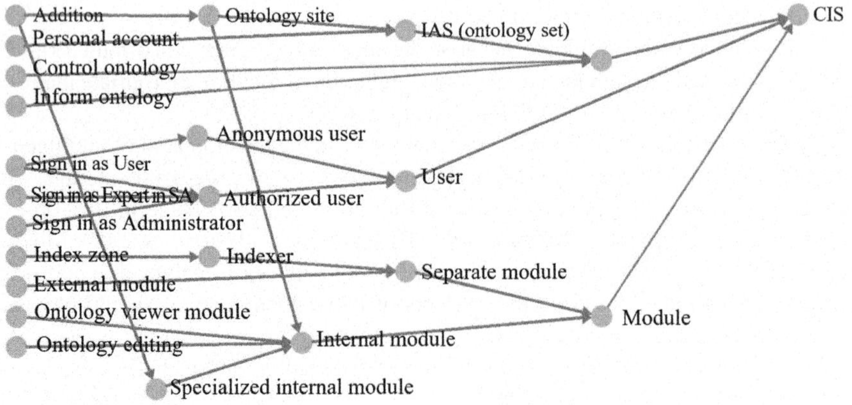

Fig. 5.2 Ontology of terms used for creation of transdisciplinary cluster of information support for development of means of destruction, appropriate technologies, and materials

Fig. 5.3 Main tasks that can be solved in transdisciplinary cluster of information support for development of means of destruction, and for monitoring of applied technologies and materials

TCIS DMD is a network-centric portal, which includes cognitive procedures of linguistic semantic analysis of network-distributed text documents contained in network repositories, knowledge bases and databases, archives, etc., and transdisciplinary procedures of their content evaluation; it provides dynamic taxonomy and aggregation of network text documents. The cluster's functioning is based on TODOS processes (Potapov et al. 2020). TCIS DMD is automatically filled in with transdisciplinary images of network-distributed documents through taxonomic indexing.

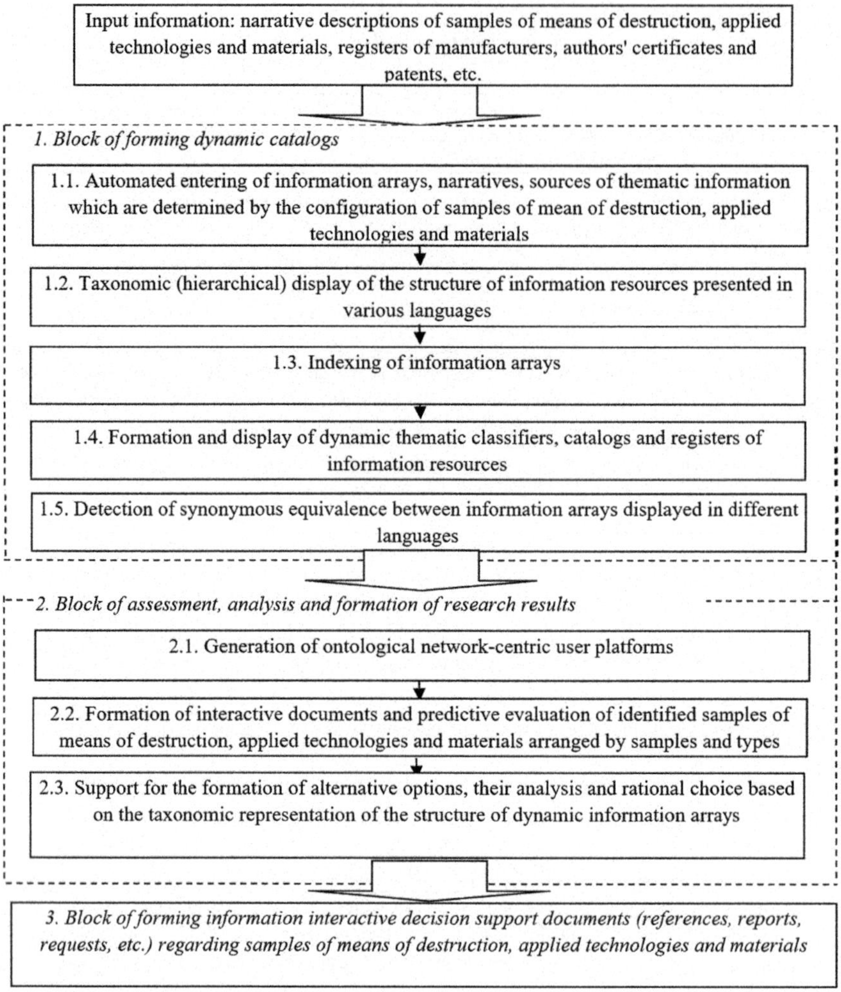

Fig. 5.4 Structure diagram of TCIS DMD algorithm functioning

The network-centric procedure of taxonomic indexing ensures the documents on a selected subject, which are located in network repositories, have a significant number of interdisciplinary relationships, and are created using various information technologies and standards, are automatically placed into the TCIS DMD environment.

The rubrics according to the content of thematic profiles and the registers of various technologies are formed in the TCIS DMD environment automatically. Each document entered into the TCIS DMD using transdisciplinary procedures is transformed into an interactive form, both in automatic and automated modes. Access to these documents is implemented based on requests in the form of natural language constructions and word combinations that are entered upon request in the TCIS

DMD environment in the user's language. Ukrainian, English, Russian, German, and French are the natural languages that provide user interaction with TCIS DMD.

Requirements for protecting confidential information in relation to those documents that are subject to the appropriate conditions for access to them and their use are implemented in TCIS DMD. At the same time, TCIS DMD is a universal scientific (and technical) means of knowledge processing. The architecture of TCIS DMD is determined based on the transdisciplinary structure of dynamic catalogs that are formed according to various thematic profiles and registers. Each user can configure his personal institutional structure, and the system components of TCIS DMD can change over time.

The thematic content and corresponding profiles of documents and registers of TCIS DMD are determined dynamically on the basis of transdisciplinary detection of intercontextual multiple interrelation of all information resources and knowledge systems used by users.

The transformation of narrative descriptions into the format of interactive knowledge systems provides the storage of natural language constructions that reflect phrases, judgments, and statements about certain facts of the subject-thematic profile. Facts are connected to each other by a set of relations, and can also be characterized by certain properties. Each narrative description of scientific and technical content reflects the conceptual system of a certain theory. The target methodological task is a competent interpretation of the conceptual system, axiomatics, rules, syntactic, and morphological foundations of the theory described. Moreover, the ability to competently interact with the network-centric system components of the information environment also depends on the completeness of the representation of the conceptual systems and disciplines described by the narrative, i.e., the conceptual systems of the thematic disciplines used are the cognitive-functional core of operationality that determines the field of their application.

The importance of scientific and technical narrative descriptions for monitoring the means of destruction, applied technologies, and materials is very high. However, the information processes that are developed today in the defense sector make certain demands on the functional properties of a narrative. One of the main requirements is interactivity and interoperability. The implementation of these requirements depends entirely on the mechanisms of transdisciplinary transformation of narratives from a passive to an interactive format.

Thus, the structural representation of narrative descriptions is implemented based on the use of cognitive procedures. Further transdisciplinary analysis solves the problem of transforming narratives into the format of interactive knowledge bases. Due to the multiplicity of taxonomies that are formed at the stage of transdisciplinary analysis, the integration of narrative descriptions is implemented. This determines the technological possibilities for the further use of tools for search, categorization, classification, and determination of the thematic equivalence of the narratives according to the content of external network information resources.

The taxonomies form the basis for the formation of ontologies of narrative descriptions. On this basis, certain conceptual schemes of scientific and technical descriptions are formed on the subjects of the disciplines, which in fact reflect a certain

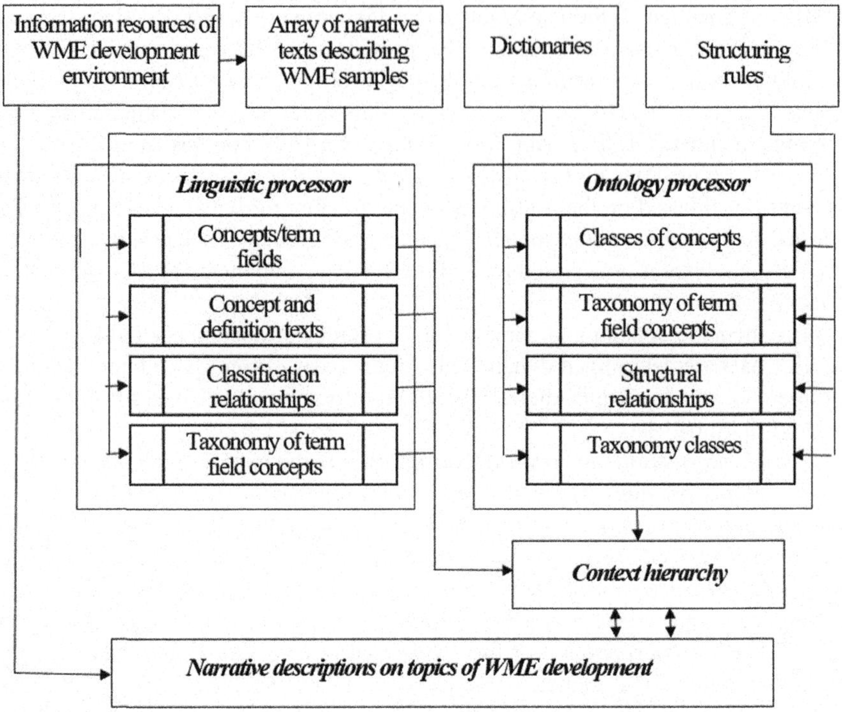

Fig. 5.5 General scheme of formation of ontological system of information support for monitoring the means of destruction

theory. The generalized diagram of the formation of ontological system of the narratives is shown in Fig. 5.5 (2019). Thus, the linguistic procedures of the corresponding processor form taxonomies based on the content analysis of the narrative texts.

The taxonomies provide the formation of classifiers of relations and connections of sets of contexts with the concepts of narrative descriptions. Semantic rules determined by the natural system of narrative ontology provide the formation of classes of concepts and their taxonomies; their connection with each other by appropriate multiple relations; and the functionality, which forms the operational environment of the ontological system of narrative descriptions (Mintser et al. 2018).

The functional diagram of narrative ontologies (Fig. 5.5) reflects the order of formation of glossaries and taxonomies, when a thesaurus and procedures of interactive interaction with each element of the ontology of the narrative are formed in the environment of the semantic processor, and multiple binary relations appear between it, concepts, and their contexts. This makes it possible to define the ontological system of the narrative description of the subject field of MD monitoring as an interactive system of knowledge. Besides, due to the selection from the set of relationships of the concepts of its field of terms, there is a connection between the taxonomy of the ontological system and network information resources selected according to certain criteria. The main feature of these resources is a more extensive contextual description of the concepts presented in the basic narrative.

The methods of analyzing narrative descriptions are diverse and depend on the goals to be achieved in the process of solving an applied problem or task. The methods of semantic processing of texts can be conditionally divided into those that carry out various "linguistic transformations", and those that are designed to solve other practical problems. In the studies of the problems of the development of artificial intelligence, interest in the text lies in the field of "extraction of knowledge", i.e., classification of messages, answers to questions, contextual translation, and understanding of discourses (Trends 2020; Golovin and Strizhak 2019). At the same time, the use of methods of transdisciplinary analysis provides the formation of interactive knowledge bases which are an important component of TCIS DMD. The basis of their formation is a mathematical model of the interactive display of MD monitoring processes, which is one of the variants of the lexicographic system (L-system) (Shapovalov et al. 2020; Shirokov 2017; Nadutenko et al. 2022) that is considered as an information environment semiotically and semantically organized according to certain rules. The acceptable L-system version is based on the use of ontologies:

$$O_{iOVT} = \langle X_{iOVT}, R_{iOVT}, F_{iOVT} \rangle, \tag{5.1}$$

where X_{iOVT} is a finite set of objects of the ontology of the development of MD, applied materials, and technologies; R_{iOVT} is a finite set of relationships between ontology objects; and F_{iOVT} is a finite set of functions for interpreting objects and technologies, which are used for this purpose and given by their attributes.

The use of ontologies for the formation of interactive knowledge bases, which are part of TCIS DMD, allows using them to describe various data structures due to their flexibility, including the system architecture. Therefore, the accepted version of the L-system is called hereafter the ontology-controlled L-system (Nadutenko et al. 2022), in which ontologies are used not only as a repository of information resources, but also to manage its own architecture, set of available functions, interface, etc. The structure of the L-system controlled by ontology is a variant of the structure of the standard L-system and can be represented as

$$L_s = \left\{E_{iOVT}, D_{iOVT}, O_{InfOVT}, O_{contrOVT}, \beta_{OVT}, \sigma\left[\beta_{OVT}\right], \Sigma_{pz}\right\}, \tag{5.2}$$

where E_{iOVT} is a set of subjects (subject area (SA) experts in the military sphere); D_{iOVT} is a set of SA, information about which is contained in the system related to MD storage, materials and technologies; O_{infOVT} is a set of information ontologies created by experts; $O_{contrOVT}$ is a set of control ontologies created by administrators or developers of the information support cluster; Σ_{pz} is an ontology-controlled software architecture of the L-system, which can change with the use of ontologies.

The use of the ontology-controlled software architecture of the L-system (Σ_{pz}) with the possibility of additionally including control mechanisms based on ontologies permits to significantly simplify its architecture:

$$\Sigma_{pz} = \{\Sigma_{konzO}, \Sigma_{zovnO}, \Sigma_{vnO}, \varphi, \xi\,(O_\Sigma)\}, \tag{5.3}$$

where Σ_{konzO} is the ontology-controlled conceptual model; Σ_{zovnO} is the ontology-controlled external model; Σ_{vnO} is the ontology-controlled internal model; φ is the operator of $\Sigma_{konzO} \to \Sigma_{zovnO}$ conversion; and $\xi(O_\Sigma)$ is the operator of $\Sigma_{zovnO} \to \Sigma_{vnO}$ conversion.

The conceptual model of the ontology-controlled L-system has the following form (4):

$$\Sigma_{cmO} = \{\{x_I, x_\Sigma\}, \{r_I, r_\Sigma\}, M\,(x_\Sigma, R_\Sigma\,(x_\Sigma))\}, \tag{5.4}$$

where x_I, r_I are the typical object and relation of information ontology; x_Σ, r_Σ are the typical object and relation of control ontology; and $M\,(x_\Sigma, R_\Sigma\,(x_\Sigma))$ is the set of functions-interpreters of control ontologies, which transform a certain object x_Σ and its relevant relations $R_\Sigma\,(x_\Sigma)$ into a certain subroutine.

Due to the use of ontologies, the model provides an extremely large degree of freedom in the data structure, but its disadvantage is the impossibility of setting integrity restrictions at this level, which are typical for a conventional L-system.

The internal model of the ontology-controlled L-system (5.5) is determined by a set of ontologies—information and control ones. Besides, the internal model includes standard subroutines that define the basic functions of information processing, and the operating environment for executing these subroutines. The set of standard subroutines, in particular, includes specific implementations of interpreter M functions. The operating environment includes all aspects of the physical functioning of the subroutines of the ontology-controlled L-system, such as programming languages, operating systems, and necessary system libraries:

$$\Sigma_{inmO} = \{O_I, O_\Sigma, G_\Sigma, E_G\}, \tag{5.5}$$

where O_I, O_Σ are the set of information and control ontologies, respectively; G_Σ is the set of standard routines of the L-system; and E_G is the operating environment.

The external model of the ontology-controlled L-system is based on the principles of a natural system (Katie Moon et al. 2020) which is formed by processing control ontologies with the help of specific implementations of $M \subset G_\Sigma$ functions. The basic component of the natural system is the vector of data processing functions, which is determined by the process ontologies; this represents the information available in the ontologies in an interactive form. An auxiliary component is a set of integrity control functions formed on the basis of ontological templates of data structures. This component is especially important for those natural systems that provide the ability to create or edit control or information ontologies:

$$\Sigma_{exmO} = \left\{O_{PR}, O_T, N\left(O_\Sigma, \bar{Q}\,(O_{PR}), Q_\Omega\,(O_T)\right)\right\}, \tag{5.6}$$

where $O_{PR} \subset O_\Sigma$ is a set of process ontologies; $O_T \subset O_\Sigma$ is a set of ontological templates of data structures; N is the dynamic natural system; $\bar{Q}$ is the vector of data processing functions $Q_i \subset G$; and Q_Ω is a set of integrity control functions.

A specific implementation of this architecture is an interactive ontological document. This document implements the internal (5.5) and external (5.6) models of the ontology-controlled L-system and can be represented by a triple

$$\langle O_I, O_\Sigma, N \rangle, \tag{5.7}$$

where O_I, O_Σ are the sets of information and control ontologies, respectively and N is a natural system that provides interactive interaction with the content O_I using the functions described in O_Σ.

The natural system that is part of the interactive document can be conditionally represented as a certain function (target function):

$$f_N(\breve{x}_1 \ldots \breve{x}_n) = \breve{Q}_n(Q_{n-1}(\ldots Q_1(O, \breve{x}^1) \ldots, \breve{x}^{n-1}), \breve{x}^n), \tag{5.8}$$

Sigma where $\breve{x}_i$ are the user commands, i.e., actions they perform in the process of interactive interaction with information; O is the set of information operated by the system at a certain moment (in the form of an ontology); Q_i are the functions of information processing; and $\breve{Q}_n$ is the representation function which forms the actual information provided to the user in text or graphic form.

In fact, the natural system N is a program module consisting of a set of subroutines (5.6):

$$N(O_\Sigma) = G_\Sigma \bigcup \sum_{x_\Sigma \in O_\Sigma} M(x_\Sigma, R_\Sigma(x_\Sigma)), \tag{5.9}$$

where G_Σ are the system kernel subroutines specified by the internal model (5.5); x_Σ are the objects of control ontologies O_Σ; and $M(x_\Sigma, R_\Sigma(x_\Sigma))$ are the interpretive functions given by the conceptual model of the L-system (5.4).

The architecture of the TCIS DMD special software is based on the principles of cognitive information technology KIT POLYHEDR (Anwar et al. 2023) and the consolidated use of information resources of cognitive services. This should provide the processes of collection, analysis, and structuring of information resources in accordance with the configurations set by administrators, and display results to the user in an interactive form with the possibility of contextual search.

The architecture of the TCIS of WME development (WMED) was implemented using the MVC (model-view-controller) template (Fig. 5.6) with levels: data level (model) represented by the ontology repository; logic level (controller) represented by the data collection subsystem; and presentation level (view) represented by the ontology interface.

A web crawler designed to collect source documents from sources selected by the system administrator collected the initial data for the TCIS DMD.

Information resources collected by a web crawler can be processed in two ways: terminological formation of SA thesauri, and structural—automated identification and structuring of information about scientific achievements of institutions.

As a result of processing information resources in the abovementioned ways, corresponding ontologies are created, and the terminological way involves the selection

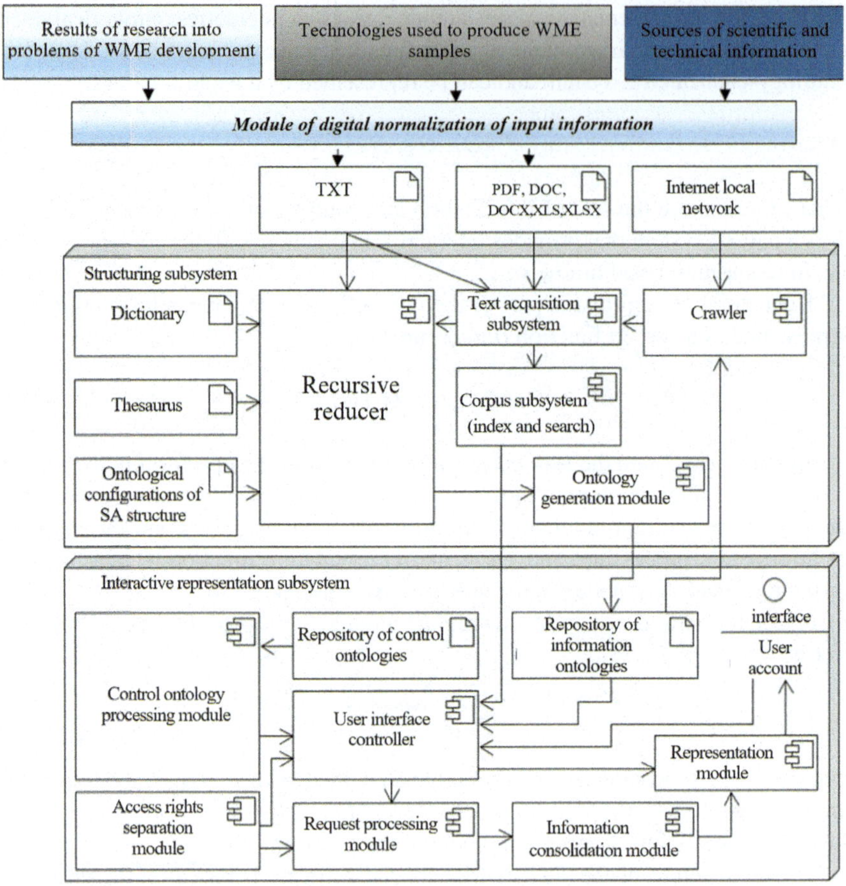

Fig. 5.6 Architecture of special software for transdisciplinary cluster of information support for monitoring means of destruction (version)

of terms present in texts using the corresponding subsystem, and the formation of ontological thesauri based on them. During their formation, terms are additionally processed in the synonym identification subsystem, which combines synonymous terms. The way of structuring involves the automated use of a recursive reducer (Podlipaev and Strizhak 2018) to identify information about the MD development, technologies, materials used for this, and other additional information resources, structuring them according to the specified configuration of the attribute set system, and forming ontologies based on this information in certain fields.

After processing information resources, an ontology database is formed, which is a specialized file storage for storing ontology XML files within a virtual file structure. The virtual file structure assumes the presence of specialized metadata which includes the names of ontologies and their other characteristics: character codes, purpose, access rights, etc. The metadata also stores a system of virtual directories that do not correspond to the actual placement of XML files within the file system.

The ontologies in the virtual directory are divided by purpose into ontological structuring descriptors, process ontologies, ontological thesauri, and ontological descriptions. According to this:

ontological structuring descriptors specify the configurations of the structuring module (recursive reducer), in particular information identification rules, sets of attributes to be picked up, etc.;

process ontologies specify application configurations within the ontological interface;

ontological thesauri are formed automatically during the operation of the information collection subsystem and are used for searching within the ontological interface; and

ontological descriptions of information about the MD development, materials, and technologies used for this, and other additional information resources are the main repository of information operated by the system. Representing this information in an interactive form is the main task of the ontological interface.

The ontological interface itself works on the basis of process ontologies. Ontology data are processed by the scenario controller that results in forming applications of the following types:

An administrative application for modifying control ontologies (process ontologies and structuring descriptors). This application should include a templated editor of control ontologies.

A user application for providing information to end users in an interactive mode: the basis of this application is the search subsystem interface.

The search subsystem operates on the basis of an index formed by processing ontological descriptions of relevant information resources in the indexing subsystem which also uses thesauri to assist the user in forming search queries.

Both types of applications provide interactive user interaction with the created ontologies; for this, user commands are processed by the corresponding subsystem.

5.3.2 Methodology for Building Interactive Knowledge Bases for Monitoring Research Results on Special-Purpose Structural Materials and Means of Destruction

Research into special-purpose structural materials is an important task. The solution of this task requires the use of various methods and techniques, and research results in the format of ontological knowledge bases according to a particular technology. An important component of the implementation of the technology is monitoring which is a dynamic component of the methodology for forming such knowledge bases. The main properties of monitoring are systematicity, dynamism, and focus on prediction. Systematicity is an important feature of monitoring, which requires determining the requirements for a set of interconnected objects based on the analysis of causal and functional random relationships, feedback loops. It should be noted that the theoretical and practical foundations of monitoring have been developed, but an information system for automating the processes of researching special-purpose structural materials has not yet been built.

The abovementioned research is also a complex process that should be foreseen in the scientific, technical, production, and technological activities, which in modern conditions have a network-centric nature. This is facilitated by such modern IT technologies as artificial intelligence (AI), cognitive computing, etc. (Konar 2005; Lee and Kirlik 2013). On their basis, a format of transdisciplinary interaction of users with sets of scientific, technical, and technological narrative (STTNr), and directly STTNr among themselves, is implemented, which characterizes the modern information space both at the global and local levels. The ontology of research on special-purpose structural materials is shown in Fig. 5.7.

In the context of scientific, technical, production, and technological activities of researching special-purpose structural materials, transdisciplinarity should be understood as a manifestation of reflexive intercontextual connectivity of their narrative descriptions, provided that they have common properties that can have different functional interpretations (Stryzhak 2023; Stryzhak and Zavalevskyi 2023; Riedl and Bulitko 2013). The transdisciplinarity of narratives is understood as the systemological, dynamic formation of classes of contextual descriptions that reflect images and have stable semantic relations among themselves according to different thematic profiles with specific terms, phrases, and word forms. The indicated semantic relations are formed on the basis of the manifestation of the hyperproperties of the narratives of all information resources used, namely, reflection, recursion, and reduction (Riedl and Bulitko 2013; Aksu-Koç and Aktan-Erciyes 2018).

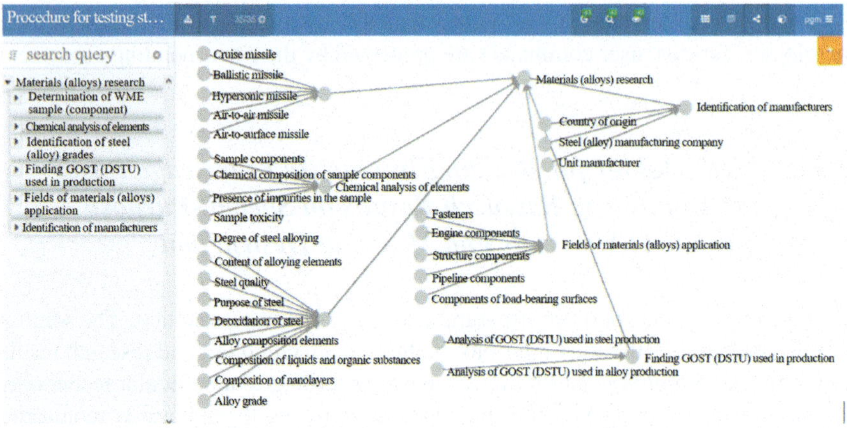

Fig. 5.7 Ontology of research on special-purpose structural materials

The category of transdisciplinarity implements the abovementioned deep interrelations and hyperrelations between various STTNr, the axiomatics of which are represented and characterized by various scientific profiles reflected in the existing profile classification of the foundations of sciences. For these narrative documents, the objectivity of their application in technological processes should be provided, especially when it is necessary to draw multi-dimensional analytical conclusions and make appropriate decisions by analyzing and studying their content. This problem has a cognitive nature, and its solution requires the implementation of the following categories of cognitive metatasks: structuring, analysis/isolation of the problem, synthesis, and choice. Platforms for their solution provide consolidated interaction of specialized experts and specialists in the format of narrative discourse (Lytvyn et al. 2019; Bogdanov et al. 2023a) with STTNr as a format for reflecting passive knowledge systems. The nature of the interaction is transdisciplinary, which provides the transformation of all documents that make up the information environment of expert interaction into an interactive format for representing knowledge systems (Riedl and Bulitko 2013; Aksu-Koç and Aktan-Erciyes 2018).

Primarily, knowledge systems can be represented in the format of an ontological system (Riedl and Bulitko 2013; Aksu-Koç and Aktan-Erciyes 2018; Bogdanov et al. 2023b):

$$O_t = \langle X, R, F, A, D, R_s \rangle , \tag{5.10}$$

where X is a set of concepts (notions) of the subject area (SA) of research on special-purpose structural materials; R is a finite set of semantically meaningful relationships between concepts in research on special-purpose structural materials; F is a finite set of interpretation functions given on concepts and/or relations; A is a finite set of axioms used to record always true statements (definitions and restrictions) in terms of the subject matter of research on special-purpose structural materials; D is a set of additional definitions of concepts in terms of the subject of research on special-purpose structural materials; and R_s is a set of restrictions that determine the scope of conceptual structures of a specific topic of research on special-purpose structural materials.

The taxonomy in reflecting the structure of STTNr in the process of researching special-purpose structural materials is the basis for implementing a systemological classification of STTNr concepts according to their properties and relationships between their contexts. This contributes to the classification and systematization of STTNr concepts, as it provides a valid reflection of the hierarchical relationships between the various elements of knowledge described by the STTNr content (Lee and Kirlik 2013; Lytvyn et al. 2019). This is especially important when building knowledge systems in the format of ontology of narrative documents—O_{nr}, which reflect scientific and technical descriptions and characteristics of samples of means of destruction:

$$O_{nr} = < X(K), p >, \tag{5.11}$$

where O_{nr} is a STTNr with a specified sequence of concepts X, defined by contexts K (descriptions), and p is a strict-order relation that determines the conditions for positioning concepts X in the text.

Therefore, O_{nr} is actually a passive system of knowledge.

The taxonomy, as a category of valid representation of connections and relationships between STTNr concepts, provides the identification of semantic relationships in the process of building an ontology (Aksu-Koç and Aktan-Erciyes 2018). An arbitrary taxonomy is a directed graph without cycles and is formed by the concepts of the corresponding STTNr which have certain hierarchical relationships between themselves. That is, the taxonomy T can be represented as the following tuple:

$$T =< X, R_t >, \tag{5.12}$$

where T is the taxonomy, X is a set of concepts, and R_t is a set of binary relationships between the concepts of taxonomy T.

Taxonomy nodes contain contexts that describe the concepts that form its vertices. They can be considered as systemological components of certain knowledge. The set of taxonomies whose concepts are elements of one ontology of the type (5.11)—O_{nr} forms a taxonomic variety.

An ontological interactive document is formed on the basis of O_{nr}. It is based on a taxonomic representation of STTNr in the format of narrative discourse, which gives specialists the opportunity to semantically process the information contained in it (Rivera et al. 2018; Trends 2020; Ansari et al. 2016; Chungora et al. 2013). Therefore, the taxonomy itself is convenient for visually displaying the content of the STTNr, but it does not yet provide full-scale dialogic interaction with the user.

For this purpose, it is advisable to use the Böhm tree format, which is a prerequisite for the formation of an ontological interactive document (OID) as a type of ontological knowledge base. Its formation is based on the categories of ontology (5.10) and natural system. The natural system provides interactive access to the information contained in the ontology (Aksu-Koç and Aktan-Erciyes 2018; Lytvyn et al. 2019; Bogdanov et al. 2023a). The class of natural systems SN is primarily given by ordered sets of pairs of the $\{< \text{action} \rightarrow \text{results} >\}$ form, which can be represented provided that there is a non-empty set of possible sets of actions—F, where F is considered as a finite set of interpretation functions, which is given over the STTNr concepts.

The use of these functions in the process of researching materials for special-purpose structures implements the process of converting STTNr into the ontological knowledge base (OKB) format, which allows determining the OID format. They also provide a whole class of conversions and transformations of the STTNr content (Lytvyn et al. 2019). This includes the following operations: clustering, classification, consolidation, hierarchization, attribution, etc.

To provide the most effective work of STTNr specialists in the OID format, the functions of processing and displaying information available within the framework of the corresponding natural system should take into account the specifics of the problem that needs to be solved. Moreover, there may be different types of data that differ radically in structure and require special approaches in more complex subject areas of STTNr, which include the study of special-purpose structural materials.

This process requires the availability of effective means of tuning natural systems, which allow changing their composition and structure in a minimum time. Such changes can be provided by addition of

- specialized functions for information representation;
- specialized functions for information processing;
- specialized functions for data preprocessing.

The specialized (control) ontologies can be used as a mechanism for determining the configuration of a natural system.

According to the defined mathematical apparatus used in the methodology for creating ontological knowledge bases for research on special-purpose structural materials, a *method of transdisciplinary consolidation* has been developed as the basis for processing information resources for research on special-purpose structural materials STTNr.

This method is intended for the implementation of a cognitive communicative act of interaction between specialists and a set of STTNr research on special-purpose structural materials based on narrative discourse. In the method, the consolidation of a set of information resources for material research is considered in the format of a unified systemically organized information environment.

The input information for the method is a set of STTNr studies of special-purpose structural materials, which is represented by documents in various formats (text in printed or electronic format, photocopies, etc.). For these formats, it is practically impossible to implement the procedure of full-scale, multi-attribute, and multi-dimensional monitoring and semantic analysis of the content of documents and STTNr.

For their full-scale use in production and technological process, their representation is required, according to the transdisciplinary paradigm of methodology in the ontological format. After this, it is technologically possible to form the corresponding OKB based on their content.

Information resources are processed in the following stages (Fig. 5.8):

At the first stage, a knowledge-oriented ontological format for representing STTNr is formed, based on their taxonomic representation. The representation of STTNr concepts is implemented in the taxonomy format. At this stage, all concepts that characterize STTNr and are connected by certain relationships based on common properties are extracted using lexicographic and conceptographic analysis. The taxonomy is a directed graph whose vertices represent fragments of STTNr knowledge.

At the second stage, semantic platforms (SP) of STTNr research on special-purpose structural materials are formed based on taxonomic variety of concepts, including their contextual descriptions. The set of STTNr can form a taxonomic variety, a hyperset of taxonomies, each of which, when forming SP concepts, is characterized by the inclusion of certain axiomatics, which can vary for different STTNr. The axiomatics are defined by the procedure of interpreting the meanings

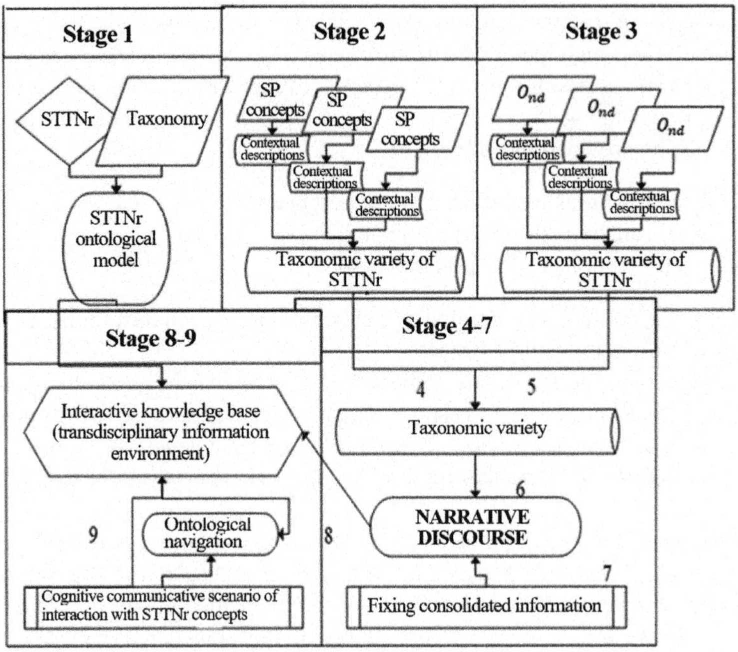

Fig. 5.8 Method of STTNr transdisciplinary consolidation

of contexts that define concepts of taxonomies. These contexts form the content of taxonomy nodes and represent certain elements of knowledge. As a result, taxonomic variety represents a certain system of knowledge according to thematic profiles of STTNr.

As an example of the formation of STTNr SP of special-purpose structural materials, the results of research on materials of missile fragments of various types used in Russia's armed aggression against Ukraine were selected. The ontology of their structure was generated based on the semantic and linguistic analysis of a set of STTNr that describes the functional, tactical, and technical characteristics, and processes of their production. These narratives were revealed during the search for relevant information (Fig. 5.9).

The ontology of materials and their properties was generated in the process of scientific and technical analysis of the fragments of *Iskander*, *Kinzhal*, R-31, R - 37, North Korean *Hwasong-11* (KN-23) missiles, etc., which were selected for research (Bogdanov et al. 2023b, c, d, 2024).

At the third stage, intercontextual relationships within the taxonomic variety of the STTNr set are revealed. An ontology of narrative discourse O_{nd} is formed as the basis of an ontological knowledge base. A verbal active consolidation function is implemented, which also interprets binary relationships between all contexts that

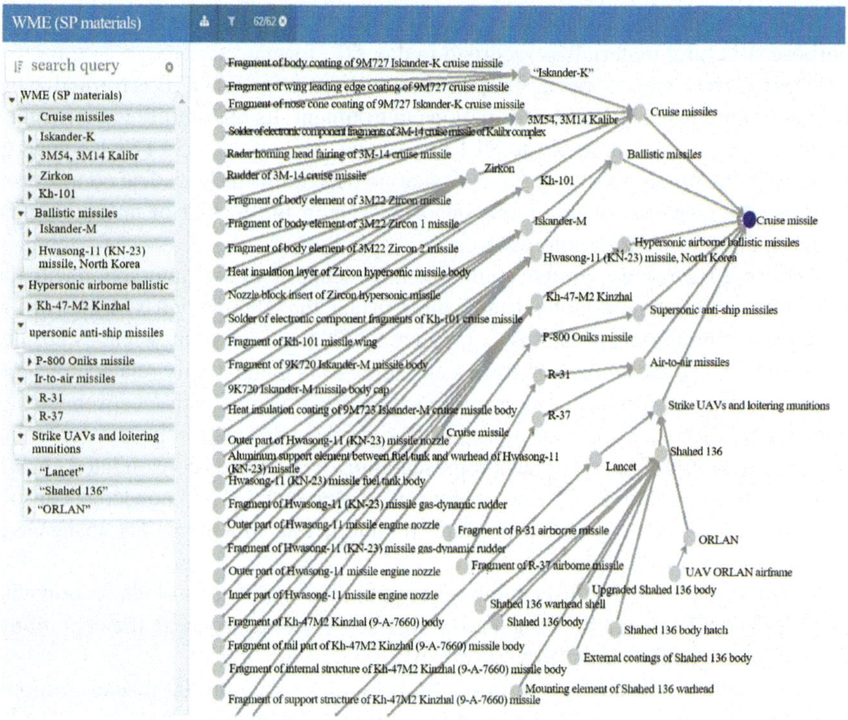

Fig. 5.9 Example of ontology of fragments of components and materials of special-purpose structures

reflect the meanings of concepts that form the content of the STTNr involved in interaction with users.

At the fourth stage, taxonomic variety is formed by encapsulating taxonomies of individual STTNr studies of special-purpose structural materials. The variety of consolidated STTNr taxonomies represents a knowledge base that unifies and reflects the set of STTNr concepts used. The encapsulation of each new taxonomy into their variety is implemented by the procedure of establishing intercontextual relationships between the concepts of the STTNr taxonomies of special-purpose structural materials.

At the fifth stage, an environment of taxonomic variety of special-purpose structural materials is formed based on taxonomies reflecting thematically defined STTNr. Similar to the previous stage, encapsulation of taxonomies is implemented, which reflect thematically defined STTNr in an environment of taxonomic variety formed by the consolidation of their concepts and taxonomies.

At the sixth stage, the taxonomic variety of materials of special-purpose structures is transformed into the format of narrative discourse based on the generation of its ontology O_{nr} which is implemented on the basis of verbal active reflection.

At the seventh stage, consolidated information on the results of research on special-purpose structural materials is recorded in the O_{nd} format.

At the eighth stage, a special-purpose construction material OKB is formed in the format of a transdisciplinary information environment. Its operability is determined by the narrative discourse of the STTNr set, and O_{nr} of each STTNr is synchronized with the functions of cognitive services to determine the logistical connections between the concepts of the entire set of STTNr in the format of an ontological representation of information resources.

At the ninth stage, a cognitive communicative scenario of user interaction with the OKB is formed using ontological navigation. In fact, the OKB, which is formed on the basis of the ontological model O_{nr} for each STTNr, is a set of knowledge capsules an ontological convolution that graphically represents contextually related concepts of the ontology in the format of a multifaceted prism where each facet reflects the contextual binary connectivity of one main concept with other concepts according to thematic profiles. An aggregated representation of connected consolidated knowledge systems based on STTNr taxonomies is implemented. The knowledge capsule consolidates the ontology with procedures for interacting with the STTNr in the OKB format (Bogdanov et al. 2023a).

Based on the above results, cognitive communicative scenarios of interaction with consolidated STTNr are implemented, whose sets of concepts are in the operational field of semantic platforms (SPs) (Fig. 5.10).

Thus, the consolidation of information resources based on the results of monitoring special-purpose structural materials is a knowledge system that dynamically combines facts from various STTNr with different thematic profiles and subject areas, and thereby forms a platform for the formation of a unified STTNr information environment. The nodes of the concepts of SP taxonomic variety by research topics are grouped into classes based on certain properties of the STTNr concepts. The names of the concepts that are updated by experts are semantically close to the contexts of the concepts of the corresponding STTNr ontologies. Therefore, the consolidation of STTNr sets, which is reflected by the variety of taxonomies, occurs on the basis of intercontextual relations established between the concepts of these taxonomies.

The efficiency of the method based on the selected initial data is demonstrated on the example of processing the results of research on materials of fragments of components of Russian missiles of various types, fragments of ontologies of which based on the results of processing the common characteristics of their properties are shown in Fig. 5.11. The formed ontologies are consolidated according to the properties of the materials used in the production of various components of missile bodies.

Based on the attribute data that define the contexts of the ontological clusters of missile types shown in Fig. 5.11, an ontology of the properties of materials used in their production is generated (Fig. 5.12). The specified attribute data are implemented into the contexts of ontology concepts, which are presented in Figs. 5.12, 5.13, 5.14, and 5.15 adapted from the relevant scientific papers and reports (Bogdanov et al. 2023b, c, d, 2024).

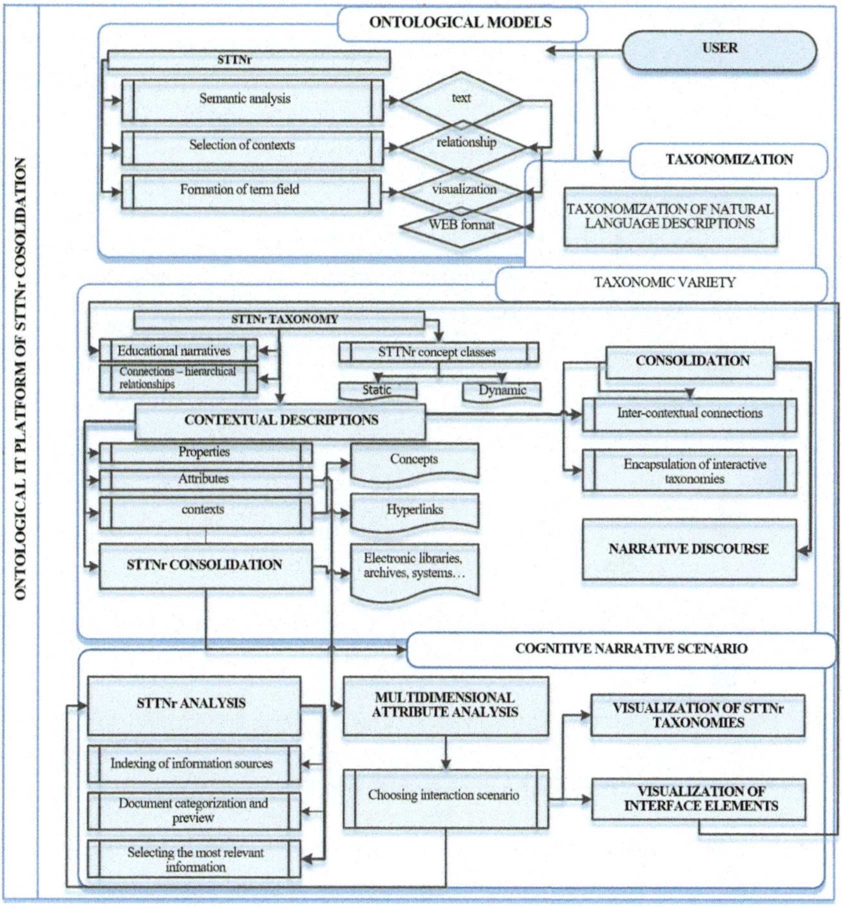

Fig. 5.10 General scheme of cognitive and communicative scenario of interaction with consolidated STTNr contexts

The indicated ontology of properties is generated due to the use of functional interpretations of transdisciplinary hyperproperties, rules for OID (ontological interactive document) forming, and ontology of the choice problem. One of the results of its generation is the implementation of information and analytical platforms that allow reflecting the presence of such chemical elements as Ti, Al, V, Mn, etc. on the basis of multi-dimensional analysis of the properties of materials used in the production of the specified samples of means of destruction. It should be emphasized that at the stage of semantic linguistic analysis of the results of research on the relevant STTNr, the cognitive services of information and analytical system to support development processes (IAS SDP) of means of destruction reveal scales for assessing the contents of specific chemical elements in the structures of means of destruction, the descriptions of which are analyzed. An example of this can be seen

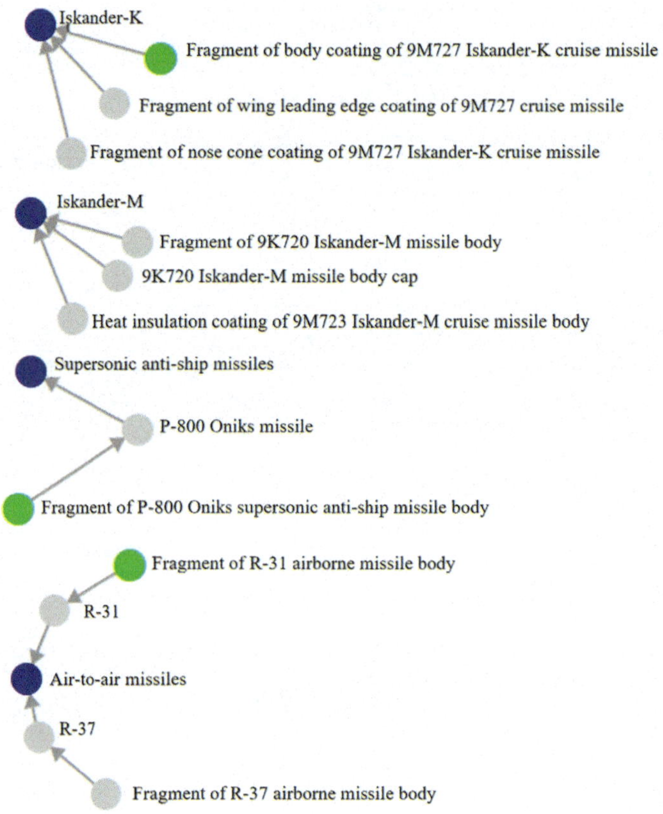

Fig. 5.11 Ontological clusters of missiles of various types

in Fig. 5.12, where the left column shows the corresponding elements and the spectra of their presence values in the structures of means of destruction.

This provides a representation of quantitative data on the contents of various chemical elements directly in each sample simultaneously. Figure 5.13 shows a bar chart of the chemical element contents in the structures of a number of missiles. Thus, specialists can evaluate the structures of means of destruction in terms of their chemical composition.

In addition, the ontology of material properties provides a reflection of the distribution of chemical elements in the special-purpose structures being analyzed.

The ontology of properties also implements a consolidated analytical assessment of the components of means of destruction and chemical elements used in their production (Fig. 5.15). Therefore, the experts have the ability to simultaneously represent the list of chemical elements both in a mean of destruction as a whole and in a specific component of its design during their analysis. This is implemented based on the method of transdisciplinary consolidation of STTNr, the procedures of which

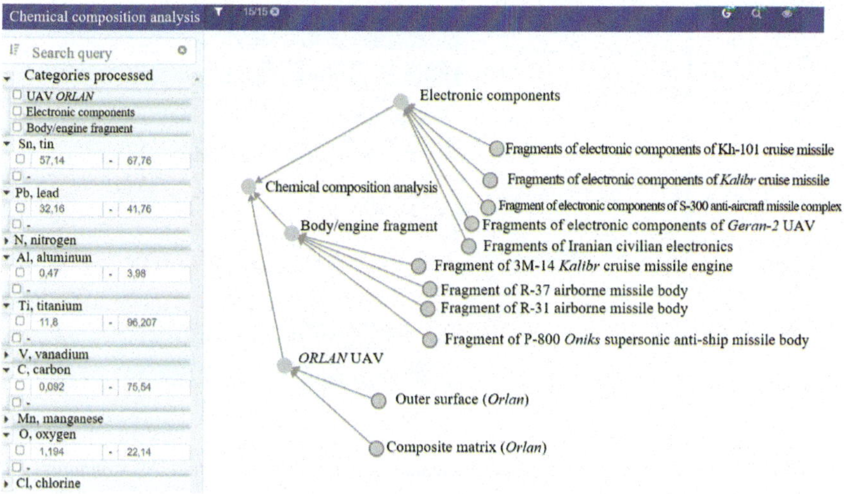

Fig. 5.12 Ontology of properties of materials of means of destruction

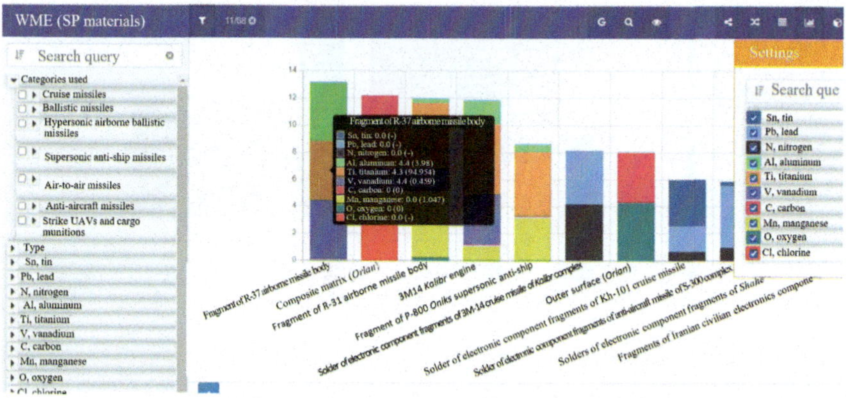

Fig. 5.13 Chemical element contents in structures of means of destruction

generate cognitive communicative scenarios of interaction of experts with a set of STTNr. Its feature is the invariance of contextual connectivity of concepts of open ontologies.

Therefore, regardless of which STTNr concept an expert begins studying the content, the ontology implements a semantic protocol for interaction with all needed STTNr contexts in the format of a cognitive communicative scenario.

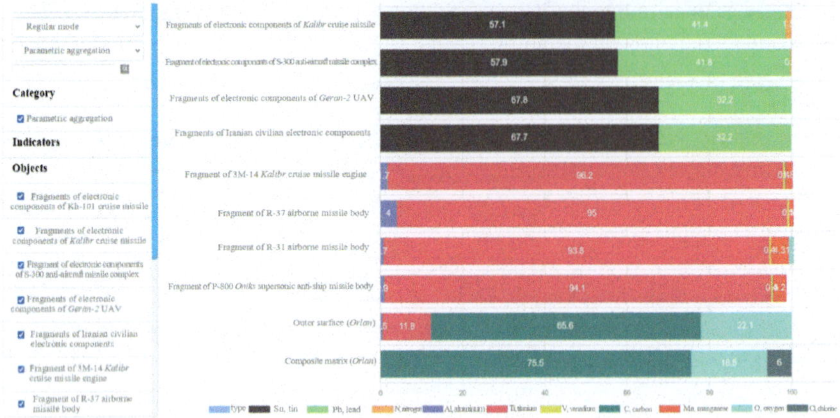

Fig. 5.14 Analytical indicators of chemical element contents in missile and UAV structures

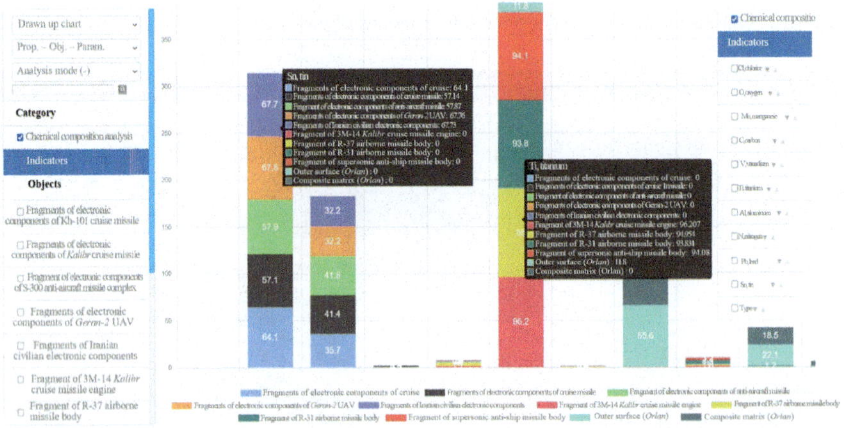

Fig. 5.15 Consolidated representation of components of means of destruction and chemical elements used in their production

5.3.3 *Scientific and Methodological Apparatus for Identifying Materials of Special-Purpose Structures Using Elements of Artificial Intelligence*

The development of weapons and military equipment (WME) is based on the use of modern scientific and technical achievements in all fields of knowledge. It is the basis of scientific and technical progress in the fields of both fundamental sciences (mechanics, solid-state physics, aerodynamics, etc.) and applied research (electronics, avionics, materials science, instrument making). The need to introduce new approaches and technologies was confirmed in Trends in science and technology

(Trends in science and technology 2020–2040 2024) it is stated there that the country's security depends on the ability to understand, adapt, and implement such technologies as artificial intelligence, autonomous control, and hypersonic systems. At the same time, the development of hypersonic systems requires the use of structural materials with increased performance, which is impossible without the use of advanced technologies for their development and production. Thus, the basis for the development of advanced technologies is the analysis of structural materials of such samples which are already used by countries around the world in producing hypersonic weapon systems.

At the same time, it is necessary to take into account the fact that the relevant information resources currently contain a large amount of information and data on technological solutions that are constantly being improved. Moreover, the vast majority of these resources contain information in natural language format. Their processing requires modern artificial intelligence tools that provide semantic, linguistic, and concept-graphic analysis of large volumes of information, namely, narratives, documents and tabular data, identification of logistical relations between them, evaluation criteria, and decision-making support in the format of the following technological chain: documents → analysis → linguistic corpus → formation of a neural network + generation of ontologies → formulation of evaluation criteria → generation of analytical decision-making platforms.

These artificial intelligence tools make it possible to implement effective research based on the results of monitoring various information resources from a wide range of open-source reports, internal assessments and research on the prospects, to develop a comprehensive understanding of the future technological environment.

Therefore, the development and improvement of the scientific and methodological apparatus for identifying materials for special-purpose structures is an urgent scientific task that requires the use of new methods and approaches (Trends in science and technology 2020–2040 2024; Nicolescu 2008; Borsato et al. 2016; Palagin 2014). Today, a promising approach is the implementation of artificial intelligence elements into scientific activities to enhance the efficiency of research and increase the reliability and validity of the results obtained.

When developing special-purpose equipment, an important stage is the analysis of the physical and mechanical properties of the materials of their structures, which primarily determine the operational characteristics of complicated special-purpose systems. Analysis is the process of imaginary or real dissection of an object and its further study; it is also a tool for studying the composition and properties of any material (Wang et al. 2013; Konar 2005).

The results of studies (Bogdanov et al. 2023b, c, d, 2024; Wang et al. 2013) and publications (Lee and Kirlik 2013; Stryzhak 2023; Stryzhak and Zavalevskyi 2023; Riedl and Bulitko 2013) indicate that considerable attention is paid to the analysis of materials for special-purpose structures, but the experience of practical implementation of the tools developed for analysis leads to the conclusion that theoretical developments in this field are insufficient. The results at each stage of the study are usually presented in a narrative format.

A characteristic feature of technologies that provide analysis of narrative descriptions of the abovementioned materials is their orientation toward multi-format and operation with heterogeneous data. At the same time, the main property here should be the ability to semantically and meaningfully operate on textual natural language information, which provides the proper level of intelligence of the technologies being created. After all, "*intelligence is a form of individualization of the system, which has linguistic status*"(Shirokov 2022). Therefore, relevant technologies and tools being developed belong to the class of intellectual information linguistics. Their implementation requires the involvement of corpus linguistics methods, theories of lexicographic systems, semantic states, and transdisciplinary consolidation already at the level of building conceptual models, and then at all other stages of design and in all elements of the information architecture of analytical research (Nadutenko et al. 2022).

The above research is a complex process that should be taken into account in the processes of scientific, technical and production, and technological activities which in modern conditions have a network-centric nature. Modern IT technologies, namely, artificial intelligence (AI), cognitive computing, etc. (Martin 2020) facilitate this. On their basis, transdisciplinary interaction of expert researchers is implemented in the format of an e-scenario with STTNr sets, and directly between STTNr, which characterizes the modern information space, both at the global and local levels.

With the help of linguistic semantic and conceptographic means of artificial intelligence, the creation and use of ontological knowledge bases is provided, based on which the process of information support of the scientific research is implemented. The use of these AI software and information tools is focused on solving many problems which include organizing the interaction of experts and their access to information sources for the formation of knowledge.

The concepts of ontology and ontological analysis have become part of the procedures and standards for modeling knowledge base systems for research purposes, since the description of such systems is the structuring of information sources for the formation of knowledge. The implementation of these technologies requires taking into account various formal and methodological requirements, criteria, and assessments. Here are the main ones:

- creation of information and functional models;
- the need to structure terms and concepts;
- rules for forming valid statements and conclusions that describe terms and concepts.

At the initial stage of building an ontological model, the following tasks should be performed:

- creating and documenting a glossary of terms;
- description of the rules according to which, based on the introduced terminology, reliable statements are formed that characterize the state of the system;
- building a model which allows forming the necessary additional statements based on existing statements.

The ontological system is characterized by unity, logical interrelations, and consistency of the concepts used.

Therefore, the ontological knowledge base has its own organizational structure that forms functional relationships between information objects of the knowledge base. Almost all information objects of modern knowledge bases are fully integrated with its functional structure, which determines the large structural dimensions of such systems. To solve this problem, it is proposed to form the structure of the knowledge base as a set of individual personalized *e*-scenarios, which, within the framework of a given ontological structure, are operationally formed in accordance with the selected object and the task set.

The process of forming an ontological model of an *e*-scenario of scientific studies consists in the possibility to consistently present them in the form of objects (Dovgiy et al. 2019).

The general formula for formalizing this ontological model has the following form:

$$S = \left\{ O_a \left\{ P_b \left\{ T_d \left\{ E_e \left\{ C_q \{ M_v \{ Z_g \} R_h \{ Z_g \} \} \right\} \right\} \right\} \right\} \tag{5.13}$$

where $O = \{O_a\}$, $a = 1, 2, 3, \ldots, m$ are the research objects; $P = \{O_a\{P_b\}\}$, $b = a1, a2, a3, \ldots, an$ are the research subjects; $T = \{P_b\{T_d\}\}$, $d = b1, b2, b3, \ldots, bn$ are the research topics; $E = \{T_d\{E_e\}\}$, $e = d1, d2, d3, \ldots, dn$ are the research stages; $C = \{E_e\{C_q\}\}$, $q = e1, e2, e3, \ldots, en$ is the research goal;pg $Z = \{C_q\{Z_g\}\}$, $g = q1, q2, q3, \ldots, qn$ are the research tools; $M = \{C_q\{M_v\{Z_g\}\}\}$, $v = g1, g2g3, \ldots, gn$ is the research route; and $R = \{C_q\{R_h\{Z_g\}\}\}$, $h = v1, v2, v3, \ldots, hn$ is the evaluation of results.

In tabular form, the ontological model of the *e*-scenario of scientific research is presented in Table 5.2.

Figure 5.16 shows an example of an ontological knowledge base formed according to the protocol of the *e*-scenario of linguistic semantic analysis of special-purpose structural materials.

An example of the tools used in the *e*-scenario of the R:$\{C_q\{R_g\}\}$ study (Table 5.2), taking into account the identification of objects $\left\{C_q\{R_g\} \middle| F\left(C_q\{R_g\}\right)\right\}$ of material analysis, is the method of scanning electron microscopy. This method was used to study the microstructural characteristics of the materials of the airframe of Shahed-136 loitering munition, the chemical composition and microstructural characteristics of the metals and alloys used to manufacture components of cruise and airborne missiles, and other fragments of munitions. One of the results of the research is the formation of an ontological knowledge base that is capable of providing the development of algorithms for detecting and implementing systems for countering loitering munitions, missiles of various types, and other means of destruction.

The *e*-scenario of forming an ontological knowledge base is a complicated process that includes the stages presented in Fig. 5.17.

Table 5.2 Ontological model of the e-scenario of scientific research

Object name	Class name	Attribute name	Conditions for attribute identification: $\{x \mid F(x)\}$*)
Research objects	O	$\{O_a\}$	$\{O_a \mid F(O_a)\}$
$\{O_a\}$	P	Research subjects	
Research subjects	P	$\{O_a\{P_b\}\}$	$\{O_a\{P_b\} \mid F(O_a\{P_b\})\}$
$\{O_a\{P_b\}\}$	T	Research topics	
Research topics	T	$\{P_b\{T_d\}\}$	$\{P_b\{T_d\} \mid F(P_b\{T_d\})\}$
$\{P_b\{T_d\}\}$	E	Research stages	
Research stages	E	$\{T_d\{E_e\}\}$	$\{T_d\{E_e\} \mid F(T_d\{E_e\})\}$
$\{T_d\{E_e\}\}$	C	Research goal	
Research purpose	C	$\{E_e\{C_q\}\}$	$\{E_e\{C_q\} \mid F(E_e\{C_q\})\}$
$\{E_e\{C_q\}\}$	R	Research tools	
Research tools	R	$\{C_q\{R_g\}\}$	$\{C_q\{R_g\} \mid F(C_q\{R_g\})\}$
$\{T_d\{E_e\}\}$	M	Research route	
Research route	M	$\{C_q\{M_v\{R_g\}\}\}$	$\{C_q\{M_v\{R_g\}\} \mid F(C_q\{M_v\{R_g\}\})\}$
$\{C_q\{M_v\{R_g\}\}\}$	O	Evaluation of results	$C_q\{O_h\{R_g\}\}$
Evaluation of results	O	$\{C_q\{O_h\{R_g\}\}\}$	$\{C_q\{O_h\{R_g\}\} \mid F(C_q\{O_h\{R_g\}\})\}$

*Conditions for attribute identification: $\{x \mid F(x)\}$—defines the set of all such x that $F(x)$ is true. *Example* $\{k \in K_a \mid k < 5\} = \{1, 2, 3, 4\}$.

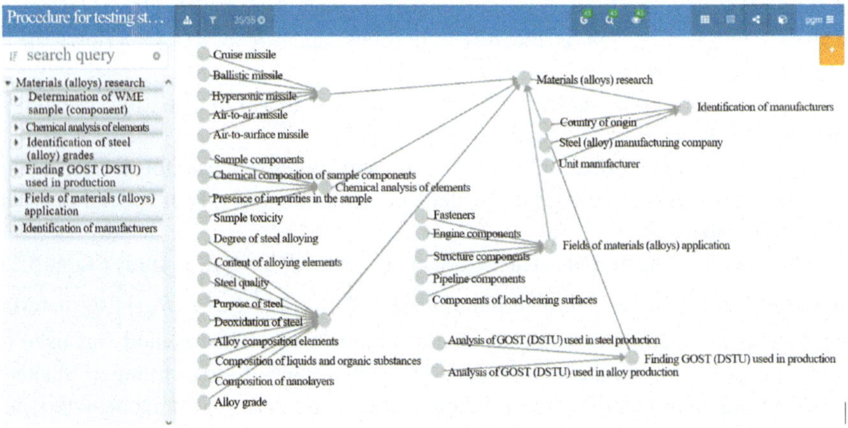

Fig. 5.16 Ontological representation of research on special-purpose structural materials (Martin 2020)

The *first stage* involves generating initial data for the analysis of special-purpose structural materials, as well as determining limitations and assumptions. The knowledge bases should include information from materials' narratives, the regulatory

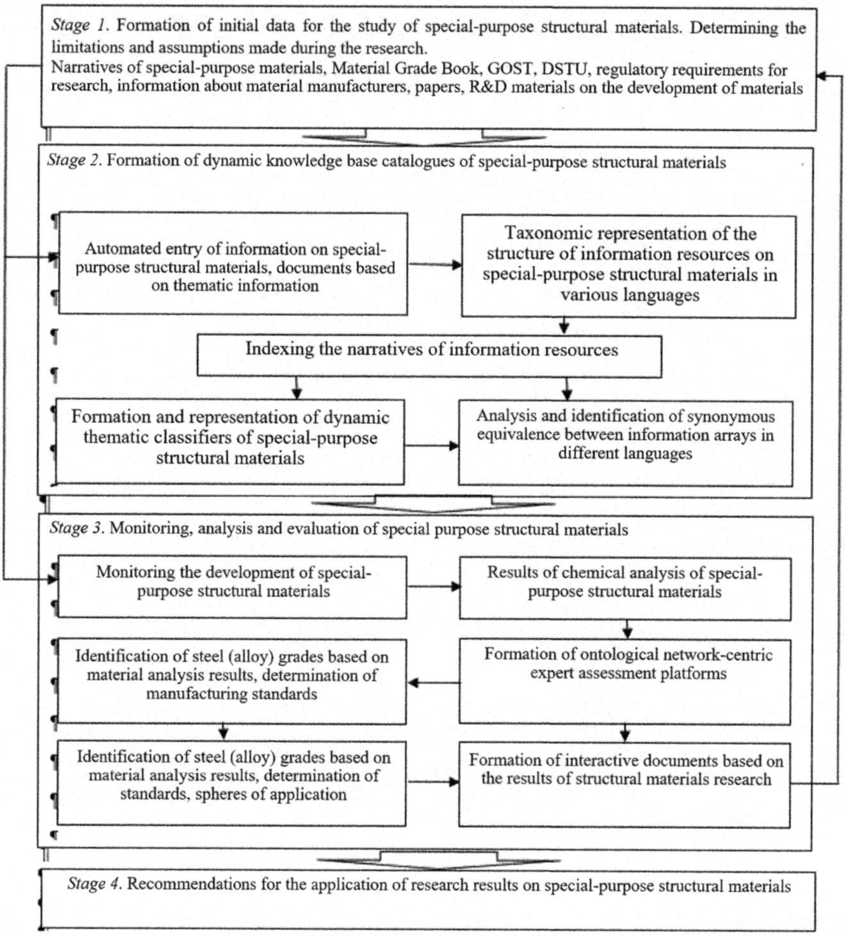

Fig. 5.17 Structure of scientific and methodological apparatus of research on special-purpose structural materials

framework for materials' production, standards for material production, papers on materials' development, etc. The ontological representation of the knowledge base on materials for special structures is shown in Fig. 5.18.

An important source of information about materials for special-purpose structures is the steel- and alloy-grade book which contains information about classification, purpose, substitutes, analogs, chemical composition, critical point temperatures, as well as other properties of alloys (Directory steel and alloys 2024). The ontology of the steel- and alloy-grade book is shown in Fig. 5.19.

At the *second stage*, dynamic catalogs of knowledge bases on materials for special-purpose structures are formed.

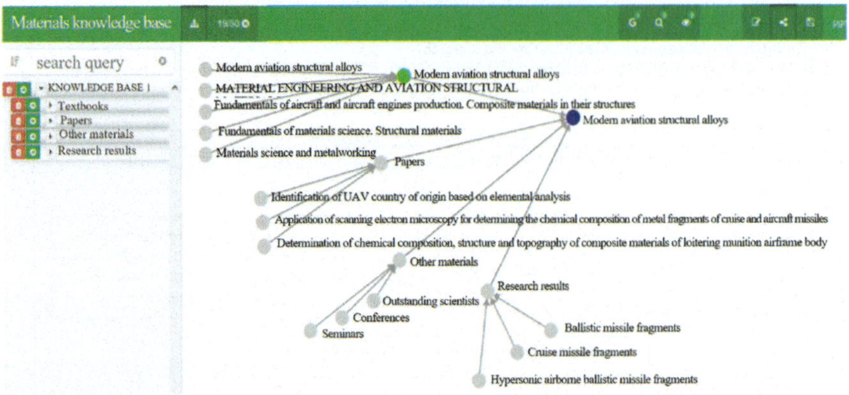

Fig. 5.18 Ontological representation of knowledge base on special-purpose structural materials

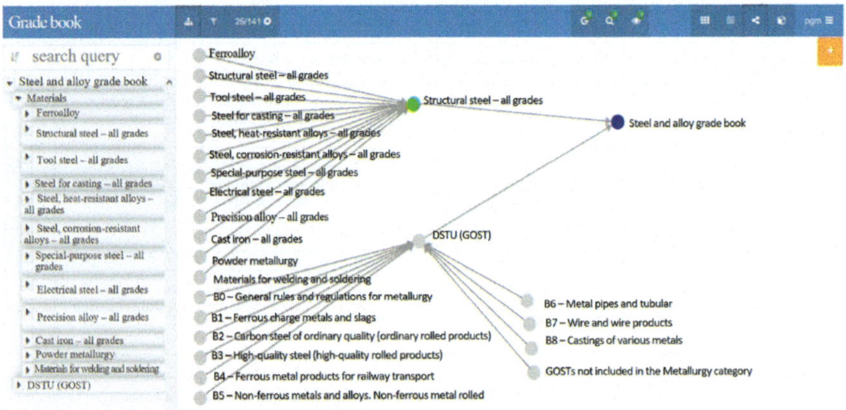

Fig. 5.19 Ontological representation of steel- and alloy-grade book

The *third stage* involves monitoring and processing research results. Based on the results of monitoring special-purpose materials, their chemical analysis is carried out to identify the microstructural characteristics of the airframe materials used in the manufacture of Shahed-136 loitering munition, the chemical composition, and microstructural characteristics of the metals and alloys used to manufacture components of cruise and airborne missiles and other fragments of WME.

For further analysis of materials, ontological network-centric platforms for expert evaluation are formed, which, based on the use of artificial intelligence tools, provide research into other characteristics and properties of materials. In their environment, special-purpose structures are identified, and task systematization is implemented (Fig. 5.20). Staying within the framework of the special knowledge of a material's expert, the study of materials that make up fragments of means of destruction, which have been exposed to significant and often alternating temperatures, pressures, and

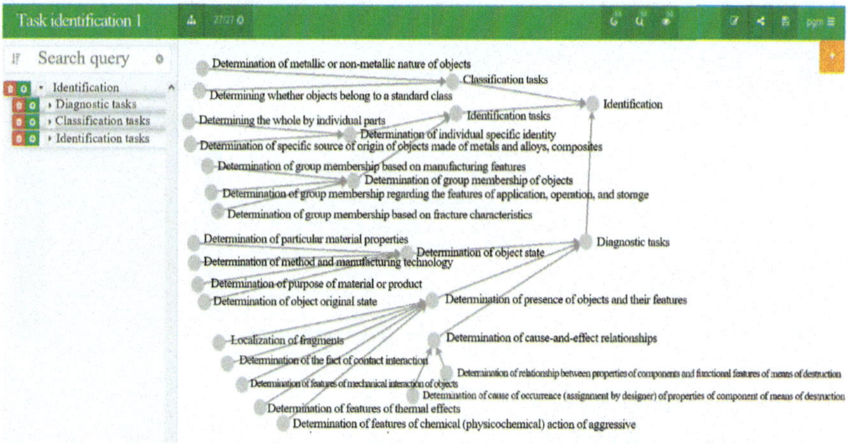

Fig. 5.20 Ontological representation of steel and alloy identification tasks (Directory steel and alloys 2024)

overloads during use, requires the use of qualitatively new methods and analysis schemes.

Further, based on the ontological knowledge base, cognitive communicative scenarios of interaction with consolidated scientific narratives are implemented, whose sets of concepts of the results of research on structural materials are in the operational field of expert assessment platforms. This creates technological conditions for the consolidation of scientific narratives that describe the analysis of special-purpose materials, based on which knowledge systems are formed, which dynamically combine facts from different narratives according to different thematic profiles. Consolidation is formed basing on intercontextual relationships established between concepts of narrative taxonomies. This consolidation is the basis for the formation of a platform of a unified information environment for researching materials.

Based on the results of the analysis, recommendations and proposals are developed for the use of research results and the formation of a knowledge base.

In accordance with the developed scientific and methodological apparatus, its use is demonstrated on the example of processing the results of research on materials of fragments of components of Russian missiles, namely, 3M-14 missile. The fragments of this missile and the results of processing common characteristics are shown in Fig. 5.21. The formed ontologies are consolidated according to the properties of the materials used in the production of various components of missile bodies.

Therefore, regardless of which concept of a scientific narrative the expert begins his research into its content with, the ontology implements a semantic protocol for interaction with all the contexts of narratives that he needs, in the format of a cognitive communicative *e*-scenario.

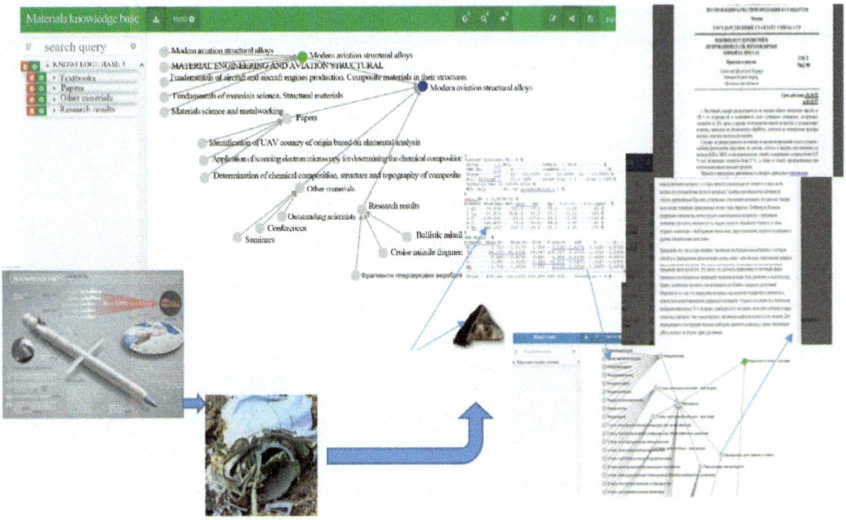

Fig. 5.21 Ontological representation of order of analysis of materials of special-purpose structures in format of active knowledge system

5.4 Methodology for Building Artificial Neural Networks for Identifying Special-Purpose Structures by Classification and Country of Origin Based on Electron Microscopic Analysis of Their Fragments

Among the known methods for automating the processes of classification and identification of special-purpose structures based on electron microscopic analysis, the use of neural networks should be considered as the most promising approach. Their current capabilities allow for the classification of images based on various characteristics (Slyusar 2022; Slyusar et al. 2021, 2022a, b; Slyusar and Sliusar 2021).

However, it should be noted that the abovementioned identification of special-purpose structures based on the analysis of their fragments using artificial neural networks is a complex task that requires an integrated approach. As a basic method for solving this problem, a multi-stage sequence of actions is proposed, which is typical for the application of neural networks (Fig. 5.22).

The first step of the proposed methodology is to collect and prepare a high-quality dataset. To do this, it is necessary to carry out electron microscopic analysis of various special-purpose structures from different countries and obtain high-quality images of fragments of structures. The collected images are further systematized and divided into classes by type of structure material and country of origin, which become tags for further learning.

At the data preparation stage, images are converted to needed size, e.g., 224 × 224 pixels, for employing a VGG16-type neural network (Qassim et al. 2018). To expand

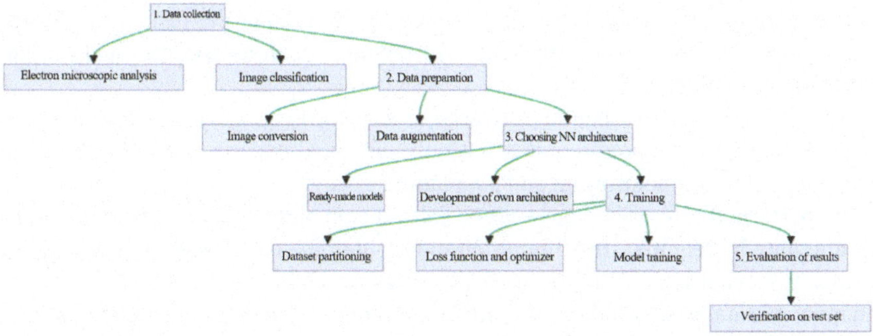

Fig. 5.22 Neural network training method for image classification

the dataset and reduce the possibility of overfitting, data augmentation should be performed, which includes rotations, shifts, zooms, reflections, partial defocusing of images, etc.

Next, the neural network architecture is selected, which depends on the specifics of the task and available resources. Ready-made models such as VGG16 (Qassim et al. 2018), EfficientNet (Tan and Le 2019), or MobileNet (Sandler et al. 2018; Howard et al. 2019) can be used, or you can develop your own architecture. It is important to emphasize that the best results are obtained when using convenient neural networks pre-trained on the ImageNet dataset (Slyusar 2022; Deng et al. 2009).

During the training phase, the dataset is divided into training, validation, and test arrays. The loss function, optimizer and its tuning parameters, training step, and batch size are determined. Then, the model is trained on the training dataset, using the validation set to tune the parameters.

Upon completion of training, it is necessary to check the effectiveness of the neural network on the test set, using the so-called confusion matrix, accuracy indicators, completeness, and other metrics to assess the quality of identification. For effective implementation, the neural network model is integrated with the interface of the analysis system or into the application, and is regularly updated by adding new data or making changes to the architecture to improve the accuracy of operation.

The described methodology is only a general approach. In real conditions, it is necessary to pay attention to many additional details, such as image quality, absence of artifacts on the surface of structures, uniform distribution of lighting, and absence of shading.

With the development of technology and the emergence of new special-purpose materials and structures, the importance of adapting and upgrading artificial neural network models will grow. To maintain relevance, the model should be flexible to introducing new data and parameters. It is also recommended to systematically analyze the identification results, examining possible anomalies or inaccuracies in the conclusions.

The effectiveness of the identification system can be increased by integrating it with other analytical tools. For example, a deeper understanding of the properties of a material and its origin can be provided by combining the results of electron microscopic analysis with other research methods, such as spectral analysis and ultrasonic scanning.

Building an effective neural network-based identification system for analyzing special-purpose structures requires not only technical expertise, but also strategic planning, constant monitoring, and adaptation.

To perform the described tasks, a specialized convolutional neural network architecture may be required. It is important to note that the effectiveness of neural networks in solving image classification problems depends largely on the quality and quantity of data on which they are trained. Moreover, each specific case may require a specialized neural network architecture. Nevertheless, the most suitable version of neural networks for solving this problem is convolutional neural networks (CNN), Fig. 5.23. In general, it is advisable to use neural networks pre-trained on the ImageNet dataset for this (Slyusar 2022). They can process two-dimensional images of surface layers of metals, responding to porosity, structural patterns, grain size, and deformations, involving both visual images in the visible range of the optical spectrum and in the infrared, or even using the results of ultrasonic scanning, etc.

Since the image sizes may exceed the scene sizes for which the pre-trained neural networks are configured, it is necessary to perform convolutional reduction of the initial images on several layers of the input segment of the neural network into which the pre-trained neural network model is integrated. The result of CNN work in the context of classification tasks will be the division of images into those containing one or another class of metal (Fig. 5.23).

An alternative option for using CNNs for image classification is to integrate them with recurrent neural networks such as LSTM or GRU. In particular, Fig. 5.24 shows the architectures of neural networks containing VGG16 and LSTM (Slyusar 2022) or GRU. However, it should be noted that such architectures do not have high accuracy: for example, it is 88.5–90 % on the CIFAR10 dataset.

According to the results of Slyusar (2022), the highest classification accuracy can be achieved by using Xception (Deng et al. 2009) or InceptionV3 (Chollet 2017) neural networks pre-trained on the ImageNet dataset with preliminary scaling of the input images toward their enlargement (Figs. 5.25 and 5.26). However, for the

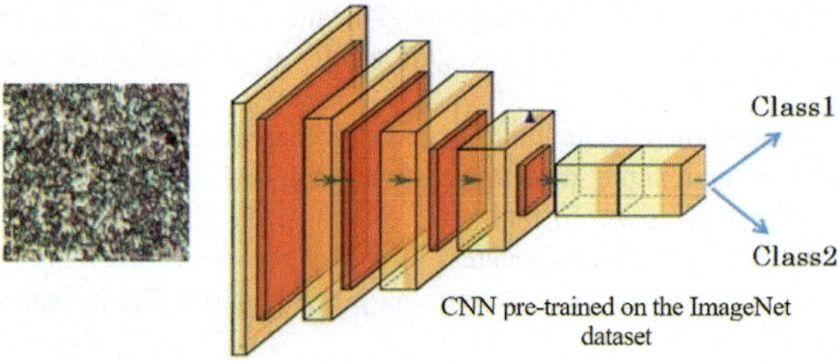

Fig. 5.23 Typical CNN structure for image classification

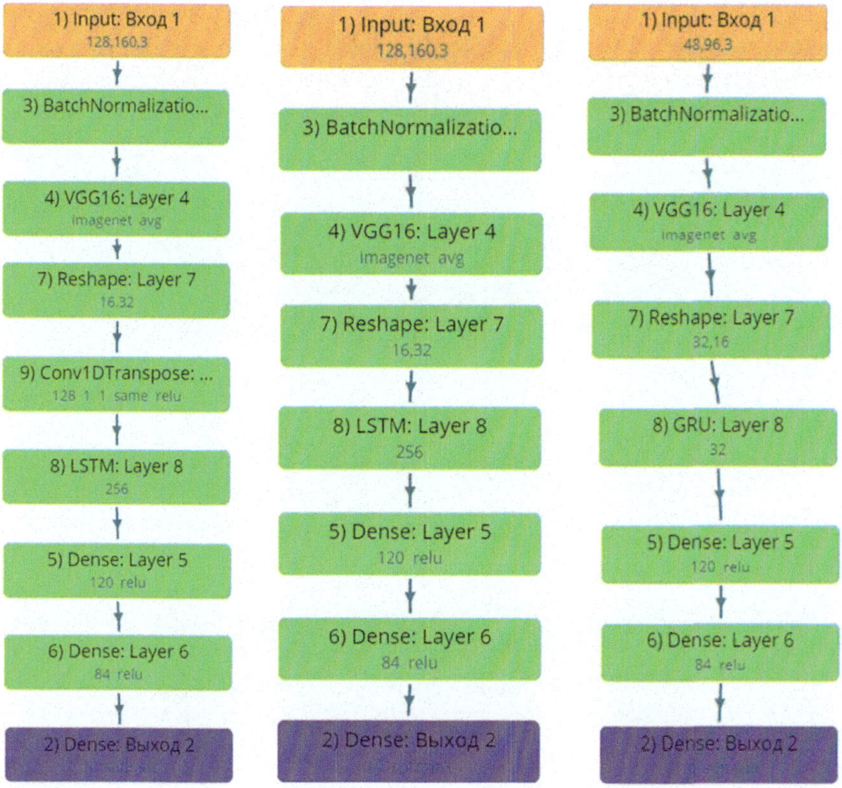

Fig. 5.24 Options for combining VGG16 with LSTM (Slyusar 2022) and GRU for image classification

Fig. 5.25 Efficient convolutional network architecture for classification (Slyusar 2022)

effective application of this approach in the context of metal type identification, it is necessary to scale the Xception (Chollet 2017) or InceptionV3 (Szegedy et al. 2016) architectures to process high-dimensional images, rather than the limited scene sizes of 299×299×3 and 229×229×3 pixels provided in them, respectively.

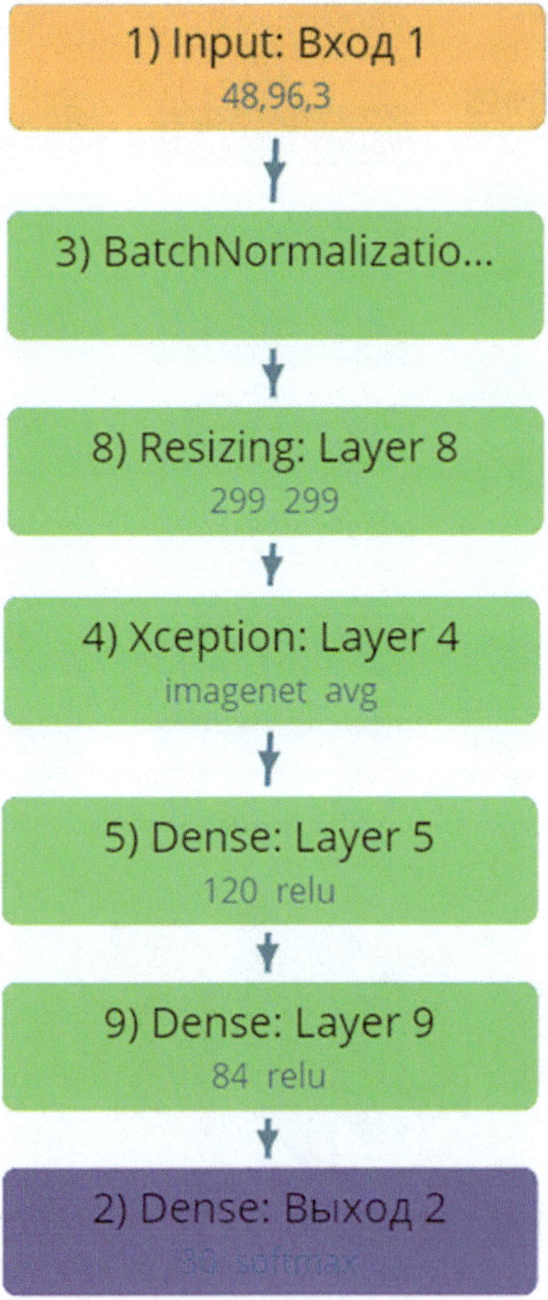

Fig. 5.26 Connection diagram of pre-trained Xception neural network

Besides, such scaled-up versions of neural networks should be pre-trained again on a dataset like ImageNet, but with larger photos. To do this, it is not necessary to accumulate millions of images with dimensions of 4288×2848 pixels or more, but it is possible to scale up the photos already available in ImageNet with dimensions of 256×256 and 500×500 pixels using neural networks for image enlargement and quality improvement.

To experimentally verify the performance of the proposed approach, the binary classification task was considered at the first stage. The authors prepared a dataset with images of two classes of metals with a total of 438 images, of which 395 were selected for training the neural network, and 43 images for testing.

246 photos were obtained for one class, of which 24 files were provided for tests, and 192 photos were assigned for the second class, including 19 photos for testing the effectiveness of learning. Thus, there was a slight imbalance in the filling of datasets of different classes, but overall the level of imbalance was within the acceptable range.

Before being fed to the neural network, the original ultra-high-resolution photos were cut into 12 pieces using Image Splitter 1.1. software (www.vovsoft.com). Further, in order to extend the dataset, these fragments were mirrored, rotated 90°C left and right, etc. Then all photos were reduced to 320×240 pixels. An example of images that were subject to classification is shown in Fig. 5.27. Figure 5.28 shows the results of applying image augmentation procedures built into the Python programming language. As is seen, the augmentation actually consisted of partially rotating the images and framing them with pixels stretched due to the apparent shift. However, due to such distortions of part of the image, the augmentation results were not included in the final dataset.

A neural network with architecture presented in Table 5.3 was used for training. It has a size of 735.19 MB and 192724946 parameters, of which only 2048 are not trainable.

The architecture of the presented neural network is sequential and consists of 19 layers, each of which performs a unique role in the learning process. The initial

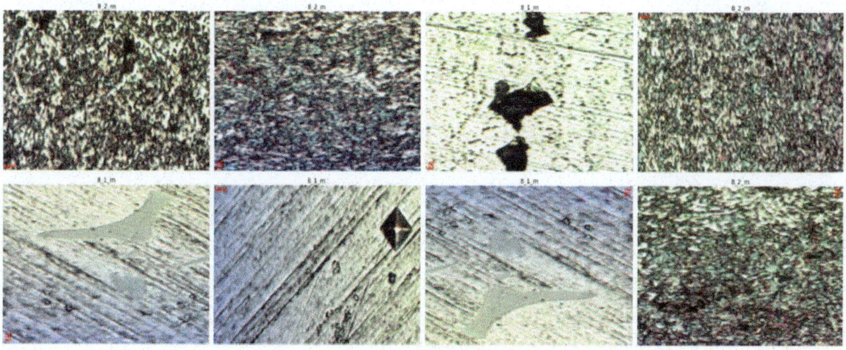

Fig. 5.27 Examples of a dataset with two classes of metals (classes "8 1 m "and "8 2 m")

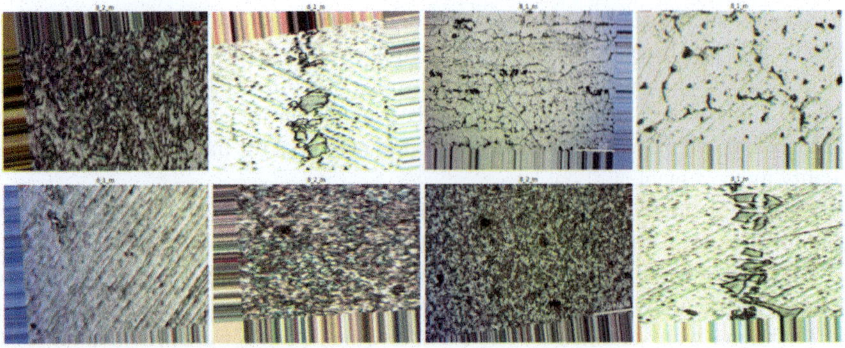

Fig. 5.28 Image augmentation results

Table 5.3 Architecture of neural network for binary classification of metals

#	Layer (Type)	Output format	Number of parameters
1	conv2d (Conv2D)	(None, 240, 320, 256)	7168
2	batch_normalization (BatchNormalization)	(None, 240, 320, 256)	1024
3	conv2d_1 (Conv2D)	(None, 240, 320, 256)	590080
4	max_pooling2d (MaxPooling2D)	(None, 80, 106, 256)	0
5	conv2d_2 (Conv2D)	(None, 80, 106, 256)	590080
6	batch_normalization_1 (BatchNormalization)	(None, 80, 106, 256)	1024
7	dropout (Dropout)	(None, 80, 106, 256)	0
8	conv2d_3 (Conv2D)	(None, 80, 106, 256)	590080
9	max_pooling2d_1 (MaxPooling2D)	(None, 26, 35, 256)	0
10	dropout_1 (Dropout)	(None, 26, 35, 256)	0
11	conv2d_4 (Conv2D)	(None, 26, 35, 512)	1180160
12	batch_normalization_2 (BatchNormalization)	(None, 26, 35, 512)	2048
13	conv2d_5 (Conv2D)	(None, 26, 35, 1024)	4719616
14	max_pooling2d_2 (MaxPooling2D)	(None, 8, 11, 1024)	0
15	dropout_2 (Dropout)	(None, 8, 11, 1024)	0
16	flatten (Flatten)	(None, 90112)	0
17	dense (Dense)	(None, 2048)	184551424
18	dense_1 (Dense)	(None, 240)	491760
19	dense_2 (Dense)	(None, 2)	482

layer, conv2d (Conv2D), is convolutional and is designed to process input images with dimensions of 240×320 pixels, using 256 filters. Each filter runs over the input image with a 3×3 convolution kernel without changing the image dimensions to detect certain features. The output of the convolution is the ReLU activation function. The number of parameters in this layer is 7168, which is the key to developing the ability to detect complex features of input data.

The second layer, batch_normalization (BatchNormalization), applies normalization process to image batches, which helps improve stability and learning speed by adjusting the distribution of input data. Normalization reduces the problem of internal covariance bias by allowing each layer to learn more independently.

The third layer, conv2d_1 (Conv2D), is also convolutional and has a similar structure to the first convolutional layer, providing an additional level of abstraction for image feature detection. The number of parameters for this layer, 590080, emphasizes its complexity and ability to detalization.

The fourth layer, max_pooling2d (MaxPooling2D), applies a maximum convolution operation, which helps reduce the size of the input data to reduce computational complexity and overhead of subsequent layers, while preserving important features of the processed scenes.

Subsequent layers continue to alternate between convolutional layers, batch normalization, and max pooling convolution, adding dropout layers to prevent overfitting by randomly ignoring some neural connections during training. The final layers of the network consist of fully connected layers (Dense) which provide classification based on detected features and abstractions.

Each layer in this architecture contributes to the progressive processing and abstraction of input data, allowing the network to effectively learn to recognize complex patterns and properties of input images for accurate classification.

The structure of the neural network under consideration was described in the Python programming language and integrated into the Google Colab Pro+ environment, within which an A100 GPU accelerator with 40 GB of graphics processor RAM, 166.8 GB of disk space, and 83.5 GB of system RAM was rented for training the neural network.

Figure 5.29 shows dependences of normalized training accuracy (vertical axis) on the number of epochs.

Figure 5.30 shows the corresponding training errors. As is seen in Fig. 5.29, the maximum training accuracy achieved on the validation sample is 100%. The same accuracy was achieved on the test sample with images that the neural networks were provided with for analysis for the first time.

Another characteristic of the quality of learning is the normalized confusion matrix. It is schematically shown in Fig. 5.31, and in the given case illustrates the absence of making incorrect decisions (zeros on the side diagonal of the matrix).

At the next stage of research, the classification task was complicated by the requirement to identify three different classes of metals, conventionally named "8_1_m", "8_2_m", and "8_5_m". The corresponding dataset was increased to 637 images, of which 575 were selected for training the neural network and 62 for testing. The first class segment "8_1_m"consisted of 246 photos, and 24 corresponding images were assigned for the test. Class "8_2_m"had 192 photos with additional 10% (??) images for the test sample. The number of images in the "8_5_m"class reached 199 photos, plus 19 images in the test sample. The photo size was limited to 320×240 pixels, and the microscope resolution was set at 50 μm. An example of a sample from the original array of images of the described variant of the dataset and the results of their augmentation is shown in Figs. 5.32 and 5.33.

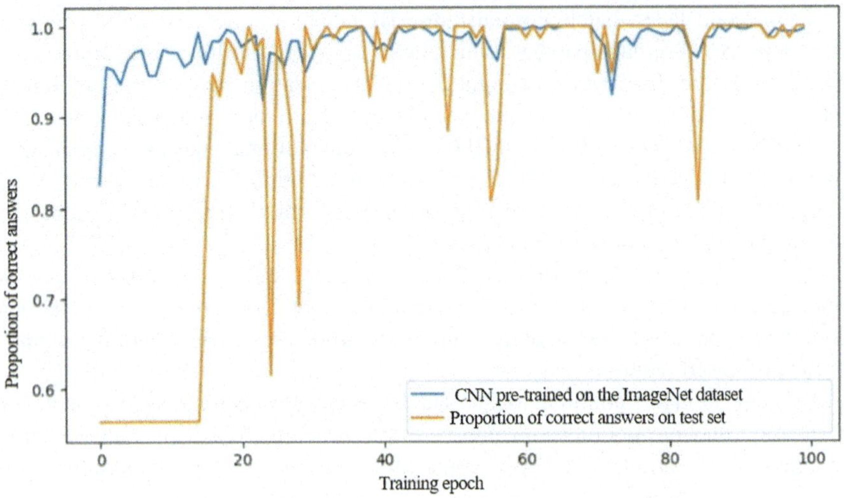

Fig. 5.29 Training accuracy within 100 epochs

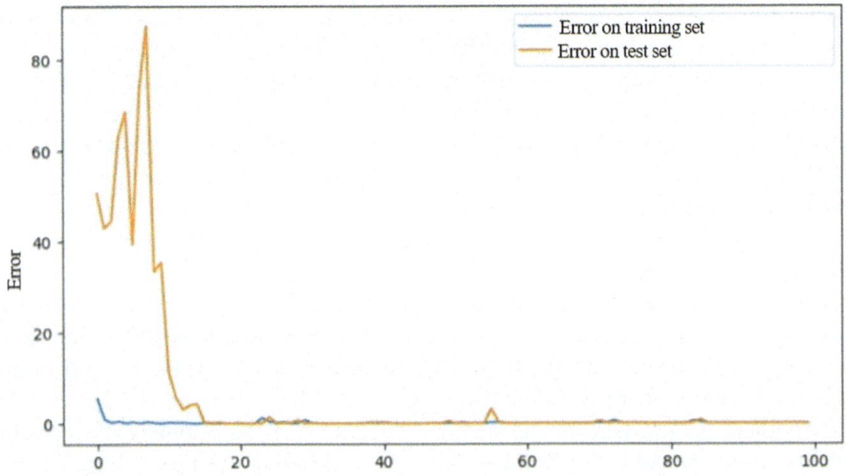

Fig. 5.30 Learning errors

The structure of the neural network was left the same as for the above-considered binary classification of two classes of metals: 19 layers combined into a sequential architecture. The only difference was in the dimension of the convolutions and the output layer, Dense, which had three outputs.

It should be noted that increasing the number of classes was expected to lead to a deterioration in learning accuracy. However, due to the limited size of the dataset, the maximum average class classification accuracy of 100% was still achieved

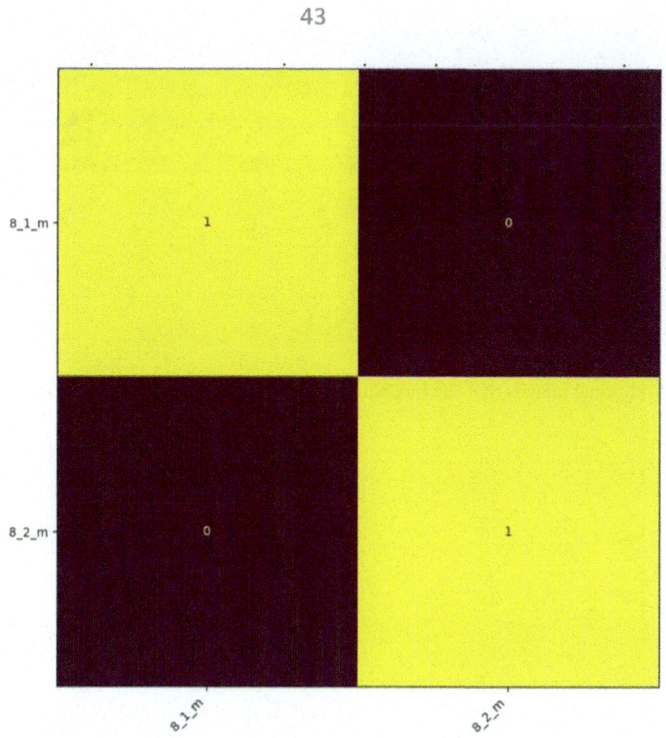

Fig. 5.31 Normalized confusion matrix

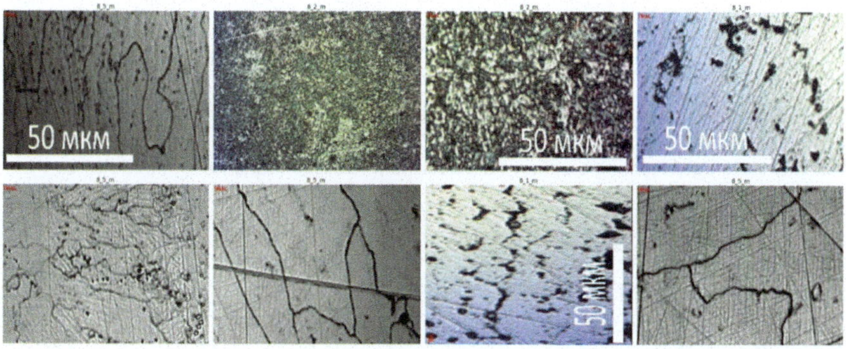

Fig. 5.32 Examples of dataset with three metal classes (classes "8_1_m", "8_2_m", and "8_5_m")

(Fig. 5.34), although in some epochs the learning error had outliers up to a relative level of 3.5 (Fig. 5.35).

At the same time, the test sample allowed determining the real classification capability. The normalized confusion matrix (Fig. 5.36), which was constructed based

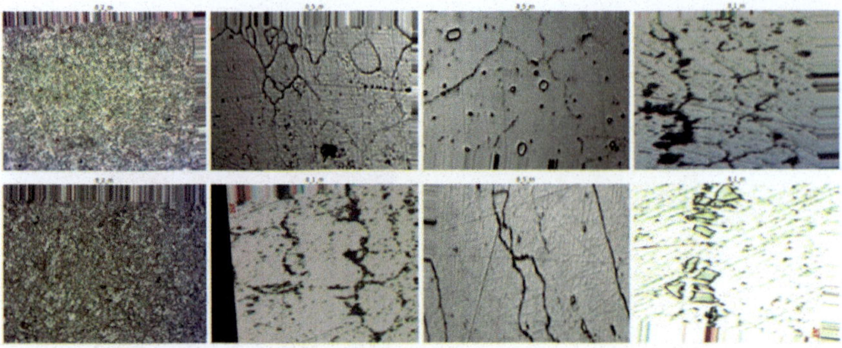

Fig. 5.33 Additional images that can be obtained from three-class dataset using image augmentation

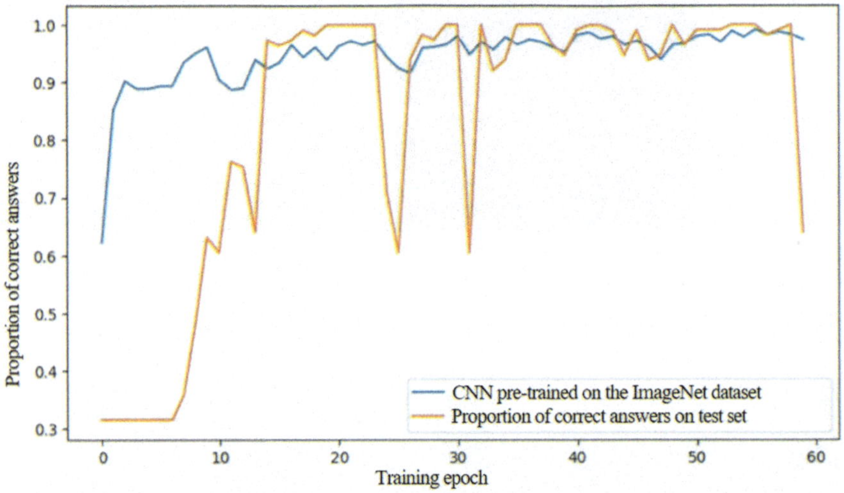

Fig. 5.34 Averaged learning accuracy on training and validation samples for three classes

on the results of using a neural network trained to process images of a three-class test sample, confirmed the presence of incorrect solutions.

In particular, the neural network incorrectly perceived 8_1_m class as metal of 8_2_m class in 75% of cases, and, conversely, attributed 63% of images of 8_5_m class to 8_1_m class. At the same time, the average accuracy of metal recognition across all classes was 54% due to the absence of errors in the identification of 8_2_m class.

Beside the need to increase the number of images in the dataset, it is also advisable to operate with larger images in order to increase the efficiency of multi-class metal identification. In this case, the deficiency of the available array of images is actually compensated by the larger area of the analyzed metal surface. To confirm

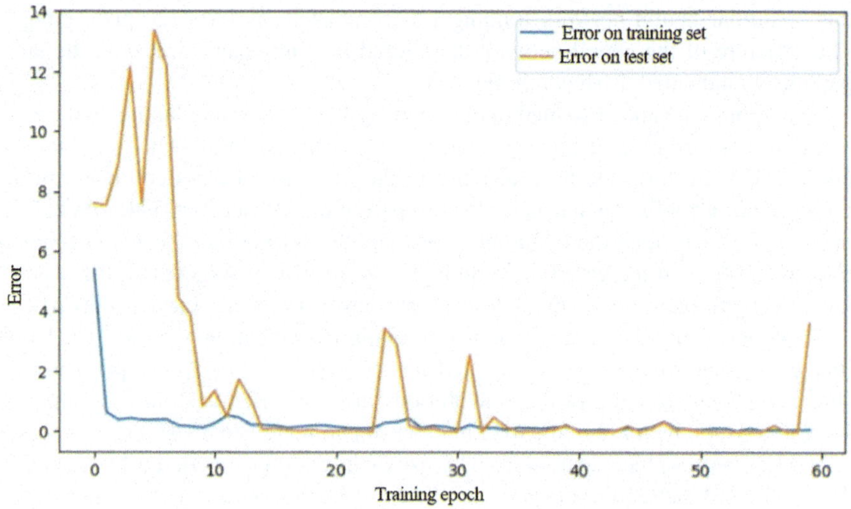

Fig. 5.35 Averaged over three classes' learning error on training and validation samples

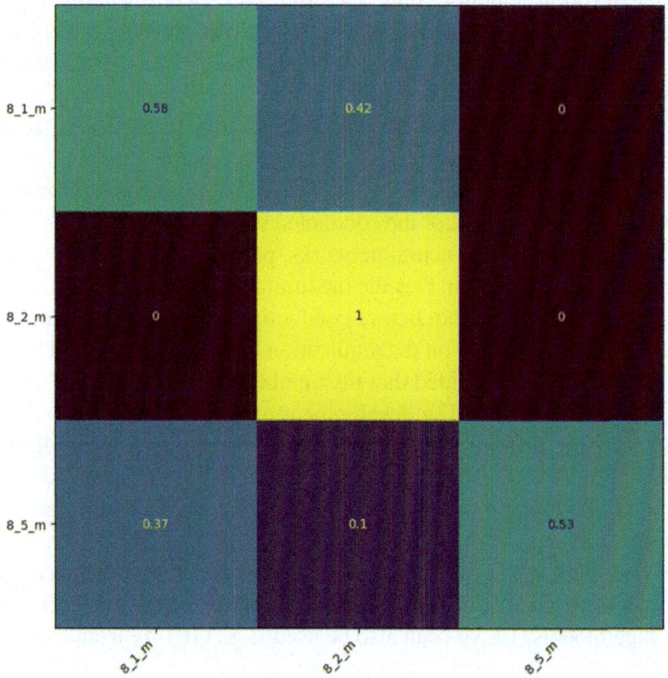

Fig. 5.36 Normalized confusion matrix for three-class test sample

this, a similar neural network training was conducted on 480×320 pixel images. The structure of the neural network configured for this image size, with the layer parameters indicated, is shown in Fig. 5.37.

The training results presented in Figs. 5.38 and 5.39 show a reduction in the level of outliers in errors, and higher accuracy of training in most of the unfavorable epochs. This, in turn, led to a decrease in the intensity of class confusion in the corresponding confusion matrix for the test part of the dataset (Fig. 5.40). As before, metal 8_2_m was identified without errors, and the correct identification of metals 8_1_m and 8_5_m increased to 58 and 53%, respectively. As a result, the average metal recognition accuracy in the test sample across all classes increased to 70%.

To confirm the obtained effect of improving classification accuracy for larger images, similar studies were conducted for a dataset with 640×480 pixel images (Fig. 5.41). The structure of the convolutional network used is shown in Table 5.4. Its feature is the increase in the total number of parameters to 836074755, of which 836072707 were subject to training. The size of the neural network reached 3.11 GB. This increased the requirements for hardware resources needed to perform calculations during training and inference. As a result, we managed to complete the stages of training and validation of its results at the same rented Google Colab Pro+ service based on the A100 accelerator (Figs. 5.42 and 5.43), while when constructing the confusion matrix, the corresponding calculations failed due to a lack of RAM.

At the same time, the behavior of the dependences of accuracy and training errors presented in Figs. 5.42 and 5.43 clearly showed the further increase in the quality of tuning the hyperparameters of the neural network. In particular, there was no decrease in the average class classification accuracy below the level of 70% in the validation sample starting from the 15th epoch and until the end of the training.

Thus, within the framework of the conducted study, visual analysis of microstructures, based on convolutional neural networks, proved to be effective in solving the problem of metal classification. Despite the limitations of this method, it can be used as a quick initial assessment tool before conducting a more comprehensive analysis. Further research should focus on the application of pre-trained neural networks of a more complex structure, provided that the number of metal classes is expanded and the dataset size is increased. The development of the proposed approach should be based on combining different neural network architectures, which will allow combining the strengths of the architectures involved and reduce the influence of the weaknesses of individual models, and lead to more accurate and reliable identification of metal structures.

It should be noted that in addition to the considered options for using convolutional neural networks to identify metals based on images of their surfaces, multimodal large language models (LLMs) can also be used, e.g., GPT-4v from OpenAI (GPT-4 2023).

A significant advantage of GPT-4v is the ability to upload a series of images to the corresponding chat for analysis. Among other well-known LLMs, Claude 3 Opus also has this capability, allowing simultaneous submitting of 20 images up to 5 MB each. However, the tests have shown that Claude 3 Opus is inferior to GPT-4v in terms of image analysis quality. In other well-known LLMs, in particular, Gemini

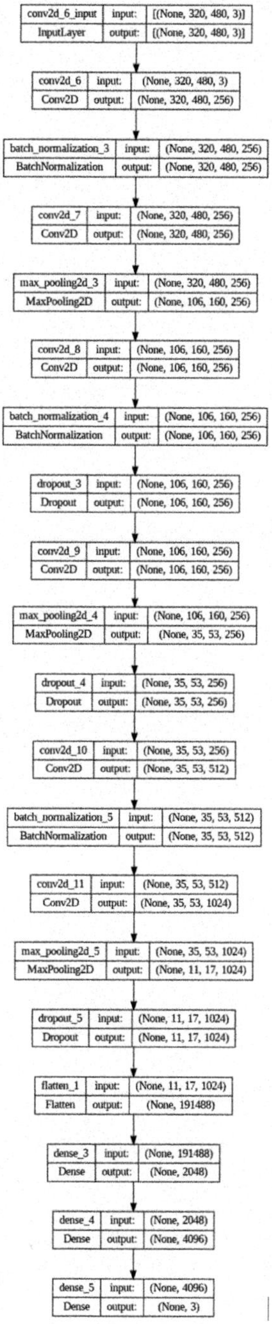

Fig. 5.37 Structure of neural network for classifying surface images of three classes of metals in 480×320 pixel format

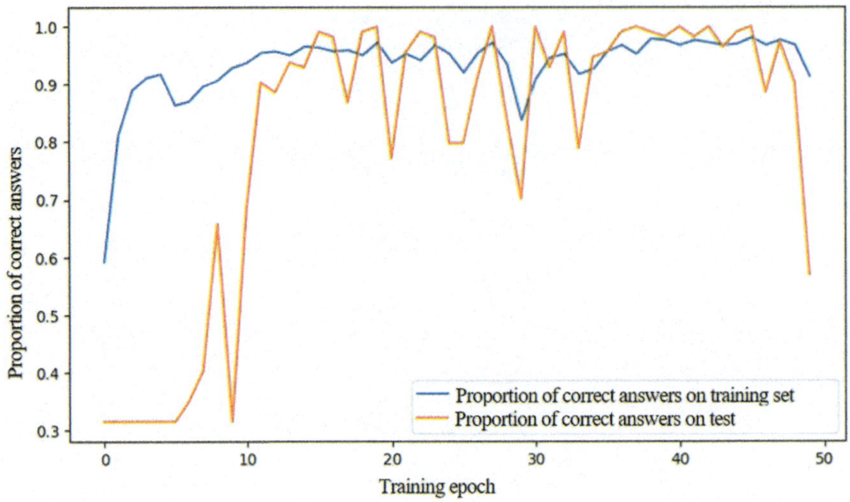

Fig. 5.38 Reduction of levels of training accuracy degradation between the 15th and 40th epochs with 480×320 pixel image format

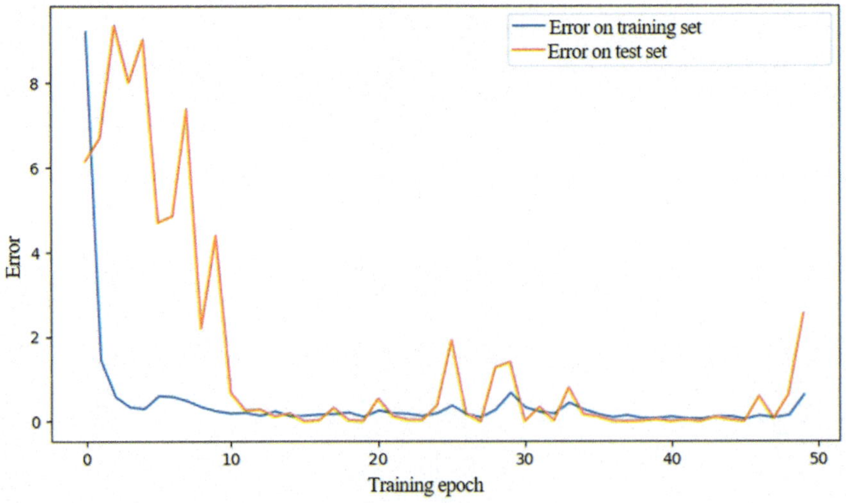

Fig. 5.39 Reduction of levels of training error outliers for 480×320 pixel image size

Ultra 1.0 from Google, LLava (Liu et al. 2023), and others, there is a limitation to loading only one image for description at a time, which makes it impossible to train such LLMs on a series of images.

In order to experimentally confirm GPT-4v ability to classify images according to the user's dataset, it was decided to return to the binary classification task. In this case, the GPT-4v configuration was carried out using the following system request:

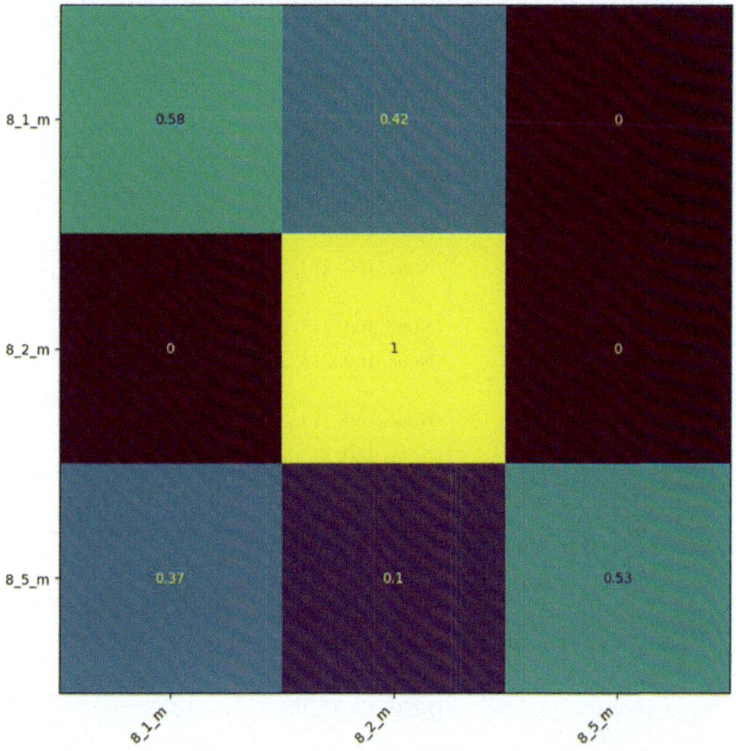

Fig. 5.40 Normalized confusion matrix for classifying test 480×320 pixel images

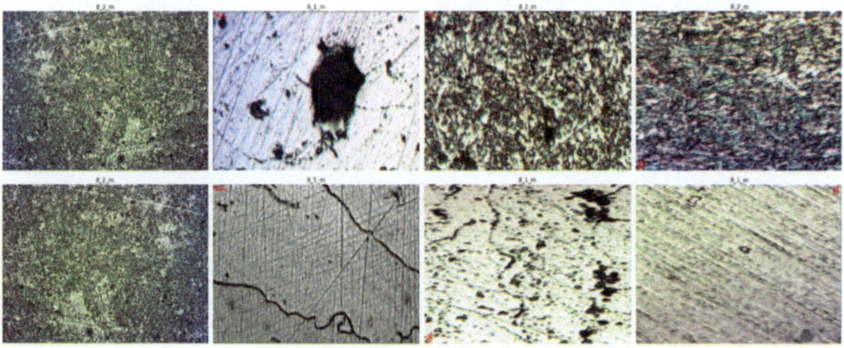

Fig. 5.41 Fragment of dataset with 640×480 pixel images

"Next, several images of metal 1 will be provided, and then an image of metal 2. It is necessary to study all the differences between these classes of metals. After that, several test images will be provided to evaluate the ability to classify metals based on images".

Table 5.4 Structure of neural network for identifying three classes of metals by images of their surfaces in 640×480 pixel format

#	Layer (Type)	Output format	Number of parameters
1	conv2d (Conv2D)	(None, 480, 640, 256)	7168
2	batch_normalization (BatchNormalization)	(None, 480, 640, 256)	1024
3	conv2d_1 (Conv2D)	(None, 480, 640, 256)	590080
4	max_pooling2d (MaxPooling2D)	(None, 160, 213, 256)	0
5	conv2d_2 (Conv2D)	(None, 160, 213, 256)	590080
6	batch_normalization_1 (BatchNormalization)	(None, 160, 213, 256)	1024
7	dropout (Dropout)	(None, 160, 213, 256)	0
8	conv2d_3 (Conv2D)	(None, 160, 213, 256)	590080
9	max_pooling2d_1 (MaxPooling2D)	(None, 53, 71, 256)	0
10	dropout_1 (Dropout)	(None, 53, 71, 256)	0
11	conv2d_4 (Conv2D)	(None, 53, 71, 512)	1180160
12	batch_normalization_2 (BatchNormalization)	(None, 53, 71, 512)	2048
13	conv2d_5 (Conv2D)	(None, 53, 71, 1024)	4719616
14	max_pooling2d_2 (MaxPooling2D)	(None, 17, 23, 1024)	0
15	dropout_2 (Dropout)	(None, 17, 23, 1024)	0
16	flatten (Flatten)	(None, 400384)	0
17	dense (Dense)	(None, 2048)	819988480
18	dense_1 (Dense)	(None, 4096)	8392704
19	dense_2 (Dense)	(None, 3)	12291

The training set of metal 1 images, which were simultaneously loaded into GPT-4v, is shown in Fig. 5.44. They illustrate various microscopic views of metal surfaces with visible defects and roughness. Next, images for the second metal (class metal 2) were also loaded into GPT-4v (Fig. 5.45). These images show different microscopic views of metal with less pronounced texture and a more uniform structure than in the previous images of metal 1 (Fig. 5.44).

After completing the above settings, the culmination of the experiment was to ask GPT-4v to classify unknown images of metals using its acquired knowledge. GPT-4v was informed that it had been introduced to two classes of metal samples and that it would then be fed individual images that it had not previously analyzed in order to get an answer as to which class they belonged to.

The first image to be loaded for analysis was the metal 2 surface image. The image, settings, and response of GPT-4v are presented in Fig. 5.46. It should be noted that

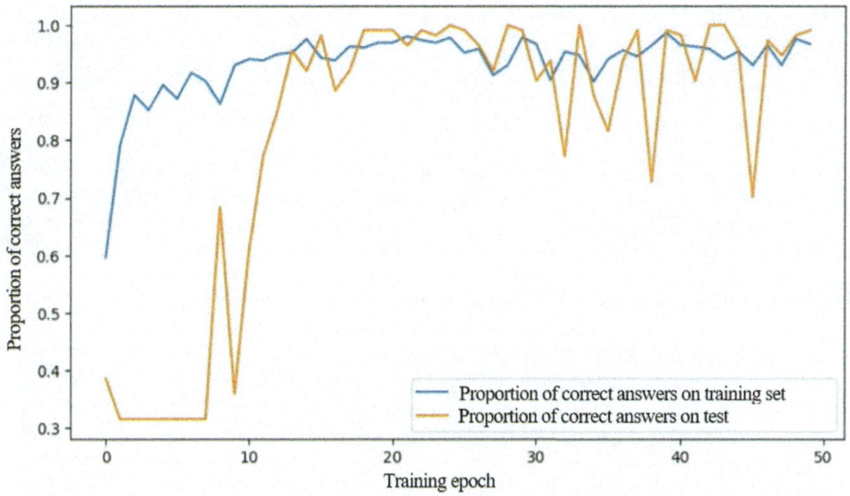

Fig. 5.42 Evolution of classification accuracy averaged on three classes for training and validation datasets

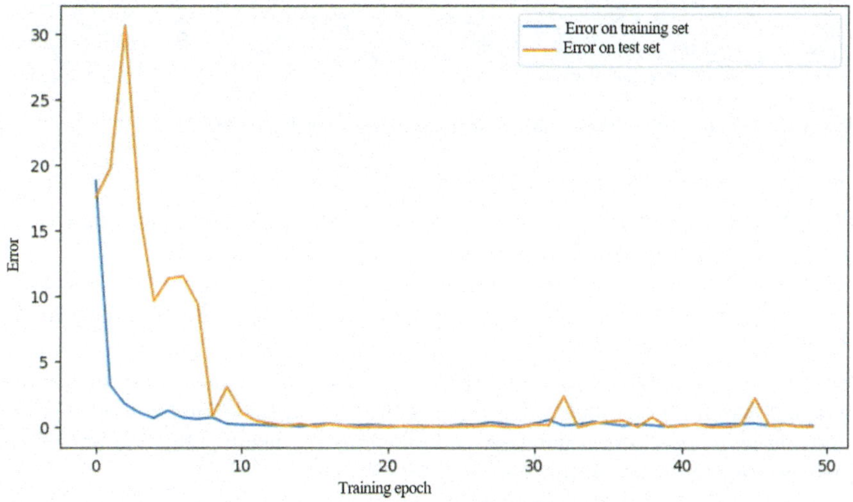

Fig. 5.43 Level of learning error depending on epoch number

GPT-4v made a clear and correct conclusion that the provided sample belongs to the metal 2 class, and justified it by features of metal surface.

Next, GPT-4v was given an image of metal 1 surface, and it answered just as convincingly and without error that it was in the case of metal 1 (Fig. 5.47).

Thus, GPT-4v coped well with the classification task and demonstrated its ability to learn and recognize different types of metal microstructures. The general principles

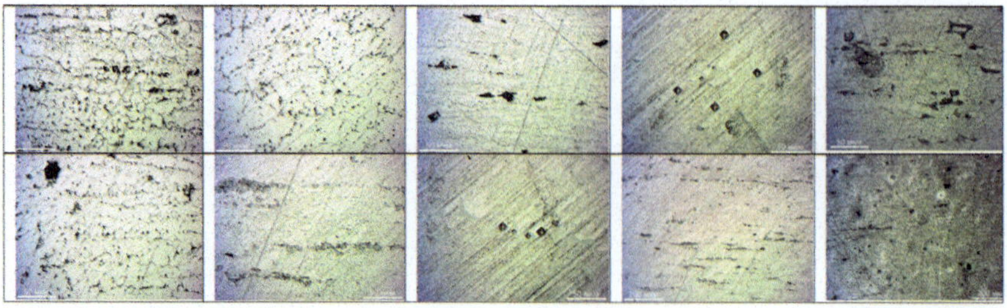

Fig. 5.44 Training set of metal 1 images simultaneously loaded into GPT-4v

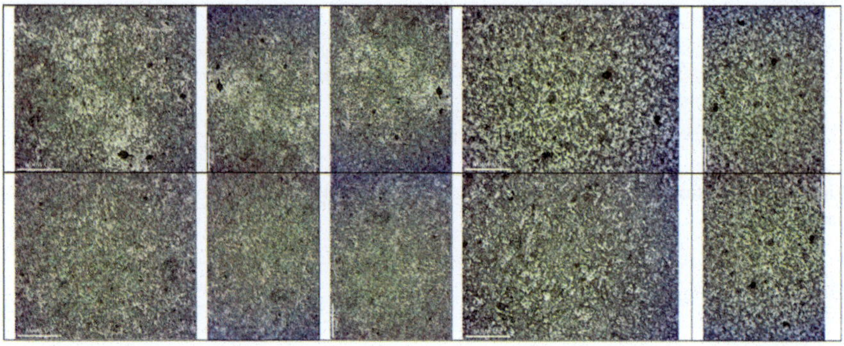

Fig. 5.45 Training set of metal 2 images simultaneously loaded into GPT-4v

of microstructural analysis and empirical data obtained from literature sources and previous experience were taken into account.

The demonstrated effectiveness of using a large language model for classifying microstructures of various metals is a new approach that allows not only determining the metal class, but also obtaining justification and description of the analysis results. Researchers can upload image arrays, and LLM will use internal algorithms to analyze and classify these images according to the training data. This process involves studying textures, grain, roughness, and other surface features that are critical for recognizing and distinguishing between different types of metals.

It is important that GPT-4v not only classifies images, but also provides an opportunity for users to test their understanding of the materials by allowing them to classify images themselves, and then check their answers against the results provided by GPT-4v. This demonstrates the potential of LLMs as a learning tool and their ability to support technical research and quality control in the metal industry. A key advantage is the ability to automate processes and provide objectivity of classification due to machine learning. The ideal strategy is to use different large language models to complement their multimodal capabilities, using the strengths of each LLM to fill the gaps of the others.

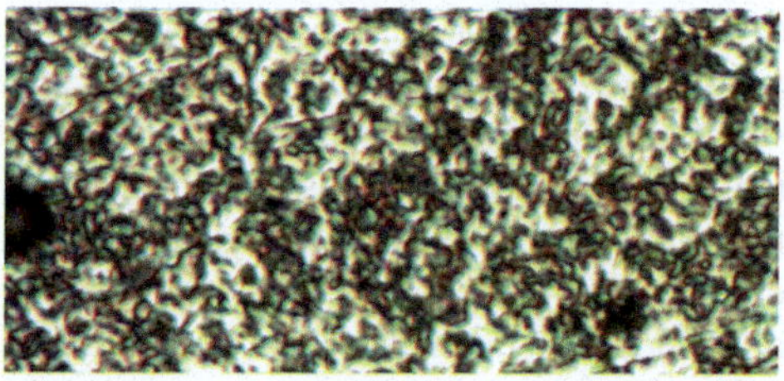

You got some pictures of metal 1 and than - pictures for metal 2. You should be study all difference between these metals classes. After this I will test you and will give few test pictures to asesment your ability to classification of metals based on pictures. Please classified what class of metal (Metal 1 or metal 2) is depicted on thus picture

ChatGPT

Based on the previously shared images of 'metal 1' and 'metal 2', I can make an informed guess about the classification of the new image.

The image you've uploaded appears to have a microstructure that is more similar to the images provided for 'metal 2'. It exhibits a fine and homogenous grain structure with a uniform phase distribution and without the clear directional texture that was present in 'metal 1'.

The consistent, fine-grained appearance and the lack of elongated grains or evidence of directional processing suggest that this is ↓ aligned with the characteristics of 'metal 2'.

Fig. 5.46 GPT-4v setup and response to metal 2 test image

SW you

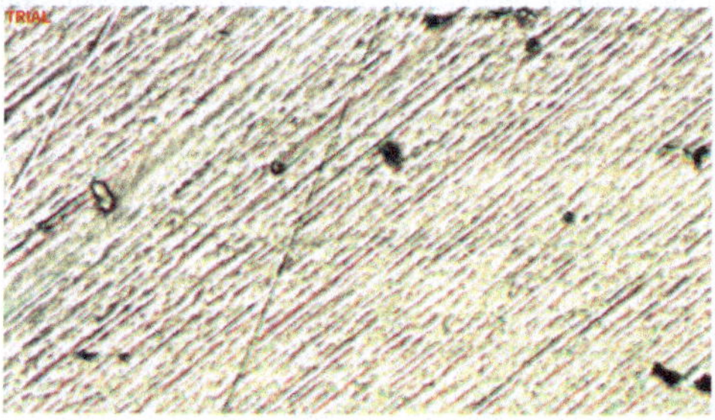

Please classifies what class of metal (Metal 1 or metal 2) is depicted on this picture

ChatGPT

The new image you've provided shows a microstructure with elongated grains and directional texture, similar to the images previously identified as 'metal 1'. The linear features and the presence of darker inclusions or precipitates along the grain boundaries are indicative of mechanical working, such as rolling or forging, which tends to elongate the grains in the direction of the deformation.

These visual characteristics are distinct from the more equiaxed and uniform microstructure of 'metal 2'. Therefore, based on the visual comparison with the earlier provided images, this image would be classified as 'metal 1'.

Fig. 5.47 GPT-4v response to image of class 1 metal

To increase the accuracy of classification, it is recommended to combine visual analysis with other material science methods, taking into account the effect of heat treatment, alloying, and other technological processes on the microstructure of metallic materials. Besides, a significant contribution to this field could be the development of machine learning algorithms that can automate the classification process based on the analysis of microstructure images.

By incorporating these strategies and methods into an intelligent system for identifying materials for special-purpose structures, basing on neural network technologies, it is possible to increase productivity and provide high accuracy of analysis. This will help reduce risks, as well as save time and resources that would traditionally be spent on this complicated and responsible process.

Further research may include the application of LLM and neural networks to X-ray diffraction (XRD) analysis to determine the crystal structure of materials, and scanning electron microscopy (SEM) to study their microstructure in more detail.

References

Aksu-Koç A, Aktan-Erciyes A (2018) Narrative discourse: developmental perspectives. In: Handbook of communication disorders. De Gruyter, pp 229–356

Ansari MH, Moradi M, NikRah O, Kambakhsh KM (2016) A hybrid recommender system for an E-learning system. In: 2nd international conference of signal processing and intelligent systems (ICSPIS). IEEE, pp 1–5

Anwar S, Ullah A, Rocha A, Sousa MJ (2023)Proceedings of international conference on information technology and applications. ICITA, Springer Nature

Araújo PC, Tennis J-T, Guimarães JAC (2024) The concept of epistemology in knowledge organization. In: Edward N, Zalta & Uri Nodelman (eds) The Stanford encyclopedia of philosophy

Bogdanov V, Grigorenko A, Chepkov I, Odnoralov I, Kremenytskyi V, Kuchinskyy A, Sperkach S (2023) Identification of the country of origin of unmanned aerial vehicles based on the analysis of the element base by scanning electron microscopy. Weapons Mil Equip Sci J 1(37):57–62

Bogdanov V, Grigorenko A, Chepkov I, Odnoralov I, Kremenytskyi V, Kuchinskyy A, Sperkach S (2023) Application of the method of scanning electron microscopy in determining the chemical composition of metal fragments of cruise and aircraft missiles manufactured by the Russian Federation. Weapons Mil Equip Sci J 2(38):18–25

Bogdanov V, Grigorenko A, Chepkov I, Odnoralov I, Kremenytskyi V, Kuchinskyy A, Sperkach S (2023) Investigation of the chemical storage, structure and topography in the composite materials in the case of the Shahed-136 boarding ammunition aerial by the method of scanning electronic microscopy. Weapons Mil Equip Sci J 4(40):53–61

Bogdanov V, Grigorenko A, Chepkov I, Odnoralov I, Kremenytskyi V, Kuchinskyy A, Sperkach S (2023) The use of scanning electron microscopy in determining the chemical composition, structure and topography of the surface of composite materials fragments of cruise missiles manufactured by the Russian Federation. Weapons Mil Equip Sci J 3(39):36–48

Bogdanov V, Grigorenko A, Chepkov I, Odnoralov I, Kremenytskyi V, Kuchinskyy A, Sperkach S (2024) Features of the chemical composition, structure and topography of the composite materials of the airframe of the modernized barrage munition "SHAHED-136" (black). Weapons Mil Equip Sci J 1(41):75–82

Borisevich VK, Vinogradsky AF, Semishov NI (1998) Structural materials science. Kharkiv Aviation Institute, Kharkiv

Borsato M, Wognum N, Peruzzini M (Eds.) (2016) Transdisciplinary engineering: crossing boundaries. Series advances in transdisciplinary engineering, vol 4

Burgot G, Jean-Louis Burgot J-L (2023) General analytical chemistry. CRC Press, Separation and Spectral Methods

Chepkov IB, Strizhak AE, Sotnik VV, Potapov GM (2022) Scientific and methodological apparatus for the study of captured samples and fragments of weapons and military equipment of the Russian Federation. Weapons Mil Equip Sci J 1(33):12–22

Chollet F (2017) Xception: deep learning with depthwise separable convolutions. In: IEEE conference on computer vision and pattern recognition (CVPR), pp 1800–1807

Chungora N, Young R, Gunendran G, Palner C, Usmman Z (2013) A model-driven ontology approach for manufacturing system interoperability and knowledge sharing. Comput Ind 64(4):392–401

Deng J, Dong W, Socher R, Li L-J, Li K, Li F-F (2009) Imagenet: a large-scale hierarchical image database. In: 2009 IEEE conference on computer vision and pattern recognition, pp 248–255

Directory steel and alloys (2024). https://www.splav-kharkov.com/main.php/ [Electronic resource]

Dovgiy SO, Strizhak OE, Velichko VYu (2019) Information and educational resources. Capsules of knowledge: collective monograph. Institute of Gifted Child, Kyiv

Dovhyi S, Stryzhak O (2020) Transdisciplinary fundamentals of information-analytical activity. In: Ilchenko M, Uryvsky L, Globa L (eds) Advances in information and communication technology and systems. Lecture notes in networks and systems, vol 152. Springer, Cham

Dubinin V, Vyatkin V (2012) Designing control applications based on the transformation of ontologies using logical programming languages. In: Proceedings of the international scientific and technical conference on contemporary information technologies, pp 6–25

Golovin OO, Strizhak OE (2019) Application of the ontology of the choice problem for describing the processes of interaction of subjects of public administration in the military sphere. Weapons Mil Equip Sci J 2(73):80–93

Gorbulin V (2020) On the protection of the sea coast of Ukraine. Bull Nat Acad Sci Ukrain 9:30–40

GPT-4 (2023) Tech Rep 1–101

Howard A, Sandler M, Chu G, et al. (2019) Searching for Mobilenetv3. In: Proceedings of the IEEE international conference on computer vision, Seoul, pp 1314–1324

Information and training resources (2019) Capsules of knowledge. In: Dovgiy SA, Strizhak OE (eds) Collective monograph. Institute of Gifted Child of NAPS of Ukraine, Kyiv

Katie Moon K, Browne Nicola NK (2020) Developing shared qualitative models for complex systems. Conserv Biol 35(3):1039–1050

Knorozov BV, Usova LF, Tretyakov AV (1987) Metal technology and materials science. M. Metallurgy

Konar A (2005) Cognitive engineering: a distributed approach to machine intelligence. Advanced information and knowledge processing. Springer, Champ

Konar A (2005) Cognitive engineering: distributed approach to machine intelligence. Advanced information and knowledge processing. Springer, Cham, Series

Kopaygorodsky A (2014) Use of ontologies in semantic information systems. Ontol Des 4(14):78–89

Lee JD, Kirlik A (Eds.) (2013) The Oxford handbook of cognitive engineering. Oxford University Press

Liu H, Li C, Wu Q, Lee YJ (2023) Vis Instr Tuning 1–19

Lyman P, Varian HR(2003) How much information. Technical report, UC Berkeley. http://www.sims.berkeley.edu/research/projects/how-much-info-2003/

Lytvyn VV, Basyuk TM, Dosyn DG (2019). Ontological engineering. Lviv Polytechnic

Martin PG et al (2020) Radiological identification of near-surface mineralogical deposits using low-altitude unmanned aerial vehicle. Remote Sens. 12(21):3562–3574

Matveev MY, Selezneva MA, Gegechkori LA (1986) Establishment of generic property of microparticles of low-alloy instrumental steels by methods of metallography and RSMD. New methods and methods of research of objects of forensic examination: expert practice 93, Moskow

Mintser OP, Strizhak OE, Prikhodnyuk VV, Shevtsova OM (2018) Transdisciplinary presentation of information using interactive documents. Med Inf Eng 1:47–52

Moylan EC, Pennington RT, Scotland RW (2002) Taxonomic account of Hemigraphis Nees (Strobilanthinae-Acanthaceae) from the Philippines. Kew Bull 57:769–825

Nadutenko M, Prykhodniuk V, Shyrokov V, Stryzhak O (2022) Ontology-driven lexicographic systems. Advances in information and communication. Lecture notes in networks and systems. Springer, Cham, pp 204–215

Nadutenko M, Prykhodniuk V, Shyrokov V, Stryzhak O (2022) Ontology-driven lexicographic systems. In: Arai K (eds) Advances in information and communication. FICC. Lecture notes in networks and systems, vol 438, pp 204–215

Nicolescu B (2008) Transdisciplinarity—theory and practice. Hampton Press, Cresskill, NJ, USA

Odinochkina TF (1973) Study of forensic objects by atomic absorption spectral analysis. Expert Tech 40

Palagin A (2014) Transdisciplinarity, computer science and development of modern civilization. Bull Nat Acad Sci Ukraine. Kyiv: NAS Ukraine 7:25-33

Podlipaev VA, Strizhak AE (2018) Integration of information resources of various nature in a network-centric environment based on the category of transdisciplinarity. Weapons systems and military technology. Kharkiv: KHNUPS 3(55):85–94

Potapov HM, Filistieiev DA, Prychodniuk VW, Gordieiev AYu (2020) Functional characteristics of network central ontographic information system of specific purpose. Telecommun Inf Technol 3:41–54

Qassim H, Verma A, Feinzimer D (2018) Compressed residual-VGG16 CNN model for big data places image recognition. In: Computing and communication workshop and conference (CCWC) IEEE 8th annual, pp 169–175

Riedl MO, Bulitko V (2013) Interactive narrative: an intelligent systems approach. AI Mag 34(1):67–77

Rivera AC, Tapia-Leon M, Lujan-Mora (2018) Recommendation systems in education: a systematic mapping study. Springer, Cham, pp 937–947

Sandler M, Howard A, Zhu M, et al. (2018) Mobilenetv2: inverted residuals and linear Bottlenecks. In: Proceedings of the IEEE conference on computer vision and pattern recognition, Salt Lake City, pp 4510–4520

Sardana MMK (2006) Information explosion, challenges and opportunities. ISID discussion notes, vision of digital India: challenges ahead for political establishments, pp 1–22

Shapovalov YB, Shapovalov VB, Tarasenko RA, Usenko SA, Paschke A (2020) A semantic structuring of educational research using ontologies. In: CEUR Workshop Proceedings, Kryvyi Rih, Ukraine, vol 2879, pp 105–123

Shcherbakovsky MG (1996) Forensic study of metals, alloys and products from them: scientific-practical. Manual for experts. Kharkiv, Osnova

Shcherbakovsky NG, Lapta SP (2009) Forensic expertise. Kharkiv National University of Internal Affairs

Shirokov VA (2017) Information system: transdisciplinarity in linguistics. Palamarium Academic Publishing, Kyiv

Shirokov VA (2022) Changing the paradigm vector of modern linguistics: some methodological considerations. Linguistics 6:3–14

Slyusar VI, Sliusar II (2021) Neural network technologies and their applications NNTA-2021. In: Scientific papers of the XX international scientific conference on neural network technologies and their applications NNTA-2021 Kramatorsk DDMA, pp 129–133

Slyusar VI (2022) Architectural and mathematical foundations of improving neural networks for image classification. Artif Intell 1:127–138

Slyusar VI, Protsenko MM, Dokuchaev AV (2022a). Methods of building neural networks for identifying weapons and military equipment. Armament Mil Equip 2(34):99–107

Slyusar V, Protsenko M, Bondarchuk M (2022b) Methodology for armaments identification using a neural network. In: International scientific-practical conference Problems of infocommunications science and technology
Slyusar V, Protsenko M, Chernukha A, Gornostal S, Rudakov S, Shevchenko S, Chernikov O, Kolpachenko N, Timofeyev V, Artiukh R (2021) Construction of an advanced method for recognizing monitored objects by a convolutional neural network using a discrete wavelet transform. Eastern Eur J Enterp Technol 4, 9(112):65–77
Strizhak OI (2014) Ontological aspects of transdisciplinary integration of information resources. Open Inf Comput Integr Technol 65:11–22
Stryzhak OE (2023) Digital university creation concept. Sci Notes Small Acad. Sci Ukraine. K.: Nat. Center Small Acad Sci Ukraine 2(27):70–80
Stryzhak OE (2023) Digital university. Creation concept. Scientific notes of the small academy of sciences of Ukraine. Kyiv: Nat Center Small Acad Sci Ukraine 2(27):70–80
Stryzhak OE, Zavalevskyi YI (Eds) (2023) The world of innovative opportunities: current issues of the development of STEM education: collective. Kyiv: Nat Center Small Acad Sci Ukraine
Stryzhak O, Prychodniuk V, Podlipaiev V (2019) Model of transdisciplinary representation of GEO spatial information. Advances in information and communication technologies. UKRMICO. Lecture notes in electrical engineering, vol 560. Springer, pp 34–75
Sturm V, Erben B, Fleige R, Wirz W (2019) Carbon analysis of steel using compact spectrometer and passively Q-switched laser for laser-induced breakdown spectroscopy. Opt Express 25(27):36855–36863
Szegedy C, Vanhoucke V, Ioffe S, Shlens J, Wojna Z (2016) Rethinking the inception architecture for computer vision. In: IEEE conference on computer vision and pattern recognition (CVPR), Las Vegas, NV, USA, pp 2818–2826
Tan M, Le Q (2019) Efficient Net: Rethinking model scaling for convolutional neural networks. In: Proceedings of international conference on machine learning, pp 6105–6114
Trends in science and technology 2020-2040. As evidenced by the NATO study. - https ://opk.com.ua/As evidenced by trends/[Electronic resource]. Accessed 25 July 2024
Trends G (2020) In the development of artificial intelligence in Ukraine information systems and technologies security. Taras Shevchenko Nat Univ Kyiv Ukraine 3–4:20–26
Tungol MW, Bartick EG, Montaser A (1990) The development of a spectral data base for the identification of fibers by infrared microscopy. Appl Spectrosc 4(44):543–549
Wang H, Fu T, Du Y (2013) Scientific discovery in the age of artificial intelligence. Nature 47–60
Zagorulko Y (2015) Semantic ontology for development of intelligent systems oriented to experts in subject domain. Ontol Des 5(1):30–46

Appendix A
Conclusions

The book presents the results of system research into physico-chemical properties of metallic and composite materials used in structural elements of the means of destruction that are employed by the Russian Federation during its large-scale aggression against Ukraine. The study also reveals the combination of forces and means involved in acquiring these data, which has made possible a more profound insight into the peculiarities of special-purpose materials.

A detailed analysis of current metals, alloys, and composite materials utilized in manufacturing special-purpose structures is performed alongside with the available methods of examining them.

The studies have been carried out with reliance on the techniques of scanning electron microscopy and X-ray microanalysis, which permits determining with maximum accuracy the chemical composition and structure of the materials in fragments of cruise, ballistic, aeroballistic and aircraft-launched missiles, loitering munitions and aircraft complexes produced by the Russian Federation and its allies (the classification of means of destruction is provided in Appendix). Specific features of employing the chosen methods to examine composite and metallic materials of structural elements of means of destruction are mentioned.

The gained experimental data permitted expanding the knowledge of the chemical composition of metals, alloys, and composites used in current means of destruction and can be implemented at missile- and aircraft-producing plants as well as other high-tech industries.

The results of the studies confirm developers' attempts to improve the characteristics of means of destruction, which, in turn, has led to wide use of composites along with high-strength and heat-resistant alloys.

At the same time, it should be noted that Russian designers of cruise, ballistic, aeroballistic, and aircraft-launched missiles restrict themselves to metal and alloy grades developed in the 1970s–1980s and launched into series production back in the days of the USSR. This results from the desire of engineers to ensure cost reduction and maximum productivity of manufacturing.

V. L. Bogdanov et al., *Identification of Special-Purpose Structures by Their Fragments Based on Scanning Electron Microscopy*, Advanced Structured Materials 233,
https://doi.org/10.1007/978-3-031-88827-4

The study substantiates the approach to computing the parameters of frequency characteristics of RF absorbent coatings and materials of means-of-destruction elements, which relies on the results of analytical research concerning electromagnetic wave interactions with composite RF absorbent materials and specific experimental data obtained with electron microscopy and X-ray microanalysis.

High-precision experimental data have been acquired concerning the internal structure of composite RF absorbent coatings of the elements of 9M727 cruise missiles of Iskander-K system and 3M-14 missiles of Kalibr system, including geometric parameters of fillers and layers, as well as quantitative chemical compositions of coating materials. This would ensure the feasibility of modeling physical processes occurring during the interaction of radio waves with RF absorbent coating material in a wide frequency range.

Taking into account the physical mechanisms of electromagnetic wave's absorption in each particular composite material, the width of the operating range of RF absorbent coatings has been determined and limit parameters of electromagnetic radiation wavelengths and frequencies of the respective range have been calculated.

The obtained results show that employing composites with fillers that have different magnetic properties allows the development of RF absorbent coatings and materials of means of destruction with variable ranges which have low probability of being detected by radar equipment of counteraction systems.

The subject and tasks of the research toward identifying means of destruction by their fragments permit forming the list of issues to be resolved, and the competence and the volume of knowledge the professionals conducting the study must possess. General principles of the methodology of the research for the identification of means of destruction have been formulated.

The reliance on the results of the conducted studies of the fragments of means of destruction has helped reveal the fundamentals of the methodology for transforming narrative descriptions of scientific and technical products, along with other technical and technological documentation, into the format of ontological knowledge bases along with the technological principles of creating them. The methodology is based on cognitive services that implement lingua-semantic and conceptographical analyses of large volumes of information, namely, natural language narratives, documents, and tabular data, reveal consolidated logistics connections among them, estimation criteria, and decision-making support. The format of the technological chain of compiling ontological knowledge base is presented: documents $\longrightarrow$ analysis $\longrightarrow$ linguistic corpus $\longrightarrow$ neural network formation $\longrightarrow$ generation of ontologies $\longrightarrow$ identification of estimation criteria $\longrightarrow$ generation of analytical decision-making platforms.

The methods and models mentioned describe the methodology of passive knowledge transformation from narrative texts into the format of ontological knowledge bases.

To improve research efficiency, the validity and reliability of the results obtained, a scientific and methodological apparatus of identifying means of destruction by their fragments has been developed. It is constructed employing the potential of cognitive information technology and elements of artificial intelligence. Ontological network-centric expert platforms that were created make it possible to aggregately identify various types of means of destruction by their fragments, where the condition of the necessary formation of an integrated information analysis space with using the principle of open taxonomies is met.

A methodology of constructing artificial neural networks for identifying means of destruction on the basis of electron microscopy analysis of their fragments has been presented. To compile the dataset, a series of microstructure photographs of various metal classes were used, and convolutional neural networks were employed for their identification. Experiments were conducted to confirm the efficiency of using neural networks for the identification of means of destruction by their fragments. Besides, the study describes specific features of involving GPT-4v Large Language Model in the classification of the materials of means of destruction fragments and demonstrates its efficiency.

Further studies advancing the present one will be aimed at the deeper integration of artificial intelligence into the processes of identifying means of destruction by their fragments and the development of interactive knowledge bases for systematizing the data obtained.

Appendix B
High-Tech Means of Destruction Used by the Russian Federation in the War Against Ukraine

Since the first days of the full-scale aggression against Ukraine, the Russian Federation has been massively using a wide range of high-tech means of destruction, such as cruise, ballistic, and air-launched ballistic, including hypersonic missiles, both ground-based and air- and sea-based, reconnaissance and unmanned combat aerial systems, as well as loitering munitions, whose characteristics are given below.

Air-Launched Cruise and Ballistic (Hypersonic) Missiles
During the full-scale Russian military aggression against Ukraine, Russian troops widely used, among others, air-launched cruise and ballistic missiles, including hypersonic ones. Depending on the relative position of the supporting and control surfaces, these missiles were designed according to the aircraft or rocket aerodynamic patterns. The general view of air-launched cruise and ballistic missiles and the main types of their carriers are presented in Fig. B.1.

The basic equipment of modern cruise and air-launched ballistic missiles, as a rule, includes the following components: an inertial navigation system, radio altimeters, route correction systems (including the global navigation satellite system), a homing head, and an onboard computer. New advances in microelectronics, engines, high-efficiency fuels, and structural materials have enabled the development of supersonic and hypersonic high-precision, stealthy missiles with a range of up to 5,500 km and an average mass of up to 2,500 kg. Their visibility (radar cross section or signature) is reduced by choosing low-reflective geometric shapes, employing radio-absorbing materials and coatings, using special devices to reduce the effective scattering surface, antenna devices, and air scoops.

The main performance characteristics of air-launched cruise and ballistic missiles used during Russia's full-scale aggression against Ukraine are given in Table B.1.

Analysis of the performance characteristics of air-launched cruise and ballistic missiles of the Russian Federation shows that they have different speeds along their paths. Subsonic missiles include Kh-55, Kh-59, Kh-101, Kh-555; supersonic missiles include Kh-22 and Kh-31; and *Kinzhal* 9-A-7660 is a hypersonic missile.

V. L. Bogdanov et al., *Identification of Special-Purpose Structures by Their Fragments Based on Scanning Electron Microscopy*, Advanced Structured Materials 233,
https://doi.org/10.1007/978-3-031-88827-4

Fig. B.1 General view of main air-launched cruise and ballistic missiles of the Russian Federation and their carriers

Table B.1 Main performance characteristics of Russian cruise and air-launched ballistic missiles

Characteristic	Kh-22	Kh-31	Kh-55	Kh-59	Kh-101	Kh-555	*Kinzhal* 9-A-7660
Length, m	11.65–11.67	5.34	5.88	5.7	7.45	6.04	7.3
Diameter, m	0.9–0.92	0.36	0.514	0.38	~0.742	0.77	at least 0.9
Launch mass, kg	up to 5780	715	1195	930	up to 2400	1500	at least 4000
Cruising speed, m/s	1100–1500	1000	200–230	250–290	200–270	200–230	up to 4080
Flight altitude, m	11500–12500 or 22500–25000	up to 16000	40–110	OS: 10–15 in CS 4–7 in TA OL: 50 in CS 10–15 in TA	40–110	40–110	
Range, km	up to 600	180–250	up to 2500	up to 290	up to 5500	up to 2000	2000–3000
Warhead mass, kg	960	110	410	320	400	410	500
Warhead types	HES or TN	GPC	SA or HEF	HEP	high explosive or nuclear	SA or HEF	nuclear or conventional
Control system	IS +ARHD (ARHH)	IS +ARHH	IS +SNS+ RA+ ARHD	IS + SNS + RA + ARHD	IS + SNS + RA + Octopus (probably) + OEHH	ARHD, GIS	IS with the possibility of correction with GLONASS, AEWS, optical HH
Homing head	Radar	Broad—bandARHH L-130	ARHD	ARHD ARGS-59E	Optical HH, radar HH	ARHD	Optical HH, radar HH

*_Abbreviations_: *AD*—attack dive, *CS*—cruising segment, *TA*—target area, *OS*—over sea, *OL*—over land, *HES*—high-explosive shaped, *TN*—thermonuclear, *GPC*—general-purpose cluster, *HEP*—high-explosive piercing, *SA*—special ammunition, *HEF*—high-explosive fragmenting, *HH*—homing head, *ARHD*—active radar homing device, *IS*—inertial system, *ARHH*—antiradar homing head, *AEWS*—airborne early warning system, *SNS*—satellite navigation system, *RA*—radio altimeter, *OEHH*—optoelectronic homing head, *GIS*—geographic information system

According to the flight range of air-launched cruise and ballistic missiles, they can be divided into the following groups: tactical (up to 600 km)—Kh-22, Kh-31, and Kh-59; operational and tactical (up to 2000 km)—Kh-555; and strategic (over 2000 km)—Kh-55, Kh-101, *Kinzhal* 9-A-7660.

The main advantages of air-launched cruise and ballistic missiles are high aerodynamic characteristics; maneuverability; the ability to choose missile trajectories at low altitudes along the contours of the terrain, which makes them difficult to detect by air defense and missile defense systems; high accuracy; and the ability to adjust the flight path using an onboard computer and automatic control system.

Sea-based cruise missiles

Among the aggressor's weapons which are most frequently used during hostilities, sea-based cruise missiles of the *Kalibr* type occupy a special place.

In the mid-1970s, the USSR defense industry was tasked with developing a general-purpose cruise missile complex, with the ability to vary the missiles on the carrier depending on the task and the specific combat situation. It was the *Kalibr* cruise missile complex (the export version is called *Club*).

The first missiles of the *Kalibr* complex were presented at the MAKS-93 air show.

The first combat use of the cruise missiles by the armed forces of the Russian Federation occurred on October 7, 2015, by ships of the Caspian Flotilla over the territory of Syria at a distance of over 1,500 km.

The *Kalibr* complex includes the following missiles:

- 3M-54T (K), 3M-54-1T (K)—cruise missiles used against surface ships (anti-ship missiles, ASM).
- 3M-14T (K)—cruise missiles for attacking ground targets.
- 91R1—anti-submarine missiles, component of the *Kalibr-PL* 3K-14K missile complex (submarine based).
- 91RT2—anti-submarine missiles, component of the *Kalibr-NK* 3K-14T missile complex (surface ship based).
- 91R1 missiles are part of the *Kalibr-PL* 3K-14K integrated missile system for submarines, while 91RT2 missiles are part of the *Kalibr-NK* 3K-14T integrated missile system for surface ships.
- 3M-14T missiles are located in a transport-launch container on surface ships, while 3M-14K missiles are located in a transport-launch tube and on submarines.

The developer classifies the cruise missiles of the Kalibr complex as high-precision weapons. The performance characteristics of the missiles are listed in Table B.2.

From the beginning of the war to the present, the enemy has been launched 3M-14 missiles on the territory of Ukraine from ships and submarines. The enemy has not yet used anti-ship or anti-submarine missiles of the complex in combat operations.

Among the main performance characteristics of the 3M-14 missile, which complicate defence, the following ones should be noted:

Table B.2 Performance characteristics of Kalibr cruise missiles [3, 12]

Missile	3M-54	3M-54-1	3M-14 K(T)	91R1	91RT1
Image					
Target	Naval targets	Naval targets	Ground targets	Submarines	Submarines
Carriers	Surface ships, submarines	Surface ships, submarines	Surface ships, submarines	Submarines	Surface ships
Length, m	8.22	6.20	7–8.2	7.65	6.20
Diameter, mm	533	533	533	533	533
Launch mass, kg	2300	1800	1770	2100	1200
Warhead	High-explosive piercing 200 kg	High-explosive piercing 400 kg	High-explosive fragmenting or cluster 450 kg	Torpedo APR-3M	Torpedo MPT-1UM
Range, km	220	300	2500	50	40
Flight speed, Mach number (M)	Cruising: 0.8 Near target: 2.9	0.8	0.8	2.5	2.0
Trajectory (flight altitude)	Cruising: 20 m Near target: 10 m	20 m	Over sea: 20 m Over land: 50–150 m	Ballistic	Ballistic
Control system	IS + Radar homing device	IS + Radar homing device	IS + Tercom + correction with GLONASS or GPS+ Radar homing device	IS	IS

- combined flight path,
- low-altitude flight,
- jamming-proof homing system,
- availability of an autonomous flight system,
- the ability of firing multiple missiles, and
- large warhead.

A combined flight path can provide a longer missile range compared to using low-altitude flight. At the same time, low-altitude flight makes it difficult to detect and destroy missiles. The missile flight route is set based on information about available air defense units. The ability to fire multiple missiles makes it difficult to shoot down missiles due to their large number. The jamming-proof homing head (according to the developer) reduces the effectiveness of electronic suppression means and increases the effectiveness of the missiles.

The missile has a large warhead that can be of shrapnel, high-explosive, cluster or nuclear type.

At the same time, a cruise missile has subsonic speed at the segments of air defense breakthrough, which increases the probability of shooting it down.

Land-based cruise missiles

The land-based cruise missiles produced by the Russian Federation include the 9M728 (9M729) guided cruise missile of the 9K720 Iskander operational and tactical missile system (9K720M *Iskander-M*).

The missile was developed at OKB Novator PJSC, Yekaterinburg, and it was put into operation in 2009.

The 9M728 cruise missile is a two-stage cruise missile with a non-separable warhead, and it is guided at all stages of its flight.

The missile has the following main performance characteristics:

Engine:

- launch engine—solid propellant;
- cruising engine—small-sized liquid-propellant TRDD-50BE engine.

Maximum range is up to 450 km.

Average speed in the guided area—240 m/s.

Average height in the guided area (at the entrance to the impact area)—0.05–0.15 km.

Strike accuracy (CEP)—up to 10 m.

Warhead mass—450 kg.

Guidance system—optical correlation-extreme.

The general view of the 9M728 cruise missile of the 9K720 *Iskander* operational and tactical missile complex is shown in Fig. B.2.

Upon the development of the 9M728 missile, great attention has been paid to decreasing the effective reflecting surface by reducing protruding parts, the extensive use of composite materials, and the use of thermal barrier and radio-absorbing coatings.

Fig. B.2 General view of 9M728 operational and tactical missile (in the launcher at left) of the 9K720 *Iskander* operational and tactical missile complex

The main advantages of ground-based tactical as well as operational and tactical cruise missiles include the ability to set an arbitrary course to the target and moving at low altitude, skirting the terrain, which makes it difficult to cover it with air defense means (ADM) as well as a sufficiently high striking accuracy. The main disadvantage of these missiles is their relatively low speed, which increases the time they spend in the air defense means (AMD) zone, and relatively high cost and technological complexity of manufacturing. To provide the effective operation of flight control and guidance systems, a special information support system is required.

Land-based ballistic missiles

During the armed aggression against Ukraine, the Russian Federation extensively used its ground-based ballistic missiles—guided operational and tactical missiles 9M723 of the operational and tactical missile complex 9K720 *Iskander* (9K720M *Iskander-M*).

The 9M723 ballistic missile is a single-stage solid-propellant missile that is controlled at all stages of flight using aerodynamic and gas rudders; it has a quasi-ballistic flight trajectory and a non-separable warhead.

The missile is produced at Votkinsky Zavod PJSC, Votkinsk, Udmurtia.

Fig. B.3 General view of 9M723 operational and tactical missile of 9K720 Iskander operational and tactical missile complex

The missile can carry warheads of the following types:

- cluster with non-contact fragmentation warheads (triggered at a height of about 10 m above the land);
- cluster with fragmenting shaped charge warheads;
- cluster with homing warheads;
- volume-detonating cluster;
- high-explosive fragmenting;
- high-explosive incendiary;
- armor-piercing;
- special (nuclear).

Upon the development of the missile, great attention has been paid to decreasing the effective reflecting surface by reducing protruding parts, the extensive use of composite materials, and the use of thermal barrier and radio-absorbing coatings deposited on the missile body.

The general view of the 9M723 operational and tactical missile of the 9K720 *Iskander* operational and tactical missile complex is presented in Fig. B.3.

The intensity of hostilities has led to a shortage of Russian ground-based ballistic missiles, and the capabilities of the Russian Federation's defense-industrial complex are unable to meet the needs of the Russian army in means of destruction.

To partially solve this problem, Russia has arranged for deliveries of KN-23 tactical missiles and self-propelled launchers for them from North Korea. On December 30, 2023, the Russian Federation launched a strike with a KN-23 missile for the first time at a distance of about 460 km from the Kursk region (RF) to the Zaporizhia region of Ukraine. Since that date, they used the KN-23 missiles against civilian infrastructure facilities in Ukraine. For instance, 24 KN-23 missiles were launched over Ukraine between December 30, 2023 and February 7, 2024.

The KN-23 is a ballistic missile, but it is designed for a descending (quasi-ballistic or aeroballistic) flight trajectory, which reduces flight time, prevents radar detection, and increases maneuverability.

The general view of the operational and tactical KN-23 missile is shown in Fig. B.4.

Fig. B.4 General view of North Korean-made KN-23 operational and tactical missile

Table B.3 Performance characteristics of supersonic general-purpose anti-ship missiles

Missile	9M723	KN-23
Image	Votkinsky Zavod PJSC	The Academy of National Defense Science, North Korea
Number of missiles per self-propelled launcher	2	2
Body diameter, mm	950	1100
Missile mass, kg	3800	5200
Speed after the initial section of trajectory, m/s	2100	–
Missile length, mm	7200	7740
Warhead mass, kg	480	500
Warhead type (options)	Nuclear, high-explosive fragmenting, high explosive, cluster, arm-piercing	Nuclear, thermobaric, high explosive, cluster
Range, km	400 (280—export modification)	690
Control system	Inertial with optical correlation-extreme homing with use of terrain images in the final segment	GPS + INS
Strike accuracy (CEP), m	up to 30	GPS + INS $\approx$ 100 m; INS $\approx$ 200 m
Angle of attack, degrees	90°	90°
Engine type	Solid-propellant rocket engine	Solid-propellant rocket engine

In general, the KN-23 (KN-23A) missile is North Korea's own development, although its design and concept are based on the 9M723 (Russia) missile, as well as the *Hyunmu-2B* (South Korea) missile.

The main performance characteristics of operational and tactical missiles used by the Russian Federation during the full-scale aggression against Ukraine are listed in Table B.3.

Supersonic general-purpose anti-ship P-800 and 3M22 Zircon missiles

P-800 *Oniks* is a Soviet/Russian supersonic general-purpose medium-range anti-ship missile designed to combat surface naval groups and single ships under intense fire and electronic countermeasures. Besides, the missile can be used against ground targets, with increased range.

The development of the fourth-generation operational and tactical anti-ship complex was started in the late 1970s. Unlike previous Soviet anti-ship missiles, which had relatively narrow sets of carriers, the new complex was designed from the very beginning as a general-purpose one: it was supposed to be placed on submarines, surface ships and boats, and aircraft and coastal launchers.

The missile was designed according to the conventional aerodynamic scheme. The solid-propellant launch engine is in the combustion chamber of a ramjet engine; when the starting stage is finished, it is separated by the oncoming air flow.

The 3M22 *Zircon* missile is a Russian medium-range maneuvering hypersonic anti-ship missile designed to strike surface ships and ground targets.

The solid-propellant engine accelerates the rocket to a supersonic speed, after which it is accelerated to a hypersonic speed by an air-jet liquid-propellant engine. The missile travels in the cruising segment at an altitude of 30–40 km, where air resistance is low. The high flight speed of the *Zircon* missile (approximately 11,000 km/h) makes it an extremely difficult target for interception by existing air defense systems; however, this does not make it impossible to intercept by modern high-speed missile defense systems.

The P-800 Oniks and 3M22 Zircon missiles have never been used in combat conditions against ships. Nevertheless, they have been repeatedly used by the aggressor to strike civilian infrastructure facilities in Ukraine, and their high supersonic speed and large mass made it difficult to shoot down them with anti-aircraft missiles.

The main performance characteristics of the supersonic general-purpose anti-ship missiles P-800 and 3M22 *Zircon*, which are used by Russia during the full-scale aggression against Ukraine, are given in Table B.4.

The P-800 Oniks and 3M22 Zircon missiles have the following advantages:

- over-the-horizon range;
- autonomous combat use (fire-and-forget);
- variety of trajectories ("low", "high and low");
- high supersonic and hypersonic speeds in all flight segments;
- fitness for a wide range of carriers (surface ships of all major classes, submarines, and land-based launchers);
- effective use under conditions of electronic countermeasures.

Loitering munitions

Before the start of the full-scale Russian aggression against Ukraine, loitering munitions (or *kamikaze* drones) have been considered exotic and specific weapons without widespread use. However, the failure of the first stage of Russia's large-scale invasion of Ukraine, and the failure to achieve the set goals at a significant expense of conventional high-precision weapons led to the transition to the next stage of the war—an attrition warfare. At this stage, the widespread use of cheap and mass-produced loitering munitions provided constant strikes, primarily on Ukraine's infrastructure facilities, and the depletion of air defense systems.

Table B.4 Performance characteristics of operational and tactical missiles

Missile	P-800 *Oniks*	3M22 *Zircon*
Image		
Purpose	Supersonic general-purpose anti-ship missile	Hypersonic general-purpose anti-ship missile
Launch engine type	Solid-propellant rocket engine	
Cruising engine type	Ramjet air-jet engine	
Missile length, m	8	10
Body diameter, mm	670	–
Missile mass, kg	3100	–
Maximum speed, km/h	3180	11000
Flight altitude, km	up to 14	30–40
Range, km	up to 600	400–1000
Warhead mass, kg	300	300
Control system	IS + ARHH	
Engine type	Solid-propellant rocket engine	Solid-propellant rocket engine

In the war against Ukraine, the Russian Federation most widely uses the *Shahed-136* (produced by the Islamic Republic of Iran) and the Russian ZALA LANCET-3M loitering munitions.

The *Shahed-136* kamikaze UAV is an Iranian loitering munition designed to strike stationary ground objects by targeting and contact detonation of the warhead. It can be launched massively from mobile launchers disguised as a dump truck body, each of which can simultaneously launch up to five UAVs.

The *Shahed-136* UAV has been developed by Iran Aircraft Manufacturing Company (HESA) (Iran, Tehran, Sepahbod Gharani Avenue 107).

The *Shahed-136* kamikaze UAV consists of a monoplane, engine, propeller, warhead, and onboard equipment which includes an onboard computer, inertial navigation system, and satellite navigation system.

The main advantages of the *Shahed-136* loitering munition include a small effective reflecting surface due to the use of composite materials, as well as the use of passive inertial and satellite navigation systems.

The main disadvantages of the *Shahed-136* UAV are the high acoustic noise of a two-stroke internal combustion engine and low speed.

ZALA LANCET-3M is a Russian loitering munition positioned as an intelligent multi-tasking weapon capable of independently finding a given target and hitting it. This loitering munition has been developed by Russian company ZALA AERO GROUP of the KALASHNIKOV group of companies.

The *ZALA LANCET-3M* loitering munition was designed to destroy moving vehicles, artillery and mortar firing positions, long-term firing positions, and UAVs at the front and in the rear. This UAV can fly autonomously, according to the flight task and along a pre-set route that can be corrected.

The *ZALA LANCET-3M* complex includes a ground control unit with a tablet, a laptop, antenna systems; power supply system for the ground control unit; a catapult mounted on a trailer; and a vehicle.

The design of the *ZALA LANCET-3M* munition is based on a cylindrical fuselage with two pairs of X-shaped wings located in the front and rear of the fuselage. It runs on electricity with an engine installed in the rear of the fuselage, and is launched with a catapult launcher.

The performance characteristics of the Shahed-136 and *ZALA LANCET-3M* loitering munitions are listed in Table B.5.

The main advantages of the *ZALA LANCET-3M* loitering munition are a small effective reflecting surface due to the use of composite materials, the passive inertial and satellite navigation systems, and a low level of acoustic noise due to the electric motor.

Table B.5 Performance characteristics of Shahed-136 and ZALA LANCET-3M loitering munitions

UAV	P-800 *Shahed-136*	*ZALA LANCET-3M*
Image		
Speed range, km/h	Up to 120	80–120
Flight duration, min	Up to 1000	40
Flight range, km,	1800–2500	48–66
Control system type	Inertial satellite	TV+ inertial satellite
Connection distance, km	–	Up to 30
Flight altitude, m	–	5000
Load mass, kg	50	3
Engine type	Piston two-stroke	Electric
Number of engines	1	
Takeoff type	Special platform	Electric
Total weight, kg	200	12
Detonator type	Contact	Non-contact/contact
Warhead type	High-explosive fragmenting	High-explosive fragmenting or shaped charge
Length, m	3.5	1.2
Wingspan, m front rear	2.5	1.5 0.62

Table B.6 Performance characteristics of *Orlan-10* UAV

Image	
Range, km	From 50 to 120
Maximum range, km	600
Flight duration, h	From 10 to 18
Payload, kg	Up to 5
Wingspan, m	3.1
Length, m	1.8
Mass:	
Without load, kg	12.5
Maximum takeoff weight, kg	18
Cruising speed, km/h	100–150
Maximum height, m	6000
Engine type	1 ICE
Fuel	Gasoline AI-95
Wind speed at takeoff, m/s	10
Operating temperature range	−30 + 400C
Deployment time, min	No more than 30
Takeoff type	Catapult
Landing type	Parachute

Unmanned aerial vehicles

The most common anti-aircraft missile system used by the Russian army for reconnaissance and adjusting the artillery fire is the *Orlan-10*.

The *Orlan-10* UAV has been developed and is produced by Special Technological Center LLC, St. Petersburg. It was put into operation in 2010.

The *Orlan-10* UAV is designed to perform reconnaissance flights to tactical depth for ground forces; it is part of the "Unified Tactical Link Control System"—a software and hardware complex designed to coordinate and adjust the actions of different types of combat arms in a certain area.

The main performance characteristics of the *Orlan-10* UAV are listed in Table B.6.

The UAV has a low-power gasoline engine that provides a speed of 75–170 km per hour, depending on tactical purposes. The I UAV can stay in the air for up to 15 h, while the maximum distance from the ground control station should not exceed 125 km. The UAV is capable of flying under difficult weather conditions.

A satellite navigation system, a raster image of the terrain with a reference to several points, or an electronic map is used for the navigation. The control is carried out both in manual mode and according to flight task. Up to 60 points and the parameters of flight there are set to specify the route.

The *Orlan-10* UAV has a *Return to Home* function and landing point determination, as well as algorithms for operating in emergencies (loss of connection or GPS signal, etc.). The UAV location can be determined due to the built-in GPS tracker or a SIM card of any mobile operator.

The manufacturer's authorised representative in the EU is Springer Nature Customer Service Centre GmbH, Europaplatz 3, 69115 Heidelberg, Germany. If you have any concerns regarding our products, please contact ProductSafety@springernature.com

Printed and bound by CPI Group (UK) Ltd, Croydon, CR0 4YY

30/09/2026

02225571-0014